UNSTEADY INTERACTION OF SHOCK AND DETONATION WAVES IN GASES

V. P. Korobeinikov
Institute for High Temperatures
USSR Academy of Sciences, Moscow

English-Edition Editor
Paul A. Urtiew
Lawrence Livermore National Laboratory
Livermore, California

HEMISPHERE PUBLISHING CORPORATION
A member of the Taylor & Francis Group
New York Washington Philadelphia London

UNSTEADY INTERACTION OF SHOCK AND DETONATION WAVES IN GASES

 Originally published as Nestatsionarniye vzaimodeystviya udarnykh i detonatsionnykh voln v gazakh by Nauka, Moscow, 1986.

Translated by Jamil I. Ghojel.

1 2 3 4 5 6 7 8 9 0   B R B R   8 9 8 7 6 5 4 3 2 1 0 9

This book was set in Times Roman by EPS Group Inc. The editor was Janine Ludlam.
Cover design by Renée E. Winfield.
Braun-Brumfield, Inc. was printer and binder.

Library of Congress Cataloging-in-Publication Data

Nestatsionarnye vzaimodeĭstviia udarnykh i detonatsionnykh voln v gazakh. English.
Unsteady interaction of shock and detonation waves in gases / [edited by] V. P. Korobeinikov; English-edition editor Paul A. Urtiew; [translated by Jamil I. Ghojel].
p. cm.
Translation of Nestatsionarnye vzaimodeĭstviia udarnykh i detonatsionnykh voln v gazakh.
Bibliography: p.
Includes index.

1. Gas dynamics. 2. Shock waves. 3. Detonation waves.
I. Korobeĭnikov, V. P. (Viktor Pavlovich) II. Urtiew, Paul A.
III. Title.
TA357.N44213 1989
620.1'074—dc19 88-39944
ISBN 0-89116-751-X CIP

CONTENTS

PREFACE

The solution of many important theoretical and applied problems in the field of unsteady gasdynamics is gaining special urgency at the present time. This is due to the need to design and construct a number of new-technology devices whose function is related to the flow of high-speed gas streams over bodies of complex shapes and with the propagation of shock waves in channels of variable profile. Such problems as the effect of blast waves on structures, the devastating effect of shock waves in mine explosions, the use of shock and detonation waves for the solution of technological questions in industry, and the launching of impulse high-temperature devices are of urgent practical importance.

Unsteady interaction of shock waves is one of the key unsolved problems of high-temperature physics. It should be noted that the processes, as a rule, are accompanied by the appearance of a developed system of oblique shock waves. In general, the shock-wave pattern is extremely complex and it has only recently lent itself to computation; therefore, the main results of the investigation of such flows have been obtained by experimental methods.

In the last decade, as a result of increasing speed and capacity of computers and the development of highly improved computational schemes, it has become possible to numerically analyze sufficiently complex gasdynamic flows, approaching practically realized situations. This allows, in a number of cases, successfully combining experimental and theoretical investigative methods.

Up to now, a large number of journal publications have been devoted to various questions of unsteady interaction of shock and detonation waves in gases and the need has arisen to combine the greater part of the developed theoretical and experimental results within the bounds of a monograph.

In this book, investigations of the interaction of shock and detonation waves in real gases with obstacles have been presented systematically. The investigations have been carried out using both theoretical and experimental methods. The techniques of acquisition and diagnostics of shock and detonation waves in gases are discussed. The results of investigations of unsteady reflection and diffraction of shock and detonation waves, interaction of shock waves with compliant obstacles, effect of boundary layer on flow structure, and propagation of detonation waves in narrow channels are presented.

Chapters 1 and 2 have been written by T. V. Bazhenova, chapter 3 by T. V. Bazhenova and L. G. Gvozdeva, chapter 4 by Yu. M. Faresov, chapters 5, 6, and 10 by L. G. Gvozdeva, chapter 7 by Yu. P. Lagutov, chapters 8 and 9 by V. N. Lyakhov, section 6.1 by L. G. Gvozdeva and V. N. Lyakhov, and sections 3.7 and 10.5 by L. G. Gvozdeva and V. P. Fokeev.

The authors are grateful to N. N. Yazeva for his help in preparing the manuscript.

CHAPTER

ONE

THERMODYNAMIC AND ELECTROPHYSICAL PARAMETERS OF GAS FLOW BEHIND SHOCK WAVES

1.1 CALCULATION OF THE PARAMETERS OF A REAL GAS BEHIND NORMAL AND OBLIQUE SHOCK WAVES

The parameters of the gas behind a normal shock wave are determined by the laws of conservation of mass, momentum, and energy [1–3]. These laws link the values of gas parameters ahead of the shock with the parameters of the gas behind the shock and with the velocity of shock travel:

$$\begin{aligned} \rho_1 \bar{u} &= \rho_0 u_0 \\ \rho_1 \bar{u}^2 + p_1 &= \rho_0 u_0^2 + p_0 \\ \bar{u}^2/2 + h_1 &= u_0^2/2 + h_0 \end{aligned} \tag{1.1}$$

where ρ, u, p, and h stand for density, velocity, pressure, and specific enthalpy, respectively; the parameters with subscript 0 are for the gas stream flowing into the shock wave, the remaining parameters denote the gas flow behind the wave.

For the determination of the gas temperature, the dependence of internal energy and enthalpy of the gas on temperature must be utilized. In the ideal gas it is given by the simple relation $h = \gamma RT/(\gamma - 1)$, and the system of equations (1.1) together with the temperature dependence of enthalpy has an analytical solution linking the parameters of the gas behind the shock and ahead of it [4].

The dependence of enthalpy on temperature in a gas at high temperatures is complicated, because when the translational temperature of the molecules is increased, part of the energy is expended to excite the internal degrees of freedom of the molecules (vibration, dissociation, ionization). The dependence of enthalpy of the gas on temperature is computed using the methods of statistical physics and quantum mechanics on the basis of experimental data on the optical constants of the molecules. The results of the calculations of thermodynamic parameters of gases are presented in the tables of [5]. Also calculated there are the values of equilibrium constants, which are needed for the calculation of concentration of the dissociated molecules as a function of temperature and pressure.

The calculation of thermodynamic parameters of a real gas at high temperatures requires the numerical solution of a system of algebraic equations linking the composition of the gas with temperature and pressure in accordance with the law of mass action through the equilibrium constant. Calculation methods are described, for example, in [6]. The calculation results of thermodynamic parameters of air are presented in [7–11], nitrogen—in [12], carbon dioxide—in [13].

For the calculation of the equilibrium concentration of free electrons from the given temperature and pressure, the law of mass action is also used.

The equilibrium equations for the reactions of ionization and formation of negative ions ("sticking" of electrons) have the general form:

$$(x_{i+k}x_e)/x_i = (1/p)K_p^{k+i}$$
$$x_i x_e/x_{-k-1+i} = (1/p)K_p^{-k-1+i}$$

Here, x_i denotes the concentration of the ionized components, index k denotes multiplicity of ionization; and the negative values of k correspond to the number of electrons captured by a neutral molecule. K_p with different indexes denote the equilibrium constants of the corresponding reactions. The concentration of the positive and negative ions are, respectively, equal to:

$$\sigma_+ = \sum_{i=1}^{5} x_{i+k}, \quad i \rightarrow O_2, N_2, O, N, NO$$

$$\sigma_- = \sum_{i=1}^{5} x_{-k-1+i}$$

The concentration of free electrons is equal to the difference $x_e = \sigma_+ - \sigma_-$.

The solution of the presented system of equations gives the following expression for x_e:

$$x_e = \left(1/p \sum_{i=1}^{5} x_i K_p^{k+i} \left(1 - p \sum_{i=1}^{5} x_i / K_p^{-k-1+i}\right)\right)^{1/2}$$

The values of x_e for temperatures higher than 4500 K in air are presented in [7–9].

In the temperature region from 2000 to 5000 K, mainly the ionization of nitric oxide affects the degree of ionization of air [7]; therefore, in the given temperature range it can be assumed that $x_i = x_{NO}$, $K_p^{k+i} = K_p^{NO^+}$

The equilibrium constant of the ionization reaction K^{NO^+} is computed from the formula [2]

$$K_p^{NO^+} = \left(\frac{2\pi m}{h^2}\right)^{3/2} k^{5/2} T^{5/2} \frac{Q_{NO} + Q_e}{Q_{NO}} e^{-E/kT}$$

where

$$\left(\frac{2\pi m}{h^2}\right)^{3/2} k^{5/2} = 6.58 \cdot 10^{-7}$$

The ionization potential of nitric oxide $E = 9.5$ eV, Q_{NO^+}, Q_{NO}, Q_e is the statistical sum of the states of molecules NO^+, NO and the electron, respectively. The value $Q_e = 2$, and the ratio Q_{NO^+}/Q_{NO} depend on the structure of molecules NO^+ and NO [5].

In the temperature region 2000–5000 K and pressures $1–10^{-3}$ atm, the effect of the formation of negative ions on the concentration of free electrons can be ignored. Therefore, when calculating the equilibrium concentration of free electrons the following formula can be used

$$x_e^2 = x_{NO} K_p^{NO^+}/p$$

Taking into account that the number of particles in a cubic centimeter at temperature T and pressure p is equal to $n = 2.7 \times 10^{19}\, p\, (293/T)$, where p is in atm, we obtain a formula for the calculation of the number of free electrons $n_e = nx_e$ per cm^3 of air:

$$n_e = 8.06 \cdot 10^{21} \frac{\sqrt{K_p^{NO} p x_{NO}}}{T} \quad (1.2)$$

Taking into consideration the thermodynamics of air, numerical calculations of the parameters of air flow behind a normal shock wave were conducted at temperatures 200–20,000 K and initial pressures from 1 to 10^{-5} atm [14, 15]. Tables were compiled for three values of the initial temperature: 200, 300, and 400 K, since the temperature of the Earth's atmosphere varies with altitude. Thermodynamic parameters of air are presented in tables for temperature behind a shock changing in 100 K increments. For each value of the temperature, the velocity of the shock wave, behind which such a temperature is established at the given initial pressure, and the thermodynamic parameters (pressure, enthalpy, entropy, speed of sound, and effective adiabatic index) were calculated. In the second part of each volume of tables of thermodynamic function [7–9], the specific concentrations of the products of dissociation of air x_i are presented.

Using the tables of [11, 14, 15], one can, from the given values of the initial pressure and velocity, determine the parameters behind a shock wave.

In particular, the concentration of free electrons behind a normal shock wave is calculated from Eq. (1.2) for the values of temperature and pressure determined from the given values of altitude H and velocity of the shock wave v_0. The results of the calculations are approximated by the formulas [16]:

$n_e = 25.2 \cdot 10^{12}\exp(-0.125\ H)\ \exp(0.142\ v_0^3 - 2.84\ v_0^2 + 19.77\ v_0 - 38.33) \pm 30\%$

$0 \leq H \leq 30$ km, $2 \leq v_0 \leq 7$ km/s

$n_e = 15.46 - 0.06\ (H - 20) + 0.463\ (v_0 - 5) \pm 30\%$

$20 \leq H \leq 80$ km, $5 \leq v_0 \leq 9$ km/s

The interaction of shock waves with obstacles include the simplest element—normal reflection of a shock wave.

The state of the gas behind a reflected wave and the velocity of the reflected wave in a real gas can be determined from the laws of conservation written in a system of coordinates, which moves with the reflected wave, when the main condition is accounted for: behind the reflected wave the flow must come to rest. The system of the laws of conservation for a reflected wave is written as:

$$u_r\rho_1 = \rho_2 u_2,\quad \rho_1 u_r^2 + p_1 = \rho_2 u_2^2 + p_2,\quad \frac{u_r^2}{2} - \frac{u_2^2}{2} = h_2 - h_1$$
$$p_2 = \rho_2 RT_2/\mu_2,\quad h_2 = f(T_2, p_2),\quad \mu_2 = f(T_2, p_2) \tag{1.3}$$

Here, u_r is the velocity of gas ahead of the reflected wave relative to the reflected shock wave; and u_2 is the velocity of gas behind the reflected wave relative to the reflected wave (subscript 2 pertains to the conditions behind the reflected wave). Since the gas behind the reflected wave must come to rest ($\bar{u}_2 = 0$, where $\bar{u}_2$ is the gas velocity behind the wave in a stationary system of coordinates), we add to the system of equations (1.3) an additional condition $u_2 = u_R$, where u_R is the velocity of the reflected wave in a stationary system of coordinates; $u_R = u_r - u_1$, u_1 is the gas velocity behind the incident wave in the stationary system; therefore, $u_2 = u_r - u_1$. Taking into account this condition, seven equations with seven unknowns are obtained for the reflected wave. These unknowns are u_r, u_2, ρ_2, T_2, p_2, h_2, μ_2, i.e., if the parameters behind the incident wave are known, a solution can be found by numerical methods similar to the computations of the parameters behind the incident shock wave. For example, the system can be transformed into form

$$u_2 = -\frac{u_1^2 + RT_i/\mu_1 - RT_2/\mu_2}{2u_1} + \sqrt{\left(\frac{u_1^2 + RT_1/\mu_1 - RT_2/\mu_2}{2u_1}\right)^2 + \frac{RT_2}{\mu_2}}$$
$$u_2 u_1 + u_1^2/2 + h_1 = h_2(T_2, p_2)$$

The computation is conducted by the method of successive approximations with the appropriate selection of the values of T [17].

For an oblique wave we add to the system of equations (1.1) one more equation linking the angle of flow deflection in the wave θ_1 and the angle of flow inclination to the wave φ_0. Let the gas flow with velocity u meet an oblique wave inclined at an angle φ_0. The Mach number of the flow is M. The normal components of the flow are u_{0n}, $\bar{u}_n$; the tangential components are $u_{\tau 0}$ and $\bar{u}_{\tau 1}$.

The ratio of the normal velocities of flow ahead of and behind the shock wave is equal to the ratio of the densities: $u_{0n}/\overline{u}_n = \rho_1/\rho_0$.

The tangential components are equal: $u_{\tau 0} = \overline{u}_{\tau 1}$.

The normal components of the velocity are linked with the values of angles φ_0 and θ_1 by: $\mathrm{tg}\varphi_0 = u_{0n}/\overline{u}_{\tau 0}$, $\mathrm{tg}(\varphi_0 - \theta_1) = \overline{u}_n/\overline{u}_{\tau 1}$. From here, it follows that: $\mathrm{tg}(\varphi_0 - \theta_1)/\mathrm{tg}\varphi_0 = \rho_0/\rho_1$.

For an ideal gas this relationship is written in the form

$$\frac{\mathrm{tg}\,(\varphi_0 - \theta_1)}{\mathrm{tg}\,\varphi_0} = \frac{(\gamma - 1)\mathrm{M}^2 \sin^2 \varphi_0 + 2}{(\gamma + 1)\mathrm{M}^2 \sin^2 \varphi_0} \tag{1.4}$$

Thus, there are several equations for an oblique shock wave (1.1) and (1.4), and an equation of state with nine unknowns: ρ_1, p_1, T_1, μ_1, h_1, u_0, u, φ_0, θ_1.

Analytical solutions of such a system are possible only for an ideal gas [4]. Numerical computations for air and CO_2 are presented in [6].

1.2 CALCULATION OF THE PARAMETERS OF AN IDEAL GAS BEHIND A SHOCK WAVE IN HYPERSONIC FLOW OVER A BODY

In hypersonic flow over blunt cones, a detached shock wave forms ahead of the body. The parameters of this wave at the front near the critical point practically coincide with the parameters behind a normal shock wave. The gas velocity behind the shock wave relative to the body falls from those values determined by the law of conservation at the front of the shock wave to zero at the body. The parameters of air behind the detached shock wave were computed by numerical methods in [18, 19]. The shock-wave detachment distance δ from the frontal critical point of a body having a radius of curvature r can be approximated by the formula: $\delta/r = 0.78\ \rho_0/\rho_1$, where ρ_0 and ρ_1 is the density of air ahead of and behind the shock. With the help of the tables of [14] the value ρ_0/ρ_1 can be expressed through altitude H and flight velocity of the body υ_0. This relationship is approximated by the following expressions [16]:

at $20 \leqslant H \leqslant 80$ km, $5 \leqslant u_0 \leqslant 9$ km/s

$$\delta/R = 0.08 \exp\,(-0.083\,(\upsilon_0 - 3)) - 3 \cdot 10^{-4}\,(H - 22.22)$$

at $0 < H < 30$ km, $2 < u_0 < 7$ km/s

$$\delta/R = (H + 1)/\upsilon_0 \exp\,(1.14 \cdot 10^{-4}\,\upsilon_0 + B(H))$$

$$B\,(H) = 1.48 \cdot 10^{-2}\,H^2 - 0.412\,H + 5.37 \text{ at } 0 \leqslant H \leqslant 15 \text{ km}$$

$$B\,(H) = 3.97 \cdot 10^{-4}\,H^2 - 0.78\,H + 3.6 \text{ at } 15 \leqslant H \leqslant 30 \text{ km}$$

The air, compressed and heated by the shock wave, flows from the region near the frontal point of the body to the lateral region. In so doing, the flow expands and cools down. The shock wave is weakened by the expansion waves and is transformed into an oblique shock wave, which deflects the incoming flow parallel to the generator of the cone. The increase of temperature and pressure behind an oblique shock is insignificant; however, the flow velocity relative to the body behind the oblique shock, unlike a normal shock, becomes comparable with the velocity of the free stream.

The temperature distribution along the normal to the axis of the cone is discontinuous, since near the wall a layer of isentropically expanding heated gas is moving from the region behind the normal shock and, at some distance from the wall, it merges with the flow which passed through an oblique shock wave.

The numerical calculation of the flow of real air over blunt cones was conducted by the method of characteristics in [20]. The results of the calculations are presented in dimensionless form, allowing transition to the needed values of the velocity and flight altitude. For the transition to dimensional values the transition coefficients, which depend on the parameters of the undisturbed atmosphere at the given altitude (ρ_0, p_0) and flight velocity (u_0), must be computed.

The distribution of the dimensionless pressure $\overline{p} = p/\upsilon_0^2 p_0$ and dimensionless enthalpy $i = h/\upsilon_0^2$ are given in tables as functions of the dimensionless distance from the body along the normal to the axis $\lambda = (y - y_T)/(y_B - y_T)$. Here, y_T is the distance of the generator of the cone from the axis at the given point x; and y_B is the distance of the oblique shock wave from the axis at the same point. The values of y_B are presented in the tables in [20]. The distance is expressed in radii of the blunt cone r. By using the tables of [20], the distribution of enthalpy and pressure is found. From the given values of the pressure and enthalpy with the help of the tables of thermodynamics of air [7], the value of the temperature and composition of air are found at every point along the normal to the body on the lateral surface of the cone.

1.3 CALCULATION OF THE ELECTROPHYSICAL PARAMETERS OF AIR FLOW BEHIND SHOCK WAVES

The principal electrophysical plasma parameters influencing the propagation of radio waves are dielectric permittivity ϵ and conductivity σ. In the simplest case, by neglecting the collision of electrons with each other and with ions and molecules, we can write the expressions for ϵ and σ in the form [21]

$$\epsilon = 1 - \frac{4\pi n_e^2}{m(\omega^2 + \nu_{eff}^2)} \tag{1.5}$$

$$\sigma = \frac{n_e^2 \nu_{eff}}{m(\omega^2 + \nu_{eff}^2)} \tag{1.6}$$

Here, n_e is the electron concentration; m is the electron mass, and ω is the frequency of the signal.

If we consider that the electron, upon each collision with an ion or molecule, loses on the average all the momentum of the ordered motion $m\upsilon$ (υ is the velocity of the electron), then υ_{eff} is the number of collisions per second.

More exact expressions for υ_{eff} upon collision with molecules and ions are obtained in the kinetic discussion. It accounts for both the dependence of the cross section of the elastic collision of electrons with molecules on electron velocity and the effect of the complex composition of the plasma on υ_{eff}.

The expressions for ϵ and σ have the form [21]

$$\epsilon = 1 - \frac{4\pi n_e^2}{m} \frac{8}{3\sqrt{\pi}} \int_0^\infty \frac{u^4 e^{-u^2}\, du}{\omega^2 + \nu^2(u)} \tag{1.7}$$

$$\sigma = \frac{n_e^2}{m} \frac{8}{3\sqrt{\pi}} \int_0^\infty \frac{\nu(u) u^4 e^{-u^2}\, du}{\omega^2 + \nu^2(u)}, \quad u = \upsilon\sqrt{m/2kT} \tag{1.8}$$

where k is the Boltzmann constant. Equations (1.5) and (1.6), obtained from the classical discussion, coincide with Eqs. (1.7) and (1.8) only when $\upsilon(u) =$ const. Otherwise, one must find the corresponding value of υ_{eff} for each frequency ω, i.e., $\upsilon_{eff} = \upsilon_{eff}(\omega)$, these being different functions for ϵ and σ.

In the special case for a single component plasma, when the cross section is dependent on velocity so that

$$q(\upsilon) = \alpha\upsilon^h \tag{1.9}$$

where α is the constant, h is an integer from -3 to $+3$, and analytical expressions for ϵ and σ are obtained and the corresponding correction curves constructed [22]. In reality, the velocity dependence of the cross section is more complex than (1.9), and one has also to deal with a plasma of complex composition. In this case, the problem is solved by numerical methods.

In this section, the results of numerical calculations of υ_{eff} for air are presented and the region of the frequency ω, at which equations of the classical theory (1.5) and (1.6) can be used, is found together with the needed corrections for the equations outside this region [23].

Weakly ionized air is discussed with the following parameters: the temperature $2000 < T < 10000$ K; pressure $0.001 < p < 10$ atm; electron concentration $10^6 \leq n_e \leq 10^{12}$, 1/cm^3; and $T = T_e$.

The discussed plasma can then be considered to be:

- **Ideal**, if the density of the potential energy will be much less than the density of the kinetic energy of plasma particles, i.e., $e^2/n_e^{-1/3} << kT_e$ or $e^2n_e^{1/3}/kT_e << 1$ (where e is the electron charge; n_e is the electron concentration; k is the Boltzmann constant; and T_e is the electron temperature). In this case the last relation is deliberately fulfilled; therefore, one can use the methods of statistical mechanics, in particular the kinetic equations. The interaction between the plasma particles is not taken into account;
- **Cold**, if the relation $\upsilon_P >> \bar{\upsilon} \sim \sqrt{kT/m}$ is fulfilled (where υ_P is the wave phase velocity; and $\bar{\upsilon}$ is the thermal electron velocity) or $\omega >> 2\pi\sqrt{(kT)/m}/\lambda$;
- **Weakly ionized**, if $\nu_{eff,i}/\nu_{eff,m} < 10^{-2} << 1$ (where $\nu_{eff,i}$ is the effective frequency of collisions with ions [21], and $\nu_{eff,m}$—with molecules); the collisions

of electrons with ions are neglected in comparison with the collisions with molecules.

Additionally, we consider that the following conditions are fulfilled:

- The plasma is continuous, i.e., grad$f$ = 0 (f is the distribution function in the kinetic equation);
- The field is "weak," i.e., $E << E_p$ (where E_p is the characteristic plasma field, mV/m; $E_p = 4.5 \cdot 10^{-3} \sqrt{\Delta T(\omega^2 + \nu_{\text{eff}}^2)}$; T is the plasma temperature in electron volts; ω is the cyclic field frequency; Δ is the average fraction of energy lost by the electron in one collision $\Delta = 10^{-3}$)

When the field is weak we can disregard the temperature change of the electrons as a result of heating by the field, and also consider that in the zero approximation the electrons are distributed according to Maxwellian law.

Since the plasma is considered to be weakly ionized, we neglect the collisions of electrons with electrons.

Let us consider the boundary case of high frequencies $\omega >> \nu(u)$. In this case we have [21]

$$\epsilon = 1 - \frac{4\pi n_e^2}{m\omega^2}, \quad \sigma = \frac{n_e^2 \nu_{\text{eff}}}{m\omega^2} \tag{1.10}$$

As seen from (1.7) and (1.8), the value of ν_{eff} is then equal to

$$\nu_{\text{eff}} = \frac{8}{3\sqrt{\pi}} \int_0^\infty \nu(u) u^4 e^{-u^2} du = \frac{2}{3\sqrt{2\pi}} \left(\frac{m}{kT}\right)^{5/2} \int_0^\infty \nu(v) v^4 e^{-\frac{mv^2}{2kT}} dv \tag{1.11}$$

Thus, in the case of high frequencies, we obtain a definite expression for ν_{eff}. Now let us consider that ν_{eff} does not vary with frequency. In this case, the expressions for ϵ and σ are convenient to write in the form

$$\epsilon = 1 - K_\epsilon\left(\frac{\omega}{\nu_{\text{eff}}}, T\right) \frac{4\pi n_e^2}{m(\omega^2 + \nu_{\text{eff}}^2)} \tag{1.12}$$

$$\sigma = K_\sigma\left(\frac{\omega}{\nu_{\text{eff}}}, T\right) \frac{n_e^2 \nu_{\text{eff}}}{m\,(\omega^2 + \nu_{\text{eff}}^2)} \tag{1.13}$$

Here, ν_{eff} is determined from (1.11), and functions $K_\epsilon(\omega/\nu_{\text{eff}}, T)$ and $K_\sigma(\omega/\nu_{\text{eff}}, T)$ are certain correction coefficients for Eqs. (1.5) and (1.6). It should be noted that in case $\omega >> \nu_{\text{eff}}$, $K_\epsilon \to 1$; $K_\sigma \to 1$, and we arrive at Eqs. (1.10). As shown in [24]

$$K_\epsilon\,(\omega/\nu_{\text{eff}}, T) = \frac{8}{3\sqrt{\pi}}(1 + \omega^2/\nu_{\text{eff}}^2) \int_0^\infty \frac{u^4 e^{-4} du}{1 + (\nu_2(v)/\nu_{\text{eff}}^2)(\omega^2/\nu_{\text{eff}}^2)} \tag{1.14}$$

$$K_\sigma\,(\omega/\nu_{\text{eff}}, T) = \frac{8}{3\sqrt{5}}(1 + \omega^2/\nu_{\text{eff}}^2) \int_0^\infty \frac{u^4 e^{-4} du}{\nu_{\text{eff}}(1 + (\nu_2(v)/\nu_{\text{eff}}^2)(\omega^2/\nu_{\text{eff}}^2))} \tag{1.15}$$

where $\nu(\upsilon) = \nu(u\sqrt{2kT/m})$.

Up to now, we have assumed that the molecules are of one kind. If we have a mixture of M components, then

$$\nu(\upsilon) = n\upsilon \sum_{j=1}^{M} q_j(\upsilon)x_j \tag{1.16}$$

n is the molecular concentration in the plasma; υ is the electron velocity; $q_i(\upsilon)$ is the transport cross section of elastic collisions of electrons with the molecules of the j-th component of the plasma; x_j is the relative fraction of the j-th component. We note that $x_j = x_j(T, p)$, i.e., the correction coefficients in this case depend also on pressure:

$$K_\sigma = K_\sigma(\omega/\nu_{\text{eff}}, T, p)$$
$$K_\epsilon = K_\epsilon(\omega/\nu_{\text{eff}}, T, p)$$

Thus, the problem has been reduced to finding the high-frequency effective collisions frequency ν_{eff} from equation (1.11) as a function of temperature and pressure, i.e., $\nu_{\text{eff}} = \nu_{\text{eff}}(T, p)$, and also functions K_σ and K_ϵ from relations (1.14) and (1.15), taking into account (1.16). In addition, the problem includes the determination of the frequency region where equations (1.5) and (1.6) can be utilized, i.e., it is assumed that $K_\sigma = 1$, and $K_\epsilon = 1$.

We rewrite Eqs. (1.11)–(1.13) in a form more convenient for numerical calculations:

$$\nu_{\text{eff}} = 0.39\,5\,(p/\sqrt{T})I_1 \cdot 10^{10} \tag{1.17}$$

where p is the pressure, mm Hg;

$$I_1 = \int_0^\infty Q(T, p)x^2\exp(-x)dx \tag{1.18}$$

T is the air temperature, eV and Q is the total cross section of elastic collisions of electrons with molecules of air;

$$Q(T, p) = \sum_{j=1}^{M} x_j(T, p)q_j(E) \tag{1.19}$$

where $x_j(T, p)$ is the concentration of j-th component of the plasma at the given temperature and pressure; q_j is the transport cross section of elastic collisions of electrons with the j-th component of the plasma; and E is the electron energy, eV. From (1.17)–(1.19) we obtain

$$\nu_{\text{eff}} = \nu_{\text{eff}}(T, P)$$

$$K_\sigma(\nu_{\text{eff}}/\omega, T, p) = (1 + \nu_{\text{eff}}^2/\omega^2) \int_0^\infty \frac{Q(T, u)u^2e^{-u}}{1 + \frac{9\pi}{16}u\nu_{\text{eff}}/\omega\left(\frac{Q(T, p)}{I_1}\right)}du$$

$$K_\epsilon(\nu_{\text{eff}}/\omega, T, p) = (1 + \nu_{\text{eff}}^2/\omega^2) \int_0^\infty \frac{u\sqrt{u}\, e^{-u}}{1 + 9\pi/16(\nu_{\text{eff}}/\omega)\, u\, Q^2(T, p)/I_1^2}\, du$$

where I_1 is determined from Eq. (1.18) and $Q(T, p)$ from Eq. (1.19).

The main components of air contributing to $Q(E) = Q(T, p)$ are N_2, O_2, NO, N, and O. The cross sections for these substances, which are used in the calculations, are given in Fig. 1.1.

The dependence of coefficients K_ϵ and K_σ and frequency ν_{eff} on pressure is weak, since the concentration of components of air changes slowly with changing pressure. As a result, the entire scale of the pressure can be divided into ranges (see Fig. 1.2) and the pressure assumed constant within the bounds of each range. Therefore, it is sufficient to plot, for each pressure range, a series of curves of $\nu_{\text{eff}}/p(T)$, $K_\sigma(\nu_{\text{eff}}/\omega, T)$, $K_\epsilon(\nu_{\text{eff}}/\omega, \text{T})$. The values of ν_{eff} were computed from Eq. (1.17). The values of $x_j(T, p)$, which are needed to calculate $Q(T, p)$ from (1.19), can be obtained from the tables in [7] and the values of $q_j(E)$ can be obtained from the graphs in Fig. 1.1.

The dependence of ν_{eff} on temperature and pressure is determined not only by the dependence of the cross section of collisions on temperature, but also by the change of air composition due to the dissociation of molecules N_2 and O_2 and formation of oxides of nitrogen. The values of ν_{eff}, thus calculated, differ by more than twice at $T = 5000$ K, $P = 10^{-3}$ atm from the results of calculation for a two-component air from Eq. (1.16) in [24] ($\nu_{\text{eff}}/P = 0.19/T^{1/6}$).

The results of the numerical calculations are presented in Fig. 1.2. The conducted calculations showed that the dependence of K_σ and K_ϵ on temperature is weak. As a result, the dependences of K_σ and K_ϵ on ν_{eff}/ω for an intermediate temperature $T = 4000$ K are presented in the same figure. These plots can be used to determine K_σ and K_ϵ at any temperature of the plasma. The error in this case does not exceed 10%.

From the obtained graphs, it is seen that formulas (1.5) and (1.6) are still valid when $\omega >> 10\,\nu_{\text{eff}}$, where ν_{eff} is determined from (1.17) or from the graphs

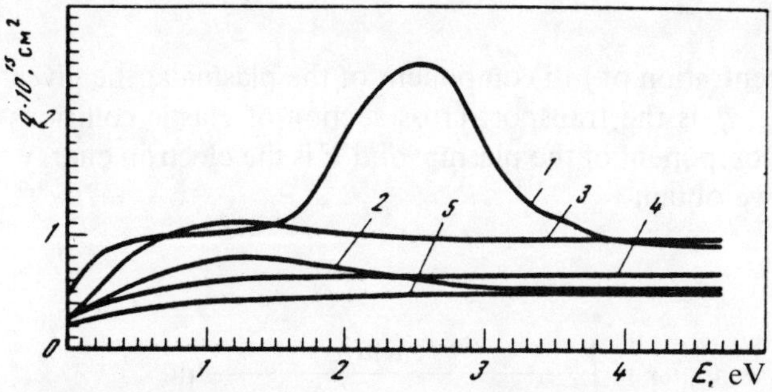

Figure 1.1. Dependence of cross sections of collisions of electrons with the components of air used in calculations *1*) N_2; *2*) O_2; *3*) NO; *4*) N; *5*) O.

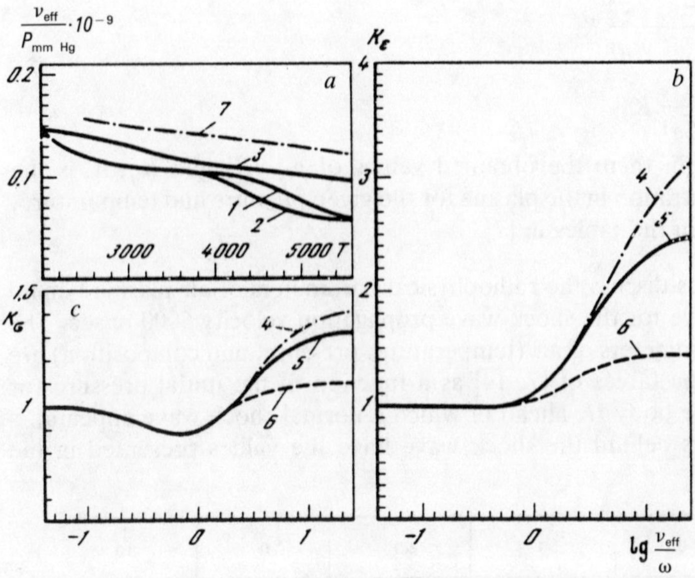

Figure 1.2. Effective frequency of collisions in equilibrium air plasma (*a*) and values of coefficients K_σ (*b*) and K_ϵ (*c*) at $2000 \leqslant T \leqslant 6000$, $10^{-5} \leqslant p \leqslant 20$ atm. *1)* $5\cdot 10^{-3} \leqslant p \leqslant 20$ atm; *2)* $5\cdot 10^{-5} \leqslant p \leqslant 5\cdot 10^{-4}$ atm; *3)* $5\cdot 10^{-4} \leqslant p \leqslant 5\cdot 10^{-3}$ atm; *4)* and *7)* calculation for air with constant composition [24]; *5)* calculation for $T = 4000$ K [23]; *6)* calculation of K_σ and K_ϵ for the case when Q is independent of υ [21].

of Fig. 1.2. When $\omega < 10\ \nu_{\text{eff}}$ the correction coefficients must be taken into account since K_ϵ, for example, can significantly affect the critical signal frequency—the frequency at which the signal is reflected from the plasma, and K_σ affects the attenuation of radio waves in the plasma.

In order to determine ϵ and σ in air behind a shock wave the temperature T and pressure p must be known. If these values are known, the procedure is as follows:

- For the given pressure the necessary graph is plotted $\nu_{\text{eff}}(T,\ K_\epsilon(\nu_{\text{eff}}/\omega),\ K_\sigma(\nu_{\text{eff}}/\omega))$
- Knowing the temperature, the value of ν_{eff}/p is found from the plot in Fig. 1.2*a*, then by multiplying the obtained value by the pressure of the plasma, the required effective frequency of collisions ν_{eff} is determined;
- The value ν_{eff}/ω is found, where $\omega = 2\pi f$ is the cyclic frequency of radio signals. If it is found that $\nu_{\text{eff}}/\omega < 0.05$, then formulas (1.7) and (1.8) are used after substituting in them the obtained ν_{eff}. If, on the other hand, $\nu_{\text{eff}}/\omega \geqslant 0.05$, then the decimal logarithm $y = \lg(\nu_{\text{eff}}/\omega)$ is found,
- Knowing y from the plots in Fig. 1.2*b* and *c*, the coefficients K_ϵ and K_σ are determined, with ϵ and σ computed from equations:

$$\epsilon = 1 - \frac{4\pi n_e^2}{m(\omega^2 + \nu_{\mathrm{eff}}^2)} K_\epsilon$$

$$\sigma = \frac{n_e^2 \nu_{\mathrm{eff}}}{m(\omega^2 + \nu_{\mathrm{eff}}^2)} K_\sigma$$

by substituting in them the obtained values of ν_{eff}, K_ϵ and K_σ; n_e is the electron concentration in the plasma for the given pressure and temperature, determined from the tables in [7].

Example. Let us discuss the radiophysical parameters of air plasma behind a normal shock wave for the shock-wave propagation velocity 5900 m/sec. The thermodynamic parameters of air (temperature, pressure, and composition) are determined from the tables of [7, 14] as a function of the initial pressure (or flight altitude of the body H, ahead of which a normal shock wave appears).

The parameters behind the shock wave have the values presented in the following table.

H, km	40	50	60	70	80
p. atm	0.94	0.38	0.1	0.025	0.0035
T. K	6300	6050	5700	5400	5000
ν_{eff}, 10^{-9} Hz	64.3	25.13	6.64	1.48	0.17
$\omega = 0.47 \cdot 10^9$ (wavelength 4 m)					
lg $\nu_{\mathrm{eff}}/\omega$	2.13	1.72	1.13	0.49	−0.44
K_σ	1.38	1.32	1.31	1.12	0.89
K_ϵ	2.80	2.54	2.44	1.75	0.994
$\omega = 18.84 \cdot 10^9$ (wavelength 10 cm)					
lg $\nu_{\mathrm{eff}}/\omega$	0.53	0.125	−0.46	−1,4	−2.2
K_σ	1.12	0.92	0.93	0.98	0.99
K_ϵ	1.6	1.2	1.0	1.0	1.0

The dependence of ν_{eff} on (T, p) can be approximated, with an accuracy up to 30%, by the formula $\nu_{\mathrm{eff}} = 4.1 \cdot 10^{12}\, p/\sqrt{T}$. In the range $1000 < T < 6000$ the temperature dependence of ν_{eff} is weak, since the decrease of the number of particles in a unit volume is compensated for by the temperature rise and rise of the total cross section of collisions $Q(T)$; therefore, as an estimate we utilize $\nu_{\mathrm{eff}} \simeq 10^{11}\, p$.

For the determination of the dependence of ν_{eff} behind a detached shock wave for a body in motion at altitude H and shock-wave velocity ω_0, the values of T and p behind the shock wave taken from the tables in [14] can be used when the initial pressure p_0 and velocity u_0 are given. The calculation results are approximated by the equations [16]

$$\begin{aligned} &\nu_{\mathrm{eff}} = 6.36 \cdot 10^6 (3.39 u_0^2 + 25.2 u_0 - 9.47)\exp(-0.12H) \pm 30\% \\ &0 < H < 30 \text{ km}, \; 2 < u_0 < 7 \text{ km/s} \\ &\lg \nu_{\mathrm{eff}} = 11.75 - 0.0608\,(H-20) + 0.122(u_0 - 3) \pm 30\% \\ &20 < H < 80 \text{ km}, \; 5 < u_0 < 9 \text{ km/s} \end{aligned} \tag{1.20}$$

The effective frequency of collisions at the front of a spherical shock wave in the atmosphere can be determined with the help of approximation (1.20). For this purpose, use is made of the analytical approximation for the dependence of pressure at the front of the blast wave on the reduced distance and blast energy [25]

$$p = 0.068(1.58 \cdot 10^{-9}\ W/R_1 + 5.4 \cdot 10^{-3}/R_1\sqrt{W/R_1} + 0.0215)$$

where $R = 3.28 \cdot 10^{-3}\ R_1$; R_1 is the distance from the center of the blast, km and W is the blast energy, kcal.

From the value of excess pressure with the help of the tables of gasdynamic functions of air [14], the shock-wave velocity at the given point and the temperature at the wave front are determined.

From the velocity value the value of ν_{eff} is computed at the given altitude from equation (1.20).

CHAPTER

TWO

EFFECT OF NONEQUILIBRIUM PHYSICOCHEMICAL PROCESSES ON THE FLOW PARAMETERS BEHIND A SHOCK WAVE

2.1 RELAXATION TIME OF PHYSICOCHEMICAL PROCESSES. FLOW BEHIND SHOCK WAVES WITH PARTIALLY FROZEN DEGREES OF FREEDOM OF MOLECULES

Analysis of the effect of relaxation processes on gas parameters behind a shock wave is presented in [26–28].

Velocity constants for a large number of reactions are presented in [26, 27, 29]. In [27], the profiles of the parameters behind shock waves in air are computed for velocities from 7000 to 10000 m/sec for an initial pressure $p_0 = 1$ mm Hg. As noted in this work, for states far from equilibrium, the principle of binary similarity is valid and the results of the calculations can be transferred from pressure $p_0 = 1$ mm Hg to another initial pressure by the linear decrease of the scale along the abscissa. Calculation showed that the vibrational temperature of oxygen molecules is close to the translational temperature, and the vibrational temperature of nitrogen molecules is considerably smaller than the translational temperature in the entire nonequilibrium zone.

The calculation of the profile of flow parameters behind a shock wave in a nonequilibrium zone is fairly complicated and, in addition, not all the values of the constants of the elementary reactions are well known. Therefore, for the assessment of the flow state behind the front of a shock wave the integral characteristic of the process is often used. This characteristic is the relaxation time, which is defined as the time it takes to reach equilibrium state for the given degree of freedom at the given temperature.

Relaxation time of the vibrational degrees of freedom and dissociation has been investigated behind shock waves in shock tubes by various methods. What is common in these methods is the investigation of variation of some parameter of the flow behind a shock wave as a function of the distance from the shock-wave front. The relaxation time is identified with the time it takes for the measured parameter to reach a constant value. Evidently, the value of this time is dependent on the degree of accuracy sacrificed by neglecting the difference between the equilibrium value of the parameter and the value of this parameter in the nonequilibrium zone, which approaches the former gradually (often exponentially). Therefore, one must give an exact definition of the approximation one is talking about. Occasionally, the relaxation time denotes the time during which the investigated quantity changes *e* times. When the value of the relaxation time differs from the characteristic gasdynamic time, such a definition gives an indication about the dimensions of the nonequilibrium zone, in which the values of the parameters differ from the equilibrium values by more than *e* times.

The methods of measuring the relaxation time include the interferometer technique of measuring density [30–32], the photoelectric shadow technique of measuring density [33], the method of temperature measurement by the transformation of the spectrum of alkali metal admixture [34, 35], the method of recording of infrared radiation [36], and the method of vacuum-ultraviolet absorption spectroscopy [37–39]. Reviews of data on vibrational relaxation times behind shock waves for various molecules are given in [39–41]. The values of relaxation time of nitrogen vibration in air turned out to be less by an order of magnitude than in pure nitrogen for some regimes [31]. This phenomenon is explained in [26] by the effect of vibrational quanta exchange between the molecules of oxygen and nitrogen. The dependence of relaxation time of vibrational excitation of various molecules on temperature, which is the mean over the relaxation zone, is depicted in Fig. 2.1*a* for atmospheric pressure behind a shock wave. For pressures differing from 1 atm, the relaxation time increases proportionally to the decrease of pressure below atmospheric. As the graph shows, the relaxation time changes relatively smoothly with temperature; therefore, for its evaluation behind the shock wave one can use both the time ascribed to the temperature, which is the mean over the relaxation zone, and to the temperature on the wave front which differs at the reduced regimes by 100–300° from the mean temperature. In the figure, beneath the temperature scale, a scale for the shock-wave Mach numbers is added, at which the corresponding temperatures are achieved. This scale is only nominal since the difference between the mean temperature and the temperature at the front is different for various gases. However, for the evaluation of the time of vibrational relaxation behind shock waves the data in Fig. 2.1*a* can be used.

For the temperature dependence of the time of vibrational relaxation of nitrogen molecules at the wave front, the following empirical formula, giving the value of τ with an error of 30%, has been proposed in [41]:

$$\tau = \frac{T}{p}\frac{1}{273} \cdot 1.26 \cdot 10^{-6}\, e^{230 T^{-1/3}}$$

where p is in atmospheres.

In [42] the following approximation is proposed for the vibrational relaxation time of nitrogen molecules:

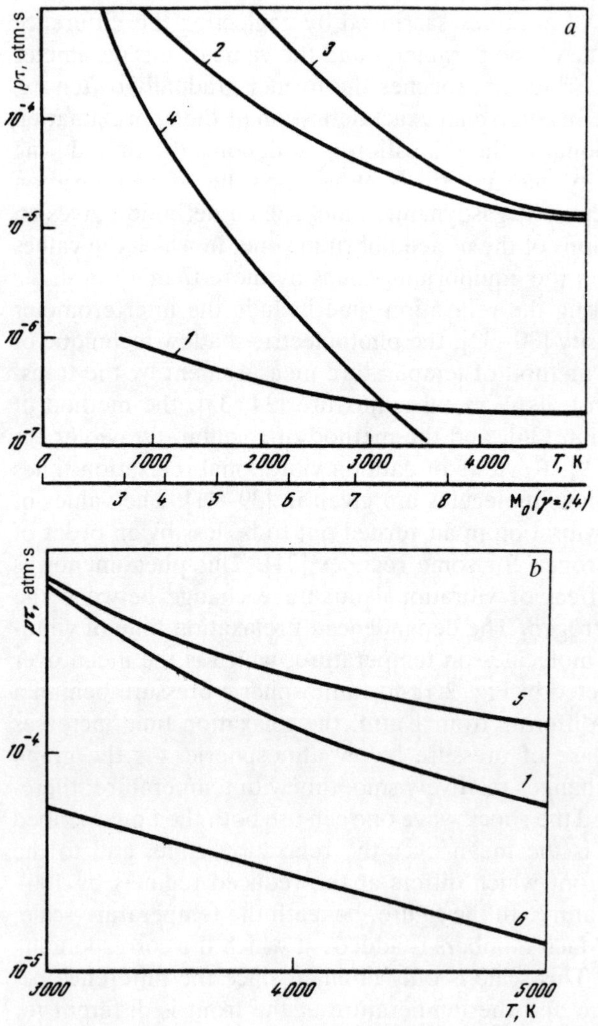

Figure 2.1. Temperature dependence of relaxation time of vibrations (*a*) and dissociation (*b*). *1*) CO_2; *2*) N_2 in air; *3*) N_2; *4*) O_2; *5*) H_2O; *6*) air constituents.

$$\lg(\tau N) = 5 + 133\,T^{-1/3}$$

where τ is in seconds and N is the number of particles per cubic centimeter.

This relaxation times of dissociation of various gases behind shock waves have been studied by the same methods as the vibrational relaxation time. The dissociation relaxation time of nitrogen molecules have been investigated in [30, 33]; of oxygen in [34, 43]; of carbon dioxide in [44–45]; of air in [27, 33, 46, 47]; and of vapors in [48].

Figure 2.1*b* shows the values of dissociation relaxation time of the components of air, carbon dioxide, and water vapor.

Analysis of the magnitudes of the times of dissociation relaxation and excitation of vibrations of molecules (see Fig. 2.1) indicates that those times can differ by an order of magnitude. This indicates the possibility of realizing such regimes, behind a shock wave, at which several of the internal degrees of freedom can be excited in equilibrium near the front, while some others, due to long relaxation time, are practically in nonexcited state at a certain distance from the front. Such states we call states with partially frozen degrees of freedom.

Calculation of flow parameters behind shock waves with partially frozen degrees of freedom of molecules is conducted the same way as the calculation of equilibrium parameters (see chap. 1). However, in this case, the temperature dependence of enthalpy and gas composition is computed by taking into account the degree of freedom which is noticeably excited during the time interval assumed in the calculations. For example, the gas, in which the dissociation time is an order of magnitude larger than the excitation time of vibrations, can be regarded as a gas with frozen dissociation up to such a distance from the wave front which is equal to the dissociation relaxation time at the given pressure by the flow velocity behind the shock wave relative to the stationary wave front.

Calculations of some flow regimes behind shock waves were conducted in conformity with the regimes realized in shock tubes. The calculation results of parameters of air behind incident and reflected shock waves were discussed in [17]. Calculations were conducted with the assumption of equilibrium degree of excitation of molecular vibrations and for frozen dissociation, and also assuming frozen dissociation at low pressure behind the incident wave and equilibrium dissociation behind the reflected wave. In [49, 6] the results of calculations are presented of the state of air, nitrogen, and carbon dioxide behind incident and secondary shock waves, as well as behind oblique shocks and triple configurations assuming equilibrium excitation of molecules and frozen dissociation.

2.2 PROPAGATION OF SMALL PERTURBATIONS IN A RELAXING MEDIUM

As it follows from acoustic equations, the velocity of propagation of small perturbations is determined by the law of pressure change with changing density.

The change of pressure in a volume element during sound propagation is adiabatic; therefore, the velocity of sound is determined by the derivative of pressure with respect to density for an adiabatic process. In the ideal gas

$$a = \sqrt{\gamma(p/\rho)} \tag{2.1}$$

In a real dissociating gas in equilibrium the velocity of sound a_e is determined by taking into account the change of the internal degrees of freedom [50].

The adiabaticity of the process of pressure change in a sound wave can be fulfilled at very low frequencies ω, at which this process slows down. At very slow vibrations the temperature of the volume element of the gas succeeds in lining up with neighboring volumes and the propagation of sound at $\omega \rightarrow 0$ must take place isothermally. However, the condition of adiabaticity is fulfilled for all the audible frequencies of sound since the heat-exchange process is sufficiently slow in comparison with the period of the majority of acoustic vibrations.

Upon transition to high frequencies the speed of sound depends on the molecular processes taking place in the medium. If energy is added to the element of a polyatomic gas, then part of it is converted to vibrational energy, the temperature of the gas decreases, and the value of its heat capacity increases. If the excitation of the vibrations takes place sufficiently slow, then dynamic equilibrium is established between the internal and external energies of the molecules. In this state, the amount of energy transferring from the external to the internal degrees of freedom for a certain duration of time is equal to the amount of energy returned by the internal to the external degrees of freedom during the same period of time.

Such slow flow of processes takes place in the gas during the propagation of sufficiently low-frequency sound. Reducing the frequency below this "zero" frequency does not bring about any change in the process down to the frequency of a few hertz when the process can no longer be considered adiabatic.

For rapid processes of compression and expansion the time for the redistribution of energy decreases. As a result of this, a shift between pressure phases and internal energy occurs during the propagation of high-frequency sound. Additionally, the energy transferred during each half of a period from the external to the internal degrees of freedom of the molecules decreases. At a certain sufficiently high frequency all the energy will remain in the external degrees of freedom without succeeding to transform into vibrational energy. This corresponds to a decrease in specific heat since all the energy is expended on the translational motion of the molecules. Increasing the frequency further does not change the value of the specific heat. The limiting frequency can be considered as infinitely large.

The frequency region, in which the sound speed in the relaxing gas is dependent on frequency, is determined by the excitation relaxation time of the internal degrees of freedom. If the period of vibration, during which there is excess energy in the volume element, is less than the time required for the transition of kinetic energy to vibrational energy of the molecules, then the process occurs as if the internal specific heat was equal to zero.

If the energy expended on the destruction of the molecule is also ascribed to the internal energy, then the presented discussion is valid for the analysis of sound propagation in a dissociating gas. Such an analysis was first done by A. Einstein, who discussed the phase shift ψ arising between the increment of density $\Delta\rho$ and pressure increment Δp in the volume element of the gas

$$\Delta p = (\Delta p)_0\, e^{[\omega(t-x/a)+\psi]}$$

$$\Delta \rho = (\Delta \rho_0)_0\, e^{i\omega(t-x/a)}$$

In this case, the sound speed $a^2 = \Delta p\ \Delta\rho$ can be expressed as

$$a^2 = \frac{V\Delta p}{V\Delta\rho} = -\frac{V\Delta p}{\rho\Delta\rho V} = \frac{1}{\rho}\left(p - \frac{\Delta p(\rho V)}{\Delta\rho V}\right)$$

By representing the specific heat of the gas as the sum of the external heat capacity c_a, which is independent of the frequency, and internal specific heat $c_i \neq 1 + i\omega\rho$, we can show that

$$a^2 = \frac{p}{\rho}\left(1 + \frac{Rc}{c_a + 1 + i\omega\beta}\right)$$

where $\beta = \omega/(k_2\rho)$; k_2 is a coefficient of attenuation.

In this expression the velocity squared is the real part and the absorption coefficient of the medium is the imaginary part [50]. Phase velocity has the form

$$a^2 = \frac{p}{\rho}\left(1 + R\,\frac{c_v + \omega^2\beta^2 c_a}{c_v^2 + \omega^2\beta^2 c_a^2}\right)$$

Instead of the coefficient γ in the equation for sound speed (2.1), the following expression appears here

$$\gamma = 1 + R\,\frac{c_v + \omega^2\beta^2 c_a}{c_v^2 + \omega^2\beta^2 c_a^2}$$

For low sound frequencies ($\omega \to 0$), γ usually has the value $\gamma = 1 + R/c_v$, for high frequencies ($\omega \to \infty$), the expression for γ acquires the form $\gamma = 1 + R/c_a$ and characterizes only the external degrees of freedom of the molecules.

Sound speed for these two limiting cases are sometimes called low-frequency (a_0) and high-frequency (a_∞) velocities:

$$a_0^2 = \frac{p}{\rho}\left(1 + \frac{R}{c_v}\right)$$

$$a_\infty^2 = \frac{p}{\rho}\left(1 + \frac{R}{c_a}\right)$$

For sound frequencies commensurate with the relaxation time the speed of sound has variable values between a_0 and a_∞.

Sound speed plays a substantial role in gasdynamics, because it is with this speed that all disturbances in the flow are being transmitted and it is this sound speed that is used to determine the structure of the gasdynamic flow. Small perturbations which promote the change of the flow state of the flow are as a rule not periodic; therefore, the problem of sound propagation of small perturbations in a relaxing medium is not as evident as the problem of propagation of a sound wave of specific frequency.

The influence of relaxation processes on the structure of gasdynamic flow is discussed in [51–60].

The system of equations of gas motion is discussed in a relaxing medium, where the ordinary equations of gasdynamics are supplemented by relations characterizing the rate of change of parameter q, which denotes the variable chemical composition of the gas or the degree of excitation of the molecules. The gas enthalpy in this case is dependent not only on p and ρ, but also on q:

$$H = H(p, \rho, q), \frac{Dq}{Dt} = -\alpha(q - \bar{q}).$$

The linearized equations have a solution in the form of an exponential curve. For the system to be solvable the following condition must be imposed on the square of the phase sound speed

$$\frac{\omega^2}{k^2} = \frac{-H_p(i\omega + \alpha) + \alpha H_q \bar{q}_p}{(H_p - 1/\rho)(i\omega + \alpha) + \alpha H_q q_p}$$

Broer, in [53], introduced two boundary values of sound speed for the analysis of the given equation:

$$a^2 = -H_p/(H_p - 1/\rho)$$

and

$$\bar{a}^2 = \frac{-(H_p + H_q \bar{q}_p)}{(H_p + H_q \bar{q}_p)^{-1/\rho}}$$

For $q = \bar{q}$, when equilibrium is reached, $\bar{a}$ becomes a. For frequencies not corresponding to condition $q = 0$ and $\bar{q} = q$, the phase and speed are complex and sound absorption takes place.

The arbitrary small perturbation, generated in the flow by small changes of state, apparently represents a set of frequencies from 0 to ∞.

In a relaxing gas it is not possible to determine the speed of the signal as the speed of amplitude motion, since sound absorption occurs. It also cannot be determined as the speed of phase motion, since dispersion takes place and part of the signal, which possesses different frequency, has different speed [57]. For the determination of the speed of propagation of such a packet we must take into account the change in its shape as a result of attenuation and dispersion.

As shown in [58, 59], the wave front propagates with a velocity a_∞ corresponding to the frozen processes of dissociation and vibrational excitation. Part of the wave packet, travelling with velocity a_∞, quickly attenuates so that the wave profile after awhile resembles the wave travelling with velocity a_0. Attenuation time decreases with decreasing relaxation time so that when the latter tends to zero, attenuation becomes almost instantaneous.

According to the estimates presented in [60], the index of the attenuation exponential curve of the high-frequency signal front is equal to $(1 - a_0^2/a_\infty^2)$ (t/τ), where t is the time from the start of travel and τ is the relaxation time. From this equation it follows that, for the regimes of interest to us in CO_2, when $\tau \approx 10^{-6}$ sec, $a_\infty = 1035$ m/sec, and $a_0 = 959$ m/sec, the forward signal front attenuates by one third in 10 μsec, which is sufficient time for observing the propagation of a signal with high-frequency (frozen) sound speed.

The disturbance propagation velocity in the flow behind a shock wave can be judged from the inclination of Mach lines at the edge of the half-wedge and first line of expansion fan.

The problem of the direction of Mach lines in the flow of a relaxing gas was discussed in a number of investigations. From [59], it follows that during the flow over a convex or concave corner, the boundary which separates the undisturbed flow from the region of disturbances is the frozen Mach line M.

References [54, 55] indicate that whereas the characteristics of the equations of motion for classical hydrodynamics determine the laws of propagation of small disturbances and coincide with the Mach lines, the characteristics of the equations of motion for relaxation hydrodynamics do not possess these properties. The direction of the characteristics is determined by the high-frequency (frozen) sound speed; however, small disturbances propagate along this characteristic only at small distances from their origin. The Mach lines further represent some region, whose width decreases with distance. The Mach line at large distances is determined from the equilibrium sound speed. Estimation of the boundary distance, at which it can be assumed that the inclination of the characteristic is determined by the frozen sound velocity, shows that it is possible to detect the propagation of the signal in the flow behind a shock wave with velocity a_∞. To determine this velocity, one must establish the processes for which the sound speed can be regarded as frozen. The process of rotational excitation of molecules, as well as the vibrational excitation, lasts for a finite duration of time. However, the rotational excitation takes place in gases at a distance of 1–2 mean-free-path legends [27]. This distance is the boundary distance for using the concepts of gasdynamic medium continuity. Therefore, we assume that any signal arising in the flow does not include components with a period commensurate with the relaxation time of excitation of molecular rotations, and this process relative to the sound signal must be regarded as instantaneous.

For determination of the flow Mach number in, for example, CO_2, the sound speed for the equilibrium case is calculated by equation.

$$a = \gamma RT\sqrt{(1 + T\sum_1^4 \alpha_i^v)/(1 + T\sum_1^4 \alpha_i^p)}$$

$$\alpha_i^{v,p} = \frac{1}{N}\left(\frac{\partial N_i}{\partial T}\right)_{v,p}; \qquad \alpha_1 = -\alpha_2; \qquad \alpha_3 = -\alpha_4/2$$

$$\alpha_4^p = \frac{Q_1}{RT^2((2/x_4) + (1/2x_3) - (1/2))}$$

$$\alpha_2^p = \frac{(2Q_1/RT^2) + \alpha_4((1/2x_3) + (1/2))}{(2/x_2) + (1/x_1)}$$

$$\alpha_4^v = \frac{Q_2/RT^2}{((2/x_4) - (1/2x_3) - ((1 - 1/T)/2))}$$

$$\alpha_3^v = \frac{(2Q_1/RT^2) + ((1/2x_3) + ((1 - 1/T)/2))\,\alpha_4^v}{(2/x_2) + (1/x_1)}$$

where x_1, x_2, x_3, and x_4 stand for the concentration of the components CO_2, CO, O_2, and O; N_i is the number of particles of the i-th component: Q_1 and Q_2 are heats of dissociation reaction of CO_2 and O_2.

The equilibrium sound speed a_e' for the case of frozen dissociation behind a shock wave is computed from equation $a_e' = \sqrt{\gamma'(RT)/\mu'}$, where γ' is assumed equal to its value at a given temperature accounting for the excitation of molecular vibrations, and μ' is taken without accounting for dissociation.

If the finite time of vibration excitation is taken into account, then one must also account for the high-frequency sound speed a_f' in the flow behind a shock wave. This velocity is computed with the assumption that the gas behind the shock wave is not dissociated but excited, and the value of γ' in the formula for the sound speed is equal to 1.4 for diatomic and linear triatomic molecules.

CHAPTER
THREE

EXPERIMENTAL INVESTIGATION OF UNSTEADY INTERACTION OF SHOCK WAVES IN A SHOCK TUBE

3.1 SHOCK TUBE FOR THE INVESTIGATION OF INTERACTION OF SHOCK WAVES

Interaction of shock waves has been investigated in shock tubes of single-diaphragm type. Shock tube UT-3 is made from a copper waveguide of cross section 72 × 72 mm. The low-pressure chamber (LPC) of this tube is composed of a continuous waveguide section 6 m long with soldered steel flanges. The high-pressure chamber (HPC) is made from steel with a square-section channel of the same dimensions. The HPC is 1.5–2 m long. Since the waveguide from which the LPC is made is 3–4 mm thick, the chamber is strengthened by additional shells of several steel plates to avoid deformation. As a result of such a design the LPC has a sufficiently long channel with good internal finish and no rough joints of separate sections, and, at the same time, acts as resistance to deformatiom which arises in the course of the experiment.

The test section (TS) is situated at the end of the LPC. In this section plane parallel glass sheets replace two opposite walls of the tube so that the shape and dimensions of the tube cross section remain unchanged. The test section also has ports for transducers and for the devices to hold the bodies, over which the flow is studied behind a shock wave. For the investigation of shock-wave diffraction a test section of special design with a convex corner is employed. In some cases, a vacuum reservoir is connected to the LPC of the tube, which eliminates the influence of the shock wave reflected from the tube end on the contact region.

Shock tube UT-6 has a square cross section 100 × 100 mm and is made from welded 10 mm thick stainless steel sections.

A shock tube with a channel of square cross section has some advantages over a tube with round cross section. In the tube with square cross section the flow is uniform in the direction of the probing ray. Optical observation can also be made behind the flow near the tube walls and the distribution of parameters across the cross section observed. The glass windows do not disturb the configuration of the channel, and it is easier to place the transducers on plane walls without distorting the surface of the channel.

In addition to the instrumentation section, one or two duraluminum sections are installed in the tube with transducers, which cannot be mounted in the thin copper waveguide of the low-pressure chamber. Diaphragms from sheet copper 0.3–1 mm thick are used in the tests with a cross precut.

The driver gases used are nitrogen, hydrogen, helium, and a mixture of hydrogen with nitrogen at a pressure of 5–40 atm, supplied by a cylinder with or without a pressure regulator. The pressure in the high-pressure chamber is measured by a standard test manometer. The apparatus has a vacuum system consisting of initial vacuum and diffusion pumps which generate a vacuum of 10^{-2} mm Hg in the tube. There is also a gas supply system which fills the low-pressure chamber with various test gases or their mixtures from cylinders, and the high-pressure chamber with the driver gas.

The initial pressure measurement system is comprised of a standard test vacuum gage, two thermocouple vacuum gages, and a U-shaped glass oil differential manometer. One bend of the latter is connected to the LPC and the other to the inlet manifold of the diffusion pump. Thus, one can measure the absolute initial pressure in the LPC from 0 to 40 mm Hg and from 0 to 1 atm. The accuracy of measuring the initial pressure is 3–5%. The leakage during the test period (2–5 min) amounts to 0.1 mm Hg. Therefore, the admixture of air in the test gas as a result of leakage amounts to 1% at an initial pressure of 10 mm Hg.

Figure 3.1 depicts a schematic diagram of a shock tube and the principal measuring devices. The high-pressure chamber (*1*) is connected to an exhaust and supply system. The low-pressure chamber is composed of sections (*2*) and (*3*). Section (*2*) is connected to the vacuum system and the test-gas supply system. Section (*3*) is equipped with transducers for the measurement of velocity with the help of a frequency meter (*4*) and a starter system (*5*). Section (*3*) is the test section. It is equipped with optical windows and mounting ports for the measuring sensors. The schematic diagram also shows the location of the shadowgraph device IAB-451 (*10*) with the light source (*7*) and a camera ZhFR (*9*), and also the voltage follower (*6*) and oscilloscope (*8*) for recording of the gauge signals.

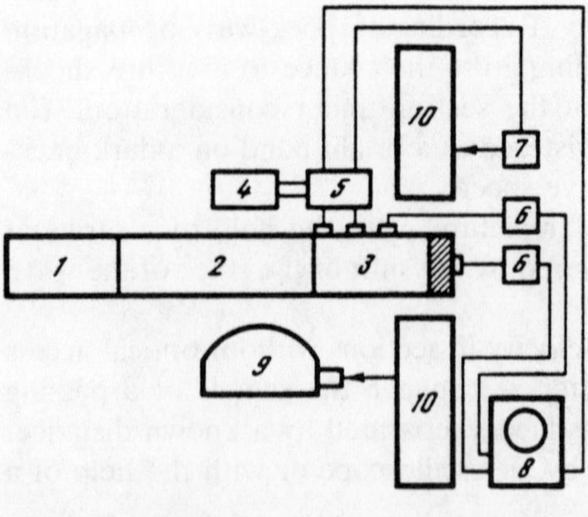

Figure 3.1. Schematic diagram of a shock tube and principal measuring devices.

3.2 OPTICAL METHODS OF VISUALIZATION AND RECORDING OF FLOWS IN SHOCK TUBES

When working on shock tubes with square or rectangular cross section, we practically deal with a one-dimensional flow in which a light ray passes basically through a uniform mass of gas. This simplifies the problem of interpretation of the shadow graphs of the process [61].

Investigations of shock wave interactions were conducted by the schlieren technique and the interferometry method using the IAB–451 device. High-speed filming was made of the processes in the test section and, as a result, streak records of the process were obtained. In the latter case, the IFK–50 impulse lamp was used as a light source and recording was made on an RF–3 film in the ZhFR camera.

The streaking method has been modified to investigate the relative location of angles between waves in triple-shock self-similar triple-wave configurations. For this purpose, the slit is located along the reflecting surface with which the interaction takes place. Formulas have been derived which allow the linkage of the inclination angles of the lines on the scan and the angles in triple configurations resulting from reflection [49].

A method had been suggested [6] in which the scanning slits were arranged along the generators of the obstacles. This allowed the simultaneous measurement of the propagation velocity of shock waves along different sides of the obstacles when investigating the diffraction of shock waves.

For the light source during the forming of the process, a special circuit was built—a thyratron TG12–400/16, giving an electric discharge in air lasting for about 1 μsec. In this case filming is made on a wide aerial film fitted into a casette. Frame sequence is obtained by exposing the same regime on frames which are shifted in time.

To obtain a series of four frames a modified Cranz-Schardin method was used [62].

Measurement of shock-wave velocity. Recording of shock-wave propagation by the Töepler method using slit scanning gives the chance to measure shock-wave velocity over the entire length of the section under consideration. The inclination of the shock-wave trace, registered as a bright band on a dark background, is determined by the shock wave speed.

During the recording of the Töepler picture, with the help of a series of exposures, shock-wave velocity is measured by the shift of the trace of the wave front from frame to frame.

The measurement of shock wave velocity in sections without optical access is conducted by recording the time interval between the signals of a passing shock wave between two pressure transducers separated by a known distance. This time interval is usually measured by an oscilloscope or with the help of a frequency meter.

The accuracy of velocity measurement by the distance method is determined by the accuracy of measurement of the time interval by a frequency meter ($\pm$ 1 μsec) and the reference distance of 10–20 cm ($\pm$0.5 mm). Generally, the error of measuring shock-wave velocity by the distance method amounts to 3–5%.

Due to blurring of the contact surface, the error in measuring the velocity of contact surface and dimensions of the contact region is greater than the error in measuring the shock-wave velocity, amounting to 5–10%.

In the area outside the viewed section, the location of the contact surface can be determined by the schlieren method through small windows.

3.3 PRESSURE MEASUREMENT BEHIND A SHOCK WAVE

Transducers based on piezoelectric ceramics TsTC-19 are used to measure velocity and pressure following a trigger of an electronic device operating in the waiting mode through a proper synchronization block. The sensitive elements of the crystals have a cylindrical form. The larger the cylinder height, the larger the electrical pulse received from the transducer, i.e., the higher its sensitivity. Increasing the diameter of the crystal leads to an increase of the power of the transducer, on one hand, and to the deterioration of spatial resolution, on the other hand. Thus, for example, when the crystal diameter is 5 mm and the shock-wave velocity is 2.5 km/sec, the pulse rise from the pressure tranducer lasts for about 2 μs, instead of being practically instantaneous. When the interaction of shock waves is investigated, cylindrical sensors of 1–1.5 mm diameter and 1–5 mm high are used [63].

The majority of designs of piezoelectric pressure transducers can be arbitrarily grouped into two principal types: with dissipating waveguide and with free suspension of the piezoelement in viscous medium, such as beeswax. In the first type, the distortion of the signal caused by the reflection of acoustic disturbances from the interface of the element and the waveguide is significantly reduced by the rational choice of a waveguide material possessing acoustic properties approaching the properties of the crystal itself [49]. These types of transducers are as a rule, considerably bulky.

The majority of the designs of pressure transducers of the second type are characterized by low mechanical strength, low stability of readings, and unsatisfactory transient characteristics. The latter leads to considerable "jitter" of the electric signal, which renders difficult the measurements directly behind the shock wave [64]. This disadvantage can be alleviated considerably by decreasing the dimensions of the piezoelement to 0.5–1 mm and by using an appropriate method for damping its natural oscillations.

To determine the shock-wave velocity in the shock tube and the pressure behind incident and reflected shock waves, and to control the continuity of flow of the air plasma behind the shock wave, a special miniature piezoelectric pressure transducer was constructed. Piezoelectric ceramic material TsTC–19 served as the sensitive element for the aforementioned transducer.

Figure 3.2 depicts a section of the schematic of such a transducer. It is composed of the housing (*1*) in which a piezoceramic element (*2*) is mounted. The latter is fixed onto a disk-shaped electrode (*3*) which has a threaded part at its aft. Pressure sealing of the transducer is achieved by a fluoroplastic gasket (*4*) when the nut (*5*) is tightened through an insulating washer (*6*). In order to smooth out the sensitive surface, which is important when measuring static pressure in supersonic flow, a close-fit cup (*7*) is mounted on the housing. A hole in the cup secures a clearance between the cup and the sensitive element. Electrical contact of the element with the casing of the transducer is achieved by a thin metallic membrane (*8*), which also serves to transmit pressure changes to the sensitive element.

The difference between this piezoelectric transducer and other transducers is in the electrode (*3*), designed as a disk, which acts as an effective dissipator of parasitic acoutic disturbances which arise when the pressure changes abruptly and distorts the signal from the transducer. The propagation of disturbances here takes place in the radial direction and not in axial direction as in the majority of known designs. Acoustic disturbances undergo repeated reflections from the surface of the electrode (*3*) during their travel, as a result of which their fronts are blurred and are generally attenuated. Additionally, the developed design is distinguished by its simplicity which makes it easy to mount the transducer and is less costly. Electrode (*3*) ensures reliable sealing of the transducer and si-

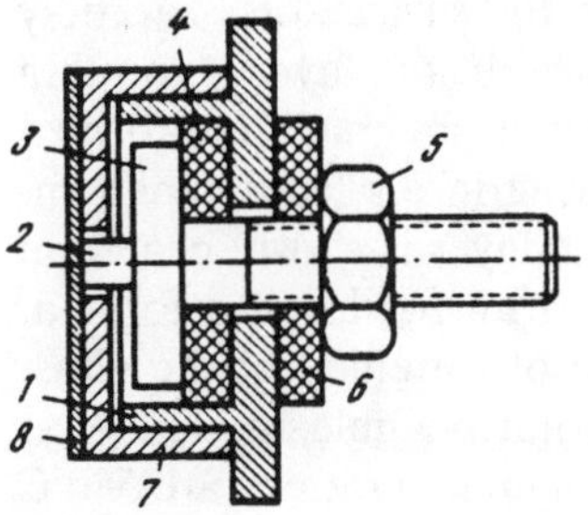

Figure 3.2. Schematic of a piezoelectric transducer.

multaneously serves as an outlet terminal for the transducer signal. The maximum dimensions of the piezoelectric pressure transducer do not exceed 10 mm and the size of the sensitive element is 1 mm; the frequency of natural resonance oscillations was about 200 kHz.

Similar piezoelectric transducers were used for triggering the synchronizing electronic device in the measuring system of shock-wave velocity.

In order to match the high output resistance of the transducer with the low input resistance of the recording device, appropriate voltage followers were used.

Voltage followers are constructed on the basis of integral operational amplifiers of type K544UD1A and are mounted in the immediate proximity of the measuring piezoelectric transducers. The gain of the indicated followers amounts to 0.995 in the frequency band of 0–1 MHz. The standard value of the input resistance is 10^9 Ohm. For matching the outlet resistance with the wave resistance of the coaxial cable, when the recording instrument is at a considerable distance (5–15 m) from the experimental device, voltage followers with additional power amplifier stages are used. The latter are based on complementary pairs of high-frequency transisitors KT315B, KT361B which ensures complete elimination of parasite "chattering" of the signal on the oscilloscope screen.

The triggering of the recording instrumentation on the UT-6 apparatus is accomplished with the help of pulse shapers with regulated operational thresholds (to prevent false triggering of the system by acoustic signals emitted by the diaphragm rupture). High-response operational amplifiers of type K574UD1 are employed in the circuits of the shapers. This allows to obtain stable amplitude of the trigger pulse ~30 V for a front duration of $\lesssim 1$ μs.

3.4 MEASUREMENT OF HEAT FLOW INTO THE WALL OF A SHOCK TUBE

To measure the temperature of the shock tube wall and to determine the heat flow in the tube wall from the time dependent temperature, thin-film resistance thermometers are used [65]. We shall describe one of the resistance thermometer types.

Resistance thermometers are usually made as follows. Following the formulation developed in [66], a polymerizing varnish is obtained by mixing platinum hydrochloride acid, lavender and rosemary oils, and absolute alcohol. Next, a layer of the varnish is applied by a small brush on the surface of a carefully polished and degreased glass. If there is too little varnish, the layer is too thin and porous. If, on the other hand, there is too much varnish, cracks are formed and the layer does not hold to the glass. After drying in air (sometimes immediately in the flame of a glass-blower), the varnish layer is slowly heated in a ventilated oven, and, at a temperature of 430°C, a mirror layer is formed. After that it is sufficient to heat to the temperature of softening of the glass. The layer by now is so strongly attached that it is not possible to remove it without damaging the glass. The recommended heating temperatures are 580°C for soft glasses and 600°C for hard glasses, and from 650 to 680°C for pyrex and quartz. The wire leads to the formed mirror film are either soldered or glued. In some cases, deposited silver strips are used as an electric terminal, and sometimes platinum wires are used as an electric terminal, and sometimes platinum wires are embedded beforehand in the glass piece and then polished with it. Chemical deposition yields very strong and thin (0.1 μm) films. However, the characteristics of the sensors (their resistance and temperature coefficient) vary within a wide range. Some researchers tend to avoid this by growing a film to the needed thickness (needed resistance) by repeated deposition of thin layers and heating every time until a mirror layer is formed. However, it is true to say that these sensors change their parameters after heating, and this change cannot be predicted. The described technique is the simplest and can be employed in laboratory conditions.

Efforts to obtain these films by the vacuum deposition techniques are well known [67]. In this case, the thickness of the deposited film is easier to control; however, the films are fragile and are easily damaged. It seems that the film, in this case also, must be heated following deposition; however, investigations in this field are insufficient. Sensors can also be made by the cathod deposition technique [68].

Using a thin-film resistance thermometer, one can measure the temperature T of the sensor's surface with reasonable accuracy as a function of time τ. For the analysis of the obtained curves and computation of heat flow q as a function of time, the following integrodifferential equation can be used:

$$q(t) = \frac{\lambda}{\sqrt{a\pi}} \int_0^t \frac{d\,T(\tau)}{\sqrt{t-\tau}\,d\tau}\,d\tau \qquad (3.1)$$

where λ and a are coefficients of thermal conductivity and diffusivity.

If the experimental curve $T(\tau)$ cannot be approximated by some simple formula, then a piecewise linear intepolation is adopted. Hence, dependence (3.1) acquires the form

$$q(t) = \frac{2\lambda}{\sqrt{a\pi}} \sum_{i=1}^{n} \frac{T_{i+1} - T_i}{\sqrt{t-\tau_{i+1}} + \sqrt{t-\tau_i}}$$

where T_i and τ_i are the values of temperature and time at the beginning of each i-th segment.

If the number of divisions n is sufficiently large, the dependence is further simplified:

$$q(t) = \frac{\lambda}{\sqrt{a\pi}} \sum_{i=1}^{n} \frac{\Delta T_i}{\sqrt{t-\tau_i^{m}}} \qquad (3.2)$$

where $\tau_i^{m} = (\tau_{i+1} + \tau_i)/2$

The most effective way of computing heat flow is by using a computer, where for direct input one uses the experimental oscillogram, the calibration relation between the beam deflection of the oscillograph and the temperature, and formula (3.1) for integration.

When the magnitudes of heat flow in the shock tube wall determined from the temperature, which is measured by the film thermometer, the coefficients of thermal conductivity and diffusivity of the material of the sensor base are used. Normally the material of the sensor base is different from the wall material. In [69] an analysis was made of the effect of the difference between the material of the sensor base and the material of the wall on the relation between the wall temperature and heat flow. This relation varies depending on the location of temperature measurement: on the lateral surface of the shock tube or at the end of the tube.

Heat is transferred between the stationary gas behind the reflected wave and the shock-tube end primarily by conduction. In this case, the dimensionless surface temperature is inversely proportional to the coefficient of thermal activity of the material of the end wall, and the heat-flow density is independent of the wall material.

When the temperature is measured by a film thermometer on the lateral tube surface behind the incident shock wave, heat is transferred mainly by convection. The temperature profile of the thermal boundary layer behind the shock wave is determined by the thermophysical properties of the walls of the channel along which the stream moves. The temperature of the sensor surface must be close to the temperature of the channel surface with the exception of the time during which a layer of gas flows over the sensor, touching its base only. Since the surface area of the sensor is small, this time is commensurate with the inertia time of the sensor (1–2 μsec). Therefore, when analyzing the results of temperature measurements on the wall surface behind an incident shock wave from formula (3.2), one must take into account the coefficient of thermal activity of the metallic wall, and not the glass base. This concept received qualitative confirmation in experiments with sensors having bases from different materials [69].

3.5 FORMATION OF A SHOCK WAVE IN A SHOCK TUBE

The shock wave in a shock tube at the moment of its formation near the diaphragm has a complex form which gradually approaches a plane wave as it moves away from the diaphragm. Indeed, let, for example, the diaphragm be burst at the center (as is usually the case). Since the time for complete opening of the diaphragm is finite and is longer than the time needed for the sound perturbations to traverse characteristic distances, the process of complete opening can be regarded as taking place in several stages [70]. In the first stage, the burst diaphragm resembles a nozzle with the smallest opening at the end (sonic nozzle). When the gas flows out of such a nozzle into a low-pressure gas, a spherical shock wave is formed. In the second stage the diaphragm is further opened, forming a supersonic nozzle. The spherical shock wave continues to propagate farther, and the parameters of the issuing gas change continuously as a result of change of the "nozzle" shape. At a certain stage of the flow a discontinuity is formed inside the nozzle, which degenerates into a sonic disturbance as the diaphragm opens more. During the opening of the diaphragm the parameters behind the developed shock wave will change. The shock wave itself impinges on the walls of the tube.

Let us consider the situation in a cross section. The shock wave is spherical or cylindrical (depending on whether the diaphragm is burst in the center with a point hole or as a slit). It interacts with solid walls of the tube and is reflected. At the very beginning, as with the incidence of a shock wave of an explosion on a surface, the wave is reflected in a regular manner. As the shock wave travels along the channel, the angle between the tangent to the wave at the incidence point and the tube wall becomes larger. The wave is reflected from the wall in an irregular manner, i.e., a triple point forms that commences to shift away from the wall. A similar situation arises on the upper wall of the tube. The irregular reflection straightens sections of the wave, located at the tube surface, making them perpendicular to the surface. When the collision of triple points occurs on the axis (assuming picture of symmetry), the result is analogous to the collision of two Mach waves which are subtended at some angle. The process of irregular wave reflection commences again and the middle section of the wave begins to straighten up during the reflection process, and the triple points move downwards. When a triple point reaches the tube surface, irregular deflection takes place again.

During this process, gradual straightening up of the shock wave takes place and it is transformed to a plane wave. This straightening up occurs due to the formation of reflected shock waves which travel across the shock tube colliding with each other, drifting apart, reflecting from the walls, and colliding once more. It must be noted that during such reflections contact entropy discontinuities are also formed, which can lead to turbulence.

Distortion of a shock-wave front can also take place in a shock tube at all nonuniformities of the tube channel [71]. Quite often, shock tubes are so designed that certain transitions exist near the diaphragm, such as from a round section to a square section and other such nonuniformities. Sometimes nonuniformities arise at the joints of optical windows with the surface of the test chambers. The presence of all nonuniformities leads to the distortion of the incident shock-wave front. Transition to a plane wave occurs by successive reflections of the wave from the walls of the shock tube. The shock wave, in this regard, is stable. It is interesting how fast the formation of a plane shock wave takes place. The question about the stability of the wave was theoretically studied in [72, 73]. Discussed was a one-dimensional problem about the change of shock-wave form after it traverses a channel with a slight convergence-divergence. The process of wave straightening was judged by the maximum axial deflection ξ of the wave front from the plane surface. In [72], an expression was obtained for the law of attenuation of disturbances as a function of distance x, which the wave traversed from the location of the obstacle. It turned out that disturbances are attenuated vibrationally. The amplitude of the vibrations decreases with distance x from the disturbance for large x such as x^{-n}, where $n = 3/2$ for weak shock waves and $n = 1/2$ for strong waves (weaker attenuation). Experiments to test this theory were conducted in [74] for weak waves and in [75, 76] for strong waves.

The experiments for weak waves were conducted at Mach numbers 1.60 and 1.41. Measured were the value of "total deflection" ξ_T, i.e., the sum, irrespective of the sign, of maximum and minimum values of ξ along the shock wave at some moment in time.

It was concluded that, for Mach numbers $M = 1.41$ and $M = 1.60$, ξ_T varies periodically, the periodic function resembling sinusoidal attenuation according to the law $x^{-3/2}$.

The attenuation of strong shock waves was investigated experimentally by Bowman [75], who showed that the oscillations are suddenly attenuated at some distance. They become comparable with the oscillations obtained due to the influence of the boundary layer. This distance is dependent on the initial pressure. Bowman obtained an empirical rule for the determination of the critical distance at which break-down of oscillations occurs: $0.6\, p_0 R^2 \leqslant L_{cr} \leqslant 2.3\, p_0 R^2$. Here, L_{cr} is the distance traversed by the wave from the disturbance, mm; p_0 is the initial pressure, mm Hg; R is the tube radius, mm. The abrupt attenuation of oscillations is explained by the effect of the boundary layer. Bowman deduced this expression by assuming that L_{cr} is the distance at which the thickness of the boundary layer displacement $\delta = (l/10p_1)^{1/2}$ amounts to 0.25–0.5 of the tube diameter.

This expression for the displacement thickness corresponds roughly to the laminar boundary layer in air and argon. The boundary layer is a fictitious wall affecting mainly the oscillations. Thus, the attenuation for strong shock waves is fairly weak. At a distance of around 50 R the value of ξ/R, according to the experiments of Bowman, is equal to 0.1. For high initial pressures the value of L_{cr} is large and a break-down of the oscillations does not take place.

Therefore, it is important in a shock tube to eliminate abrupt flow disturbances existing along the tube length, since any transition from a round cross section of the tube to a square or similar sections causes disturbances which affect the shock wave for a long time by distorting it.

Lately, the question of the front structure of strong ionized shock waves has been given great attention. The specific features of the wave form, arising in strong nonequilibrium conditions of the gas, have been found ahead of and behind the front of the shock wave [77, 78].

It is interesting to note that under the same conditions the shock wave experiences an increasing effect of the disturbances resulting from nonconformities on the surface along which the shock wave propagates. The attenuation of the disturbances depends on the type of gas. It appears that the stability of the shock wave decreases as the adiabatic index decreases. In carbon dioxide and polyatomic gases, shock waves retain their nonlinear form longer. As an example of a nonuniformity on the surface, we shall discuss a cylindrical segment.

The investigation of the interaction of shock waves with cylindrical segments was carried out in shock tube UT-3 with a square section 72 × 72 mm. Mach numbers were varied in the range of 1.5–6 and the initial pressure was from 10 to 30 mm Hg.

Segments of cylinders were used as models with a diameter of 123 mm. They differed from one another by the angle of interface α_0 between the flat and cylindrical surfaces. These angles varied from 71 to 4°. For an angle $\alpha_0 \simeq 4°$, a linear approximation can be used in the analytical solution of such a problem.

Wave configurations for two consecutive moments in time corresponding to the location of the incident shock wave (propagating from left to right) on the front (*I*) and rear (*II*) slopes of the segment are presented in Fig. 3.3. If the conjugate angle λ_0 between the cylindrical surface and the flat plane is less than critical, then Mach reflection takes place—a triple point forms which begins to move along the trajectory OA_1. When the Mach wave propagates along a curved surface its diffraction takes place. Further development of the process leads to the formation of a steady flow over the surface by the incoming stream behind the shock wave travelling at Mach number M_1. A new triple point A_3 appears on the passing wave the moment it leaves the segment as a result of reflection of the diffracting wave from the horizontal plane. Thus, as a result of the interaction with the segment, the shock wave is deformed and breaks appear on it (at the location of triple points), from which reflected shock waves and contact surfaces originate.

When experiments were carried out, consecutive shadow photographs and interferograms of the interaction region were obtained. Argon, nitrogen, and carbon dioxide were used to simulate the various depths of physicochemical processes in the gas behind the shock wave. For the range of velocities and times, in the course of which shock waves interact with the segment, no excitation of molecular vibrations and dissociation takes place and nitrogen behaves like an ideal gas ($\gamma = 1.4$). In carbon dioxide, excitation of molecular vibrations takes place without dissociation, and if the concept of effective adiabatic index

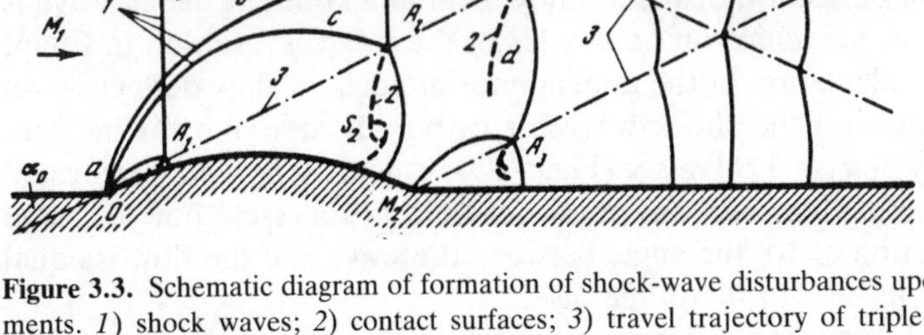

Figure 3.3. Schematic diagram of formation of shock-wave disturbances upon interaction with segments. *1*) shock waves; *2*) contact surfaces; *3*) travel trajectory of triple points; *I*) shock wave reflected as a Mach wave on the front slope of the segment (A_1-triple point in Mach configuration); *II*) wave is diffracted on the rear slope of the segment (A_2-triple point, S_2-vortex); *III*) reflection of wave section A_2M_2 from the lower surface of the channel (A_3-new triple point).

is used, the parameters of the hot gas behind the shock wave correspond to $\gamma = 1.18$. For argon the value $\gamma = 1.67$ was used.

The effect of the physicochemical processes on the interaction of shock waves can be described as follows. Endothermal physicochemical processes, occurring in the gas behind the shock-wave front at high temperature, lead to a decrease of the adiabatic index and increase of the density ratio at the shock-wave; as a result, the Mach number of the incoming flow increases. Hence, for example, when the shock wave travels in the shock tube in carbon dioxide and the Mach number of the incident shock wave is increased to 10, the Mach number of the incoming flow reaches the value of 1.8 if the observation time is small and the physicochemical processes have not been completed. For the same Mach number of the incident wave the Mach number of the incoming flow becomes equal to 3 if the time is sufficient to reach complete equilibrium. In the meantime, physicochemical processes are known to have little influence on the pressure behind the front of the incident shock wave because an increase in density leads to a simultaneous sharp decrease in temperature.

Many processes of shock-wave interaction can be studied on the basis of the decay of an arbitrary discontinuity. Here, the shock polars are plotted in the pressure–velocity plane, and the compatibility conditions of equal pressure and normal velocity along contact surfaces are assumed. Since physicochemical processes have little influence on pressure and significantly alter the gas velocity, the influence of the physicochemical processes on the configuration of shock waves resulting from their interaction is considerable.

The following elements can be singled out in the flow pattern (see Fig. 3.3): the head wave at the leading edge of the segment (*a*), inclined section of the reflected shock wave (*b*), triple-wave configuration (*c*), and lines of contact discontinuity (*d*). The effect of the physicochemical processes on the separate elements can be observed. For the head wave, increasing the velocity of incoming flow leads, evidently, to a decrease of velocity of the reflected wave at the initial stage and decrease of detachment distances at the steady state. The wave appears to be pressed to the body by the incoming flow. The effect of decreasing the adiabatic index on the configuration of the wave is analogous to the effect of increasing the Mach number of the incident wave. The inclined shock wave is pressed to the body at the stage of regular and Mach reflections the same way. Indeed, the angle of deflection of the gas flow behind an oblique shock wave is dependent on the gas velocities on both sides of the wave. Presented in Chap. 5 are the results of calculation of the dependence of angle of flow deflection on the angle of incidence on the shock wave for carbon dioxide and for incident shock-wave velocity of $u_0 = 1.9$ km/sec (Fig. 5.8*b*). Calculations were performed with and without taking into account physicochemical processes. For the same angle of flow deflection of 10° the angle between the wave and the flow is equal to 30° in one case, and 60° in the other case.

The effect of physicochemical processes on the location of waves in triple configurations are particularly significant. In Chap. 5, Fig. 5.8*a* shows the dependence of the inclination angle of the reflected wave ω_2 on the inclination angle of the incident wave ω_1 (the angles are defined with respect to the trajectory of the triple point) for carbon dioxide, $u_0 = 1.9$ km/sec, as computed for various values of the effective adiabatic index. It appears that angles ω_2 change not only in magnitude by 30–40°, but also in sign, which leads to an abrupt change in the configuration as a whole. Thus, upon reflection from a wedge, the effect of the physicochemical processes leads to the appearance of new Mach configuration, the so-called double Mach configuration.

Experiments show that when a shock wave interacts with a cylindrical surface, the change in the flow pattern as a result of physicochemical processes is analogous to the observed change on a flat wedge. The head wave and the inclined part of the shock wave are further pressed to the body, the stage of regular reflection is tightened, and a double Mach configuration appears instead of a simple Mach configuration, i.e., an additional disturbance occurs on the reflected wave. With the appearance of a Mach reflection the shock waves intersection point begins to move along a curvilinear trajectory on the cylinder surface The trajectory of the triple point in a gas in which physicochemical processes occur lies nearer to the body than in a gas with undisturbed degrees of freedom.

Contact surfaces, arising as a result of Mach reflection, are turned into large-scale vortexes (S) (Fig. 3.4). In a gas with excited internal degrees of freedom the vortexes reach large dimensions and are quickly removed from the surface of the body. Figure 3.4 depicts Schlieren photographs of a shock wave in nitrogen interacting with a segment. The photographs show all the indicated features of the process. Figure 3.4*a* shows the stages of Mach-wave reflection, and Fig. 3.4*c*—the diffraction stage. Figure 3.4*b* shows the manner in which large-scale vortex formations appear in the flow behind the shock wave.

Generally, physicochemical processes occurring in a gas behind a shock wave lead to a noticeable change in flow pattern, particularly to sharper breaks on the passing wave, more intensive reflected waves, and higher flow turbulence. Any curvatures on the surface of the channel can be represented as a set of several cylindrical surfaces. If the discontinuity of the surface of the channel is small, i.e., the conjugate angle α_0 is small (see Fig. 3.3), then the disturbances in the flow arise the same way as upon the interaction with thicker cylindrical obstacles. Triple points cause the distortion of the passing wave, and the contact surfaces which turn into a vortex cause flow turbulence. Further movement of the shock wave along the tube leads to the restoration of the plane form. This takes place as a result of repeated reflections of the wave from the tube walls.

The greater the progress of physicochemical processes, the lower the trajectory inclination of the triple point. Therefore, the disturbance of the shock-wave form in a gas with excitation is maintained longer and the intensity of the appearing vortexes is increased. We note that the restoration of the plane form of the wave takes place in such a manner that the wave becomes more of a plane wave at some distance from the disturbance, then its form is distorted, then it again becomes a plane wave, and so on. Usually, the test section, in which observations are carried out, is located at a fixed distance from the dis-

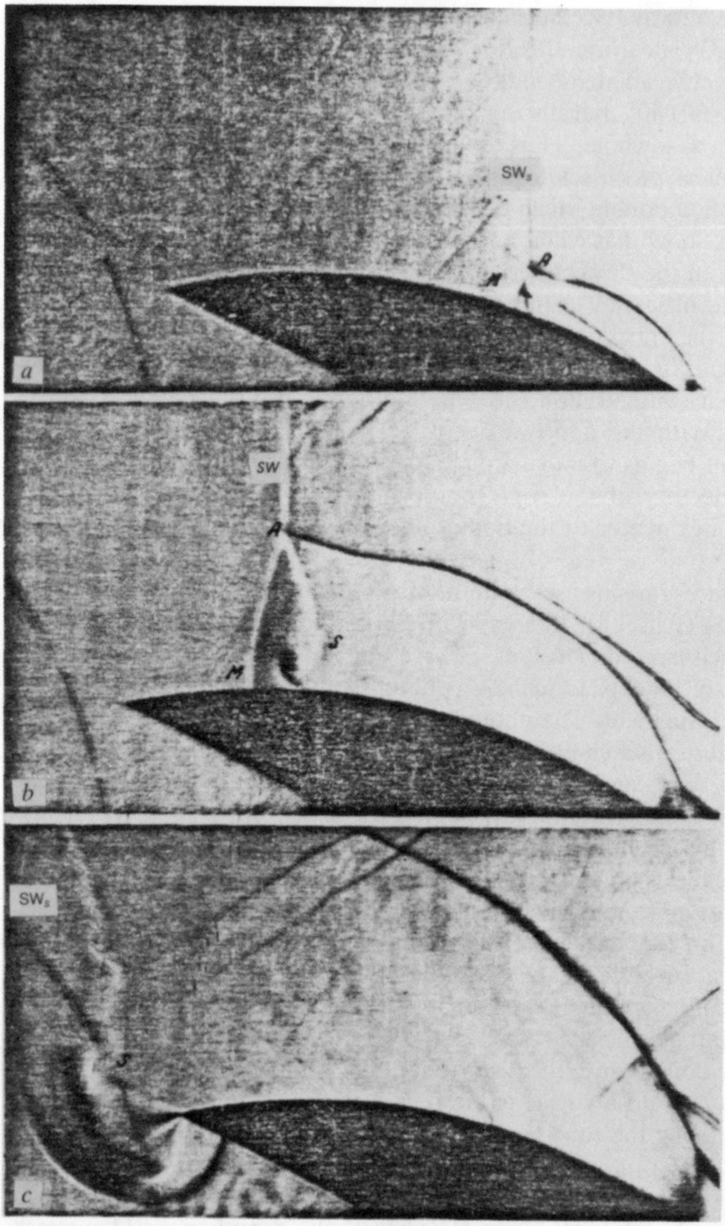

Figure 3.4. Consecutive stages of shock wave interaction in nitrogen with curvilinear surface ($M_0 = 4,5, p_0 = 10$ mm Hg, $\alpha_0 = 34°$) SW—incident shock wave; AM—Mach wave; AS—tangentional surface; S—vortex.

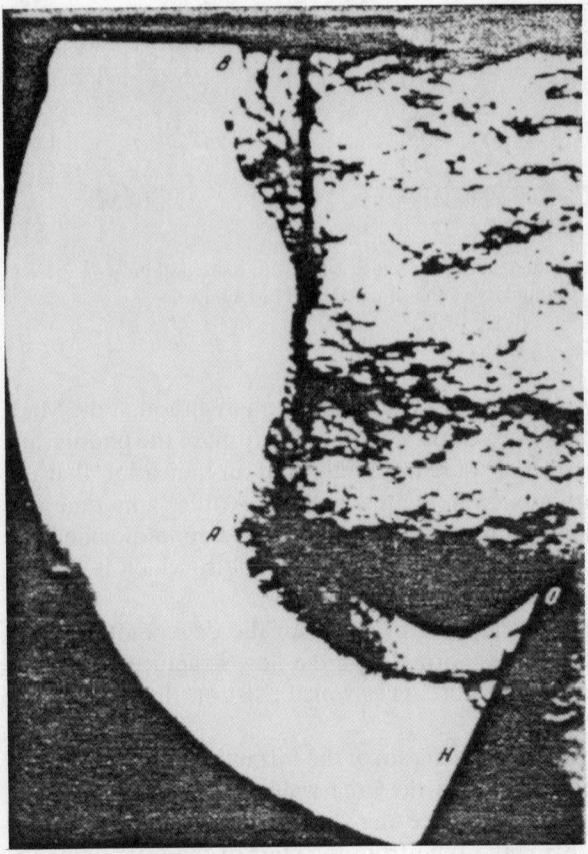

Figure 3.5. Shock wave in freon.

continuity of the tube at, for example, the diaphragm assembly. Therefore, with increasing Mach number of the incident shock wave it is observed that the front in the test section is alternately either more distorted or plane. The process bears a "threshold character" i.e., the distortion of the front in the given tube section can be regarded as the result of an apparent instability of the shock wave. The flow is noticeably distorted and complicated at high Mach numbers, when the degree of progress of physicochemical processes taking place in the gas behind the shock wave increases. Thus, for example, Fig. 3.5 shows a Töepler photograph of a shock wave diffraction at a convex corner. Investigation was carried out in freon. The region disturbed by the shock wave diffraction is AOH. Above line OA we see the front of the shock wave AB, which is sharply distorted as a result of interaction of the shock wave with the joints on the side wall between the test section and the low-pressure chamber. The Mach number M_0 of the incident wave was around 3. In Chap. 6, Fig. 6.2*d* shows a photograph of diffraction of a wave, moving in freon with a Mach number $M_0 = 3.64$, at

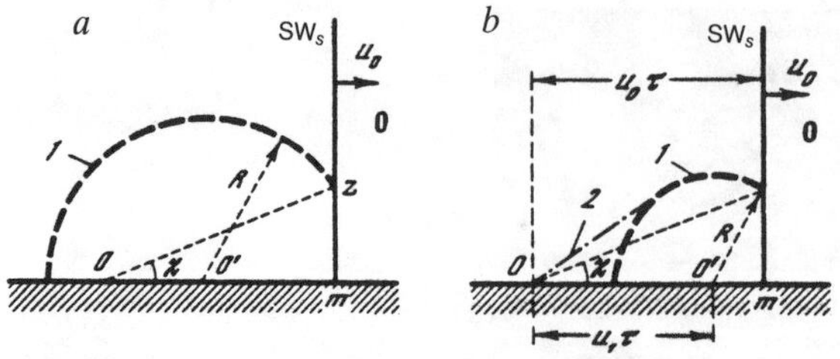

Figure 3.6. System of waves from the interaction of a shock wave with a discontinuity on the wall at point 0. *a)* $u_1 < a_1$; *b)* $u_1 > a_1$; *1)* boundary of the sonic circle; *2)* Mach line.

an identical corner. The wave front in this diagram is normal, although the Mach number M_0 is larger. Thus, the observed distortion of the front in the photograph in Fig. 3.4 is not a manifestation of wave instability, but an indication that not all disturbances have been attenuated in the test section. At the same time, the shock waves in the test section in argon, air, nitrogen, and carbon dioxide were normal within the bounds of the investigated Mach numbers, which is evident from the photographs in Chaps. 5 and 6.

At a large distance from the diaphragm, and with the degeneration of all disturbances, the wave front becomes normal, and the flow structure behind the wave is defined by the same microdisturbances which exist on the walls of the shock tube.

Presented in Fig. 3.6 is a schematic diagram of the formation of a disturbance system from separate nonuniformities on the tube wall. The center of the disturbed region is shifted to point O'. Since the disturbances travel at a sonic velocity, then $R = a_1\tau$, where τ is the time from the start of the interaction; a_1 is sound speed behind the shock wave. The point of intersection of the sonic circle with the incident wave travels at an angle χ_1 to the wall surface, which is easily found from geometrical considerations: $\mathrm{tg}\chi = \sqrt{a_1^2 - (u_0 - u_1)^2}/u_0$, where u_0 is the shock-wave velocity; u_1 is the flow velocity behind the wave.

Since there is always a certain degree of roughness on even very smooth walls of the shock tube, then as a result of superposition of such patterns from many random disturbances on the wall the flow structure behind the shock wave travelling in the shock tube will be made up from the following disturbance mesh (Fig. 3.7): lines of tangential discontinuities (*I*), transverse sonic waves (*II*), and the mesh of Mach lines for the case of supersonic flow (*III*). By photographing the Schlieren patterns in a shock wave one or another disturbance system becomes prominent depending on the location of the knife edge. If the knife edge is vertical, the wave itself and the Mach-line mesh become prominent, and when the knife edge is horizontal, the transverse waves stand out clearly. By fine tuning of the Töepler device one can render visible the tangential surfaces.

In the photograph of Fig. 3.5, the disturbances behind the wave front are multiple contact surfaces which have turned into vortexes. It is likely that the

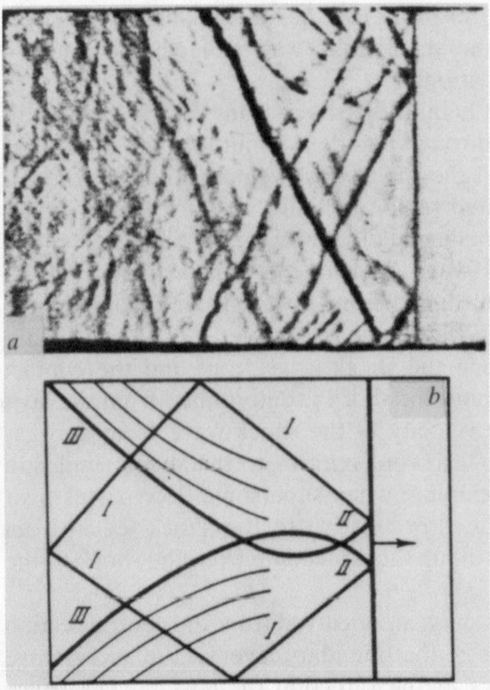

Figure 3.7. Wave pattern of the flow behind a shock wave travelling in a channel. *a*) shadow photograph of the disturbances behind the shock wave; *b*) schematics of the flow.

presence of these small-scale vortexes leads to the development of flow turbulence behind the shock wave.

3.6 EFFECT OF BOUNDARY LAYER ON FLOW PARAMETERS IN A SHOCK TUBE

From the ideal shock-tube theory it follows that the flow parameters behind the shock-wave front are constant up to the contact surface, which moves with a velocity equal to the gas velocity behind the shock wave. A closer inspection of the profile of flow parameters behind the shock wave shows that they vary from one point to another along the heated region behind the shock front. Indeed, the parameters determined from the laws of conservation for an ideal gas are realized only at the shock-wave front. Then follows the relaxation region, in which the state of the gas changes due to the physicochemical processes. However, even after reaching the equilibrium state, the flow parameters do not remain constant. This is so because a boundary layer growth develops on the tube wall over the length of the heated region behind the shock front. This phenomenon leads to a change of state in the inviscid core of the flow [79–81]. It is noted in [82] that the flow velocity increases gradually from the shock to the contact surface. In subsequent works [83–113], the question about the effect of the boundary layer on flow in the shock tube was subjected to detailed experimental and theoretical investigations.

The growth of the heated region behind the shock front ceases when equilibrium is reached between gas flow through it and gas input through the shock-wave front. This results in equality of the shock-wave velocity and the contact-surface velocity. The size of the heated region behind the shock front reaches its maximum value and remains unchanged. Flow becomes steady, not unlike the establishment of a steady shock-wave detachment from a blunt body in supersonic flow, when the gas input through the shock-wave front becomes equal to the flow through the sonic lines in the flow passing over the lateral surface of the body. Flow of the gas between the shock-wave front and the contact surface is isentropic. Gas velocity behind the shock is determined from the laws of conservation as a function of the velocity of the shock-wave front and gas velocity at the contact surface is equal to zero relative to the shock similar to the one at the body on the zero streamline when supersonic flow passes over it. The isentropic deceleration of the flow relative to the shock leads to an increase in temperature and pressure along the streamline from the shock to the contact surface.

In a shock-tube flow losses of gas mass are incurred from the central section to the boundary layer. The thickness of the boundary layer at the shock-wave front is equal to zero, then it grows in the direction of the contact surface accompanied by an increase in gas losses to the boundary layer. Flow in the core of the stream is equivalent to subsonic flow in a nozzle, the cross-sectional area of which increases from the shock to the contact surface. The relation for the increase in cross sectional area A is determined by the relation for the growth of boundary-layer thickness δ with the distance from the shock l. This relation is different for laminar and turbulent boundary layers: $\delta = \beta\,(1/\mathrm{Re})^n$, where n is equal to 1/2 and 1/5 for laminar and turbulent layers, respectively; β is the coefficient of proportionality.

The transition from laminar to turbulent boundary layer is achieved when the Reynolds number Re^* of the flow reaches 10^6. The Reynolds number Re^* for the transition in the flow behind the shock wave is determined as $\mathrm{Re}^* = \tilde{u}\tilde{l}/\nu_e$, where $\tilde{u}$ is the flow velocity; ν_e is the viscosity in the core of the flow; $\tilde{l}$ is the length along the contact region; $\tilde{u} = u_1(w - 1)$; $\tilde{l} = l^*(w - 1)$; $w = \rho_1/\rho_0$; and l^* is the distance from the shock-wave front along which the transition from laminar to turbulent boundary layer is achieved.

The thickness of the boundary layer and transition distances were determined by optical methods in [112, 113]. Good correlation with the theory of Mirels was noted.

The boundary layer significantly influences the parameters of the heated gas.

The boundary layer affects gases with various adiabatic indexes γ in different ways, since the effect is manifested according to the theory of Mirels by the appearance of an isentropic expansion wave. The influence of boundary layer is most pronounced in monatomic gases, and least apparent in gases with low values of the adiabatic index. If, at Mach number M_0 of the order of eight, the maximum possible deviation of the relative Mach number in Ar reaches 30%, then in N_2 it reaches 20%, and in gases with excited vibrations ($\gamma = 1.3$) it amounts to several percentage points.

Changes in flow parameters are most pronounced at low Mach numbers M_0. When $M_0 > 5$, the value of the deviation becomes practically independent from the Mach number of the shock wave.

It is experimentally verified [82] that the flow parameters behind the shock wave in a shock tube change from the shock to the contact surface, and that this change can be determined for a known length of the contact region by estimates presented in Mirels' work [91]. For the shock-wave velocities M_0 = 3–6, the assumption on which these estimates are based are fulfilled with an accuracy of at least 4%.

In [105, 106], methods are proposed to compensate for the effect of the boundary layer on flow parameters behind a shock wave by changing the shape of the high-pressure chamber. The high-pressure chamber is so designed that the walls reflect the unsteady expansion wave converting it to a compression wave or to a large-amplitude expansion wave. Upon propagation downstream of the flow, these waves catch up with the shock wave and interact with it in the needed manner. If there is a need to keep the shock wave intensity constant, then an additional compression wave is created to compensate for the attenuation of the shock wave. In order to maintain constant gas parameters along the contact region, one must decrease the temperature and pressure which rise from the shock to the contact surface. The high-pressure chamber has, therefore, a decreasing cross section.

3.7 EFFECT OF BOUNDARY LAYER ON THE SHAPE OF THE REFLECTED SHOCK WAVE

When a shock wave is reflected from a solid wall it propagates in an inhomogeneous flow. In this case the reflected shock wave propagates in a gas which can be regarded as being comprised of two layers—a core and boundary layer, created by the incident wave. Similar phenomenon is observed when a strong incident shock wave propagates along the shock-tube walls heated by radiation from the shock-wave front [114].

Griffith [115] predicted and Mark [116] observed experimentally that in this case such regimes might develop when wave bifurcation takes place with the formation of λ-type configurations. The characteristic dimensions of the bifurcation are one order of magnitude larger than the thickness of the boundary layer.

Mark observed that when a shock wave is reflected from the end of the shock tube a configuration of the type depicted in Fig. 3.8 is obtained. In the figure, *IA* is the reflected wave; *AO* is the oblique wave; *AB* is the third wave of the configuration; AT is the tangential surface of the discontinuity; and OB is the boundary of flow separation. An analogous situation exists in the steady interaction of a shock wave with the boundary layer when flow separation occurs. Two questions arise. First, what is the criterion for the occurrence of bifurcation and, second, how to compute the entire wave pattern of the flow. Up to the present time, an answer to the first question has been found, but not a satisfactory answer to the second.

The principal idea about the bifurcation criterion was stated by Griffith and formulated precisely by Mark. It was proposed to compare the stagnation pressure of the gas stream passing through the shock wave in the boundary layer with the pressure behind the reflected shock wave. If the pressure behind the reflected wave is greater than the stagnation pressure of the gas stream near the wall, then the result is flow separation, its accumulation in region OBE, and formation of wave OA. Then, accordingly, wave AB and tangential surface AT are formed.

Mark computed the separation criterion as follows. Let the flow velocity of the gas behind the incident wave be u_1, and the velocity of the reflected wave u_R in the laboratory coordinate system. Then with respect to the reflected wave the Mach number of the flow in the core is $M_R = (u_R + u_1)/a_1$ and the Mach number of the flow in the boundary layer is $M_w = u_R/a_0$. It is assumed here that the gas temperature at the wall does not change when the gas passes through the shock wave, that it is equal to the temperature of the cold gas, and that the gas velocity in the boundary layer in the laboratory coordinate system is equal to zero. For the perfect gas the number M_w is calculated from the relation

$$M_w = [2(\gamma + 1) M_0^2 - (3 - \gamma)]/(\gamma + 1)M_0$$

where M_0 is the Mach number of the incident wave. When $M_w < 1$, the ratio of the stagnation pressure p_{st} to the pressure behind the incident wave p_1 is determined from the relation

$$\frac{p_{st}}{p_1} = \left(1 + \frac{\gamma - 1}{2} M_w^2\right)^{\gamma/(\gamma-1)}$$

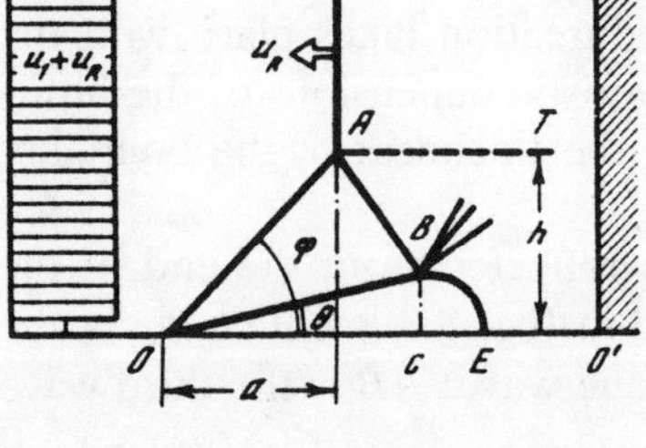

Figure 3.8. Schematic of bifurcation of a reflected shock wave upon interaction with the boundary layer.

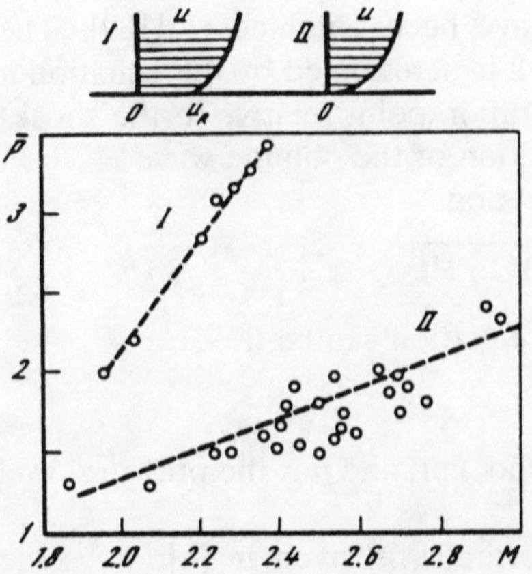

Figure 3.9. Dependence of the pressure of flow separation on the Mach number. *I*) separation upon bifurcation; *II*) separation of the boundary layer by a steady shock wave in a nozzle.

When $M_w > 1$, the pressure rise is determined from the equation

$$\frac{p_{st}}{p_1} = \left(\frac{\gamma+1}{2} M_w^2\right)^{\gamma/(\gamma-1)} \left(\frac{2\gamma}{\gamma+1} M_w^2 - \frac{\gamma-1}{\gamma+1}\right)^{1/(1-\gamma)}$$

The pressure behind the reflected wave for the perfect gas is equal to

$$p_5/p_1 = (2\gamma M_R^2 - (\gamma-1))/(\gamma+1)$$

The value of M_R is easily found from the relation $M_R \overline{M}_2 = 1$, where $\overline{M}_2$ is the Mach number of the flow behind the incident wave in the coordinate system moving with this wave. The ratio p_5/p_{st} is easily determined from these relations. When this ratio is greater than unity, then bifurcation exists and when $p_5/p_{st} < 1$, there is no bifurcation.

This criterion was qualitatively confirmed in tests by Mark and in other experimental works [117–130].

In [129], experimental comparison was carried out of the separation criterion upon bifurcation of the shock wave in a shock tube and upon steady separation in a nozzle (Fig. 3.9). In this figure, $\bar{p}$ is the critical pressure rise on the wave causing separation of the flow, which enters the wave, and M is the Mach number of the flow. In the nozzle, the shock wave is stationary relative to the wall; upon reflection the shock wave travels along the lateral surface, i.e., separation appears to take place on a surface moving in the direction opposite to that of the wave. In this case the profile of the boundary layer is fuller than for a separation in a nozzle. The velocity profiles are depicted in Fig. 3.9. Evidently, the reason behind this is that the separation pressure during bifurcation is almost twice as high as the pressure for steady separation in a nozzle.

As regards the computation of the wave structure resulting from bifurcation, this problem has, as yet, to be solved completely. Experimental data exist regarding the trajectory of the motion of the triple point, the height of which increases with the movement of the wave from the wall and then reaches its steady value.

Approximating formulas for nitrogen have been obtained in [129]. The height of the triple-point zone above surface h is determined by the equation $h = a \operatorname{tg}\varphi$, where a is the location of the separation point relative to the shock-wave front, mm; and φ is the angle of inclination of the oblique wave.

From Mark's theory φ is found from equation

$$\varphi = \arcsin \sqrt{(p_{st}/p_s + (\gamma - 1)/(\gamma + 1))(\gamma + 1)/2\gamma M_R^2}$$

For $\gamma = 1.4$, the following approximate equation can be used

$$a = 0.7x^{0.8} \lg M_0 \, p^{-0.1} \pm 2$$

where x is the distance from the shock-tube end, cm; and p is the pressure, mm Hg.

The experimental validation of these formulas is given in [6]. By some transformation of the path travelled by the reflected wave, one can match trajectories of the triple points in the bifurcation for various gases for the initial stage of reflection [127]. The knowledge of the trajectory of the triple points allows one to determine the gas parameters in their vicinity on the basis of calculations using the triple-wave theory (see Chap. 5).

All the well-known works investigating bifurcation are conducted within the limits of a plane flow pattern. In this case, the changes occurring as a result of the interaction of the bifurcation zones, forming on the adjoining channel walls, are not taken into account. Also not considered are the regimes of reflection with joining bifurcation zones travelling in opposite directions from the side walls of the rectangular channel. The indicated phenomena are particularly significant for narrow channels, where they play a determining role during the formation of the flow pattern.

Experimental investigation of the structure of such flows was undertaken in [131, 132] with shock-wave reflection from the end of rectangular channels. The channels were mounted in a shock tube with a square section 72 × 72 mm with the help of two wedge-shaped plates about 140 mm long in the field of view of the observation windows (dimensions of the window—180 × 72 mm) at the far end of the test section of the low-pressure chamber of the shock tube. In one case, the half height of the channel $H = 11$ mm, which is comparable with the value of maximum bifurcation height for the investigated reflection regimes. In the other case, $H = 4$ mm, which amounts to fractions of the maximum bifurcation height. Experiments were conducted in carbon dioxide, nitrogen, and air at the initial pressures of 10–50 mm Hg and Mach numbers 2–4. The study of the reflection process in the shock tube and channels was carried out up to the moment of interaction of the reflected shock wave with the contact surface separating the driving gas. The visualization of the flow pattern was realized with the help of a four-channel device of the Cranz-Schardin type, constructed on the basis of the shadow system IAB-451 and attachment RP-463 [62, 133].

In three channels of the device, shadow patterns were photographed of the entire field of view of the device (dimensions of the image on the film are 10 × 25 mm). In the fourth channel a streak record was recorded in the axial section of the channel. The photorecorder was of type ZhFR. To obtain shadowgraphs, xenon lamps ISP-1500, were used with a duration of the light pulse about 0.5 μsec. To obtain proper streak record the IFK-50 pulse lamp with a burst duration of 800 μsec.

Figures 3.10 and 3.11 show shadowgraphs of reflected shock wave in carbon dioxide in channels of various sections where the end is located on the right. When reflected from the shock-tube end, the configuration of the waves in the bifurcation zone corresponds, generally, to the known flow pattern, although even here one can see the specific features of both the bifurcation zone itself and the flow behind that zone. First, the separation line of the boundary layer (*L* on the lateral channel walls, and at the projection of point *O* on the upper and lower walls, Fig. 3.10*a*, engulfing the entire perimeter of the channel, is not

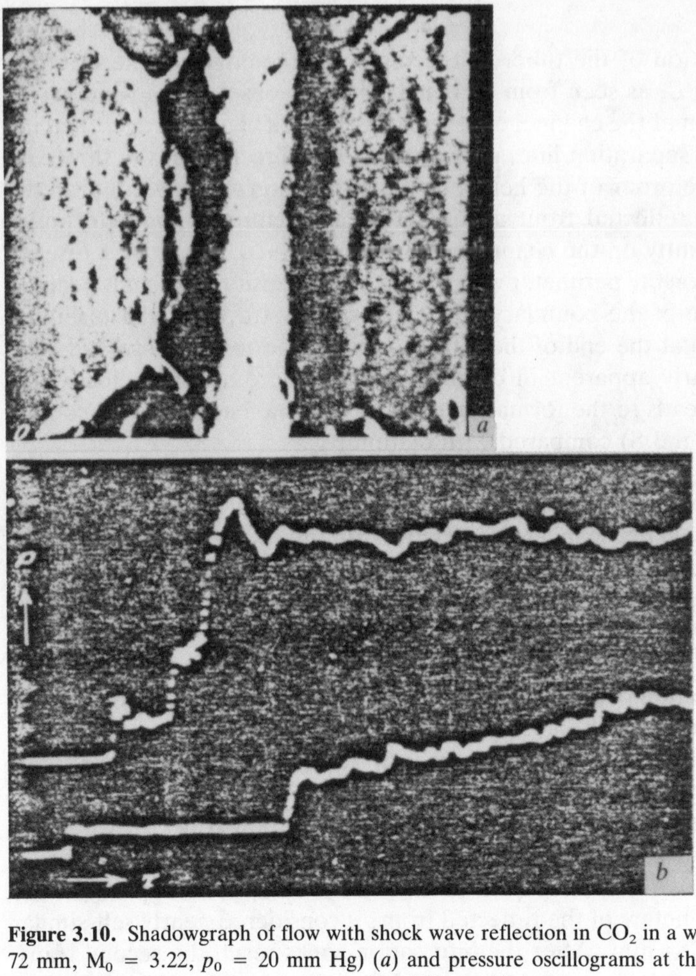

Figure 3.10. Shadowgraph of flow with shock wave reflection in CO_2 in a wide channel (S = 72 × 72 mm, M_0 = 3.22, p_0 = 20 mm Hg) (*a*) and pressure oscillograms at the lateral surface of the channel (CO_2, M_0 = 3.2, p_0 = 10 mm Hg, 100 μsec/div) (*b*).

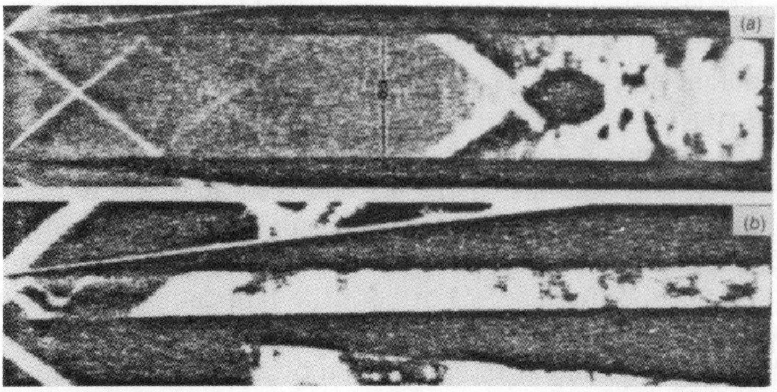

Figure 3.11. Shadowgraphs of flow behind a reflected shock wave in carbon dioxide at $p_0 = 20$ mm Hg in rectangular channels of various sections. *a*) $M_0 = 3.04$; $S = 22 \times 72$ mm; *b*) $M_0 = 3.09$; $S = 8 \times 72$ mm.

located in one section of the tube. In the region of the edges of the channel, the separation line L, as seen from its behavior in the observation windows, is bent in the direction of the channel end. Such an effect of channel edges, leading to the delay of the separation line, propagates in the direction of the flow core at a distance on the order of the height of the bifurcation zone. Second, in the region behind the reflected front a number of characteristic nonuniformities situated predominantly on the channel walls, are observed. Thus, the presence of a region located on the perimeter with considerable density gradient is peculiar to the joining place of the boundary layer. Farther toward the end a turbulent zone is located and at the end of the tube vortex formations are observed. The latter are particularly apparent in carbon dioxide, where a large velocity discontinuity, which leads to the formation of vortexes, is a feature on the contact surface AT (see Fig. 3.8) compared with diatomic gases.

The reflection process with strongly developed bifurcation is characterized by various behaviors of the pressure change on the lateral channel surface at various distances from the channel end. Figure 3.10*b* shows pressure oscillograms which are obtained with a piezoelectric transducers located at distances of 25 and 85 mm from the shock-tube end from the top and the bottom, respectively. For the transducer near the end, according to the bifurcation model discussed earlier, a pressure rise is observed as the incident wave passes followed by a rise when the separation point O reaches the transducer. The pressure is constant during a time interval of about 50 μsec, corresponding to the passage of the bifurcation zone, then a second rise follows up to the reflection pressure. For the transducer away from the end, one might expect that the change of pressure profile would correspond to the increase in the bifurcation zone. However, in reality the variation in pressure here differs qualitatively from the variation discussed earlier. This occurs despite the fact that the reflection process, described by the parameters of the reflected front, is considered nearly self-similar up to a distance of 85 mm. After the bifurcation zones reach the second transducer, a continuous pressure rise is observed during 1 msec and more. This is linked, as shown by individual frames and streak-records, with the deceleration of the trailing edge of the bifurcation zone [117] which does not reach the section where the transducer is located during the experiment. This feature of the behavior must be taken into consideration when the state of the gas behind the reflected wave is determined by the readings of the lateral transducers.

The characteristic configuration resulting from the reflection in carbon dioxide in a channel of height 22 mm (H = 11 mm) is shown in Fig. 3.11*a*. Here, the phase which is observed following the moment of joining of the triple points of the bifurcation zones is presented. The effect of corners, as in wide channels, is apparent on the retardation of the separation lines L in the narrow sides of the channel (observation windows) up to their disappearance from the field of view. After that it is difficult to say anything definite about the character of the flow in the vicinity of the entire narrow side of the channel. For the flow in the stream core it is characteristic for the features to appear which are inherent in the reflection process in a sufficiently narrow channel. When the triple points draw together and finally collide, the gas in the flow core moves away from the reflected shock wave and into the region enclosed by the shock which causes the final retardation of the gas. Thus, the retardation of the gas flowing into the reflected wave is caused by a system of waves of a pseudo-shock type. The latter were observed before during steady flows in channels and following the interaction of the reflected shock wave with the contact surface in shock tubes [134, 135].

When the channel is further narrowed (Fig. 3.11*b*), the possible asymmetry of the V-shaped front leads to its degeneration, at some moment in time, into one inclined front F which occupies the entire section of the channel. The effect of the corners of the channel, which cause the delay of the separation point, appears in the visible (in projection) expansion of the oblique front of the reflected wave and in the corresponding deformation into a cylindrical surface.

CHAPTER

FOUR

INTERACTION OF SHOCK WAVES WITH A POROUS COMPRESSIBLE MEDIUM

4.1 NORMAL REFLECTION OF A SHOCK WAVE FROM A WALL COVERED BY A POROUS COMPRESSIBLE MATERIAL

Experimental investigations of the interaction of air shock waves with a solid wall covered by a layer of compressible material, such as polyurethane foam, indicate a considerable increase of peak pressure on a wall under that layer of material as compared with the case without a coating [136, 137]. A study of the interaction of an air shock wave with a porous screen from polyurethane foam was conducted in [138], in which a considerable decrease in peak pressure on the wall was recorded in the presence of air gap between the wall and the screen. To theoretically explain the propagation process of the shock wave in a porous compressible medium, a computational method [139], which was initially developed for the description of gasdynamic flows with solid particles, was used in [138].

It should be noted that strongly compressible porous materials such as elastic polyurethane foam are similar to a certain degree to disperse gas–liquid systems (water–mechanical foams, liquid with gas bubbles), in which wave pressures have been widely investigated both experimentally and theoretically [140–145]. The results of experimental investigations are presented on the interaction of a steady air shock wave with a wall covered with layers of porous plastic materials of various thicknesses. Elastic polyurethane foam of type PPU-EM-1 and foam plastic of type PKhV − 1 were used as coating materials. The density of the indicated materials was 33 and 50 kg/m^3, respectively, with a porosity coefficient (ratio of volume of pores to the total volume of the specimen) $\alpha = 0.95{-}0.98$. The stiffness of foam plastic is relatively high; noticeable deformations in this material are observed at stresses higher than $(4{-}7) \cdot 10^5$ N/m^2. Typical values of the compression strength of polyurethane foam at 40% deformation is less than 10^4 N/m^2.

It is important to note that the pressure of normal shock-wave reflection from a rigid wall was varied in the experiments in the range of 1–15 atm which considerably exceeds the characteristic value of the compression strength of polyurethane foam and is comparable with the corresponding value for foam plastic. This allows revealing the role of the mechanism of acoustic pressure transmission through the rigid structure of the coating. A schematic diagram of the experimental layout and instrumentation is shown in Fig. 3.1.

Investigations were carried out in a single-diaphragm shock tube with a square-section channel 0.1 × 0.1 m. The length of the low-pressure chamber (LPC) and the test section (TS) with optical windows was 8 m, the high-pressure chamber (HPC) was 1.5 m long. Flat layers of the investigated materials of various thicknesses were placed on the end wall of the test section, completely covering the cross-section of the shock tube channel. A pressure transducer was also mounted on the end wall. The pressure-measuring piezoelectric transducers used in the tests were of the design described in [63]. The measurement of the

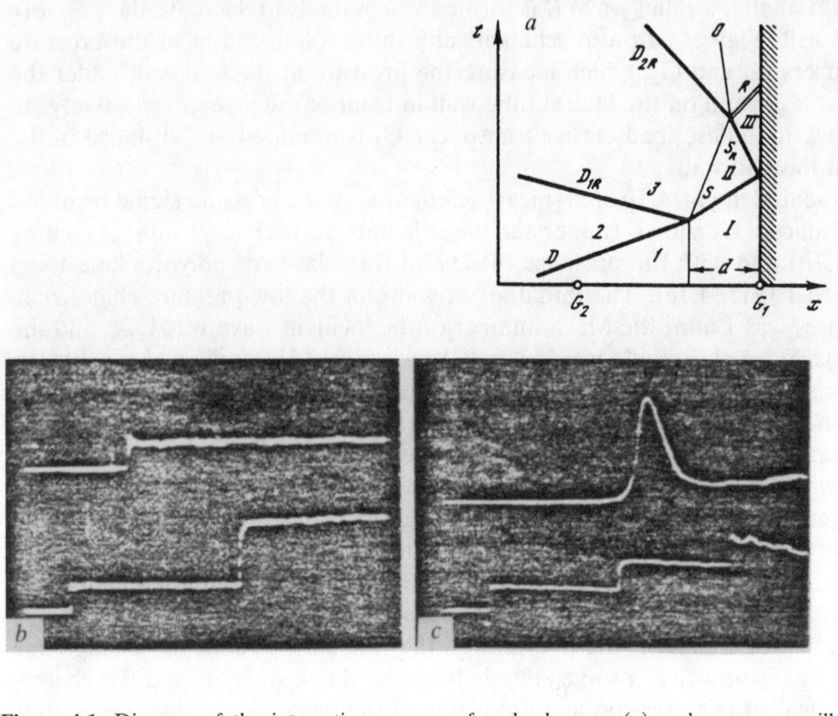

Figure 4.1. Diagram of the interaction process of a shock wave (*a*) and pressure oscillograms (*b*, *c*).

velocity of the incident shock wave was made with the help of pressure transducers mounted in the lateral walls of the channel and also with special circuits of pulse shapers for triggering the digital frequency meter ChZ–33 and oscilloscope "Tektronics 453A." To match the high output resistance of the piezoelectric transducer with the resistance of the line voltage, followers were used based on an operational amplifier and push-pull transistor power amplifier. The visualization of the deformation process of the porous material was carried out with a pulse light source, a shadow system IAB-451, and a camera ZhFR working in the slit-scanning mode. Air at a pressure of 5–760 mm Hg served as the test gas, and helium and nitrogen as the driving gas. The Mach number of the incident shock wave was varied between 1.3–5.

Figure 4.1*a* depicts x–t diagrams of the interaction process of air shock wave D in the shock tube with a plane layer of porous compressible material laid on a rigid wall. The parameters of the undisturbed gas ahead of the incident wave are denoted by number *1*; behind the incident wave D—by *2*; and behind the reflected wave from the layer surface D_{1R}—by 3. The Roman numerals *I*, *II*, and *III* denote, respectively, the parameters of the undisturbed porous medium, parameters of the medium behind the wave refracted in material S, and the wave reflected from the rigid wall S_R. The interaction of the reflected wave from the wall with the contact surface leads to the formation of a secondary reflected wave in air D_{2R} and to a change in the direction of motion of the material surface O. Additionally, a relief wave R is formed which tends to decrease the pressure on the wall. Figure 4.1*a* also schematically shows the location of the pressure transducers G_1 and G_2, which measure the pressure at the end wall under the material layer and on the lateral tube wall in front of the layer, respectively. In the shock tube described earlier transducer G_2 is mounted at a distance of 0.2 m from the end wall.

Presented in Fig. 4.1*b* are typical oscillograms of the pressure signal recorded by transducers G_1 and G_2 (upper and lower beams, respectively) without coating (Fig. 4.1*b*) and with the presence of 0.08 m thick layer of polyurethane foam PPU-EM-1 (Fig. 4.1*c*). The initial air pressure in the low-pressure chamber in both cases was 1 atm; the Mach number of the incident wave was 1.74; and the time sweep on the oscillograms was 0.2 msec/div. It is noticed that with the presence of a layer of the porous material, instead of the usual step-like pressure pulse on the wall, a pulse of complex shape is obtained. The maximum amplitude of this pulse p_3 is considerably greater than the amplitude of the transducer signal without a coating p_{20}. The streak-records of the compression process of the layer of porous material give an indication about the coincidence of the moment of time, corresponding to the maximum compression of the layer, and the time for reaching maximum pressure p_3. Following the compression process, the unloading phase sets in and the pressure gradually approaches the value of p_{20} (that is, of course, if the pressure on the free surface does not change due to the expansion waves propagating behind the shock wave during the characteristic time of compression and unloading of the layer).

The second pressure jump (the lower beam on the oscillograms) has a characteristic rounding-off near the top which is an indication of the bifurcation of the shock wave when it interacts with the boundary layer.

It is important to note also the presence of a secondary reflected shock wave in the air D_{2R} (Fig. 4.1*c*) which is recorded by the lateral transducer G_2 as a third pressure jump (the first jump corresponds to the incident shock wave D; the second corresponds to the shock wave reflected from the surface of the material). It was indicated earlier that this wave is generated when the shock-compressed layer of the porous material is unloaded.

The superimposed pressure oscillograms on the wall, obtained in tests with layers of polyurethane foam of various thicknesses, are illustrated in Fig. 4.2*a*, *b*. In the first series of tests (Fig. 4.2*a*) the Mach number of the incident wave was $M_0 = 5$ for an initial air pressure of $p_0 = 12$ mm Hg; in the second series (Fig. 4.2*b*) $M_0 = 2$ for $p_0 = 1$ atm. In each series of experiments, the maximum amplitude of the pulse first increases with an increase in thickness of the layer material, and at values of layer thickness $d \gtrsim 80$ mm it stabilizes at an approximately constant level which depends on the material type and initial conditions of the experiment. With increasing d the duration of the front of the pressure pulse increases somewhat and the front can be conventionally divided into two regions: in the first region a slow increase of pressure is observed; in the second a relatively sharp increase in pressure, which is almost linear, takes place. The moment of appearance of the signal, from the transducer covered by a layer of the polyurethane foam, noticeably lags behind the signal from the open trans-

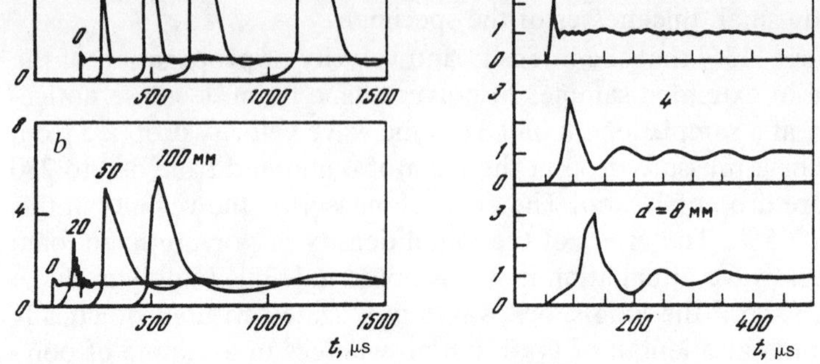

Figure 4.2. Pressure profiles at the end wall for various layer thicknesses of polyurethane foam (*a*, *b*) and foam plastic (*c*).

ducer. The time delay is proportional to the thickness of the coating. The maximum value of the coefficient of pressure increase on the wall $R_1 = p_3/p_{20}$ (p_{20} is the pressure of normal shock wave reflection from the rigid wall) which was recorded during the tests amounted to approximately 10 (obtained at $p_0 =$ 10 mm Hg, $M_0 \simeq 5$).

When pressure p_0 is reduced to values below 1–2 mm Hg at the same Mach number, coefficient R_1 decreases to 2–3. Variation of pressure p_0 within a wide range at constant Mach number $M_0 = 5$ was not feasible due to the design features of the shock tube. Oscillograms of pressure signals of the end transducer, covered by layers of foam plastic of various thicknesses, at $p_0 = 12$ mm Hg and $M_0 = 5$ are shown in Fig. 4.2*c*. In this case, the time delay of the pressure signal for all values of the layer thicknesses is practically not noticeable, which is an indication of relatively high sound speed in the material ($\geqslant$1 km/sec). The pressure amplitude at the wall under a layer of foam plastic is less than in the case of polyurethane foam (for similar initial conditions of the experiment); however, it is still noticeably larger than the amplitude of the signal from the transducer without a coating. The distinguishing feature for the given series of experiments is the oscillation of the signals of the pressure transducer, with the period of oscillations being directly proportional to the thickness of the coating. Measurements of the time interval from the moment the shock wave reaches the free surface of the specimen to the moment maximum pressure is reached on the wall for different thicknesses of material layers indicates the constancy of the velocity of propagation of the compression wave in the porous medium (for thickness of specimens of polyurethane foam up to 100–150 mm and of foam plastic up to 10–15 mm). The absolute values of the velocity of the compression waves in polyurethane foam at $p_0 = 12$ mm Hg ($M_0 = 5$) and $p_0 = 1$ atm ($M_0 = 2$) are equal to 110 and 215 m/sec, respectively, and in foam plastic they are equal to 90 m/sec. It should be noted, however, that the velocity of compression waves in polyurethane foam, computed from the time delay pressure response, is 1.2–1.5 times higher than the given values due to the presence of sufficiently extended "pedestal" of the pressure pulse. The assertion about the constancy of velocity of propagation of compression waves in the investigated materials is true for sufficiently small thicknesses of the specimens.

It was found out [136] that the intensity and velocity of propagation of the compression wave in extended samples of polyurethane foam decrease noticeably; in particular, at a sample length of 0.34 m the wave velocity dropped from 320 to 280 m/sec for a pressure drop at the front of 4 atm and from 380 to 280 m/sec for a pressure drop of 12 atm. The error of measuring the velocity in this case amounted to ~5%. The effect of the initial density of polyurethane foam on the rate of shock wave attenuation in it was noted in [138]. In polyurethane foam of density 20 kg/m^3 the long shock waves propagated without practically noticeable weakening at a length of up to 0.8 m, whereas in a sample of polyurethane foam with large initial density (35 kg/m^3) a twofold attenuation was observed over the same length. The decrease of wave intensity in denser samples is linked with the increase in energy expended on the deformation of the skeleton of the porous material.

Measurement of the displacement velocity of the free boundary of the porous material, conducted by streak-record, indicates that this velocity is independent of time with the exception of a small initial section. When $p_0 = 12$ mm Hg and $M_0 = 5$, the absolute value of displacement velocity of the layer boundary of polyurethane foam amounted to 95 m/sec, and that of foam plastic 75 m/sec. The layer of polyurethane foam of initial thickness of 65 mm is compressed 13–15 mm followed by the unloading phase. The residual deformation in the samples of polyurethane foam, subjected to the action of shock waves, is small ($\leqslant 5\%$). The structure of the upper material layer becomes more spongy, and at low initial pressures and high Mach numbers of the incident wave the free sample surface becomes noticeably darker and partially charred. The analysis of the results confirms the hypothesis presented in [136] stating that the mechanism responsible for the abnormally high pressure on the wall under the porous compressible material consists of the drawing-in of the solid phase of the material into motion behind the front of the penetrating wave. Direct measurement of the momentum, transmitted to the wall through coatings of various types and thicknesses, indicates the absence of the momentum loss. The momentum of the air shock wave, initially distributed in the gas mass behind the front, partially localizes in a relatively compact but sufficiently dense mass of porous compressed material, where interaction with the wall leads to the appearance of a pressure peak. Analysis of pressure profiles on the wall beneath the layer of porous material allows one to assume [137] that the compression wave in polyurethane foam, which is noted for its high skeleton elasticity, is a shock wave. In relatively rigid foam plastic, on the other hand, superposition of the shock and acoustic waves takes place when the latter propagates across the skeleton of the porous material. The conventional division of the front of the pressure pulse on the wall beneath a layer of polyurethane foam into two regions can be qualitatively substantiated as follows. The elasticity of the material structure and its permeability lead to a slow pressure rise. Relatively sharp pressure jump follows, which corresponds to the shock wave front being formed in the porous medium whose thickness is equivalent to several diameters of the pores of the material.

Measurements of gas pressure in the shock wave reflected from the free surface of the material layer p_{21} which, evidently, is equal to the pressure behind the shock-wave front in the porous medium, showed that in the majority of cases p_{21} differs insignificantly from the pressure of normal reflection of the shock wave from the rigid wall ($p_{21}/p_{20} = 0.6$–0.8). This is explained by the relatively low velocity of motion of the free surface of the material (due to its considerable initial density) in comparison with the velocity of the reflected air shock wave. The decrease of the coefficient R_1 when the initial gas pressure p_0 is decreased to 0.5–1 mm Hg can be interpreted as being the result of decreasing the pressure p_{21} to values close to the elasticity of the skeleton of the porous material.

In conclusion, we shall briefly discuss some experimental results obtained in [138] by measuring shock-wave intensity in air, which passes through a block of elastic polyurethane foam 0.3 m thick. The initial air pressure during the test was 1 atm. It is revealed that when a steady shock wave of intensity $\delta_1 = p_1 p_0^{-1} = 3.5$ impinges on the block, the pressure piezoelectric transducer, mounted in the side wall of the shock tube behind the block, registers a shock wave which is refracted in air and has an intensity $\delta_3 = p_3 p_0^{-1} = 1.9$. As a result of the subsequent motion of the compressed sample behind the air shock wave, a pressure rise is registered during a time interval of 1.8 m/sec up to $\delta_0 = 3.4$, which is close to δ_1.

When a shock wave of triangular profile having intensity $\delta_1 = 4.1$ (direction of the positive phase 1 m/sec, wavelength 0.5 m) impinges on the block of polyurethane foam a noticeable decrease in intensity of the refracted shock wave is observed behind the block ($\delta_3 = 1.5$).

Analysis of the experimental results allowed the authors of [138] to conclude that when the length of the porous screen is comparable with the length of the incident unsteady shock wave, the screen acts as an effective attenuator of shock waves.

4.2 CALCULATION OF SHOCK WAVE PARAMETERS IN A POROUS COMPRESSIBLE MEDIUM

In this section an approximate mathematical model describing the processes of propagation and reflection of shock waves in a porous compressible medium will be discussed [146]. Elastic forces of the skeleton of the porous medium for finite deformation can be assumed negligibly small compared with the values of the static and dynamic gas pressure at an arbitrary point of the medium. This assumption strongly limits the scope of porous media for which the described computational method is applicable; however, it allows one to derive simple analytical expressions which completely describe the kinematic and dynamic characteristics of the shock wave in similar media. Since all the following qualitative calculations are conducted in relation to the case of elastic polyurethane foams, we shall briefly discuss the properties of these materials.

Elastic polyurethane foams, based on simple and complex polyethers, relate to a class of materials which are characterized by high porosity (the total volume of pores can reach 98% of the overall volume of the sample) and by low values of skeleton elasticity. The characteristic size of the pores of polyurethane foam depends on its brand and manufacturing technology and lies within the range from fractions of a millimeter to several millimeters. We note that, unlike the majority of brands of foam plastics, polyurethane foam has, as a rule, an open lattice structure, i.e., it is permeable material. The apparent density of polyurethane foam of different brands varies within a wide range (20–250 kg/m^3); however, for the most widely used brands (PPU-EM-1, PPU-ET, PPU-E-25 and others) the value of the apparent density amounts to 25–60 kg/m^3.

Henceforth, we shall regard strongly compressed porous materials with sufficiently large linear dimensions of the sample (considerably exceeding the characteristic dimensions of the pores) as a mixture of solid particles and gas. This representation in regard to polyurethane foam is justified as follows. The elasticity of the skeleton of the majority of brands of elastic polyurethane foam is noticeably less than the value of the atmospheric pressure at normal conditions.

Thus, for example, for the widely used polyurethane foam PPU-EM-1 with an apparent density of 25–45 kg/m^3 the compression strength at 40% deformation is equal to $(0.4–1.0) \times 10^4$ N/m^2. This indicates that it is possible to neglect the elasticity of the skeleton of the said materials in comparison with the gas elasticity in the pores at an initial gas pressure of $p_0 \approx 1$ atm, and simulate the real porous material by a liquid with gas bubbles or by a homogeneous suspension of solid noninteracting microparticles in a gas. The dynamics of propagation of acoustic waves and shock waves in analogous media were analyzed in [139–145] on the basis of the representation of a two-phase medium as a homogeneous mixture. As noted in work [143], the homogeneous mixture represents the simplest model of a two-phase medium in which all the effects linked with discrete structures are ignored. Only the volume fraction of the solid phase, which influences the compressibility of the medium as a whole, is taken into consideration. The speed of sound in a homogeneous mixture simulating a real porous medium with quasi-elastic skeleton can be determined from Wood's equation [142]

$$a^2 = \frac{1}{(\alpha\rho^0 + (1-\alpha)\rho^1)(\alpha/\rho^0 (a^0)^2 + (1-\alpha)/\rho^1 (a^1)^2)} \tag{4.1}$$

where ρ^0 and ρ^1 are the densities of gaseous and solid phases; α is the volume gas content; a^0 and a^1 are the sound speed in gaseous and solid media. (Here and later, the superscripts "0" or "1" of any parameter pertain to the gaseous or solid media. Subscripts are used to denote the averaged parameters of the medium in various flow regions).

At normal conditions the density of air in the pores of the material $\rho^0 \approx 1.2$ kg/m^3. The density of monolith polyurethane $\rho^1 \approx 1.2 \cdot 10^3$ kg/m^3, and the speed of sound in it $a^1 \approx 2 \cdot 10^3$ m/sec. Assuming additionally that in the case of polyurethane foam $\alpha \approx 0.5–0.98$, we can simplify relation (4.1):

$$a^2 \approx (a^0)^2/(\alpha(1-\alpha)\rho^1/\rho^0) \tag{4.2}$$

In accordance with Eq. (4.2), when α equals 0.5 the sound speed in the medium has a minimum value. For the complete thermodynamic description of the two-phase porous medium, one must assume an exact or approximate law of phase heat transfer as applied to a specific type of porous material. If the radius of the pores is small, interphase thermal equilibrium in the medium can be established in less time than the characteristic duration of propagation of acoustic disturbances. In this case, it is admissible to regard the medium to be in equilibrium thermodynamically and, even, isothermally (as a result of high specific heat of the solid phase). Similar approximation has been used by the authors of [140–143]; however, in [141] an estimate of the cooling time of an isolated gas bubble of radius r in water, which is initially compressed adiabatically from pressure p_1 to pressure p_2 ($T_0 = 21°C$), is also proposed:

$$t_1 \approx 43r^2 (p_2/p_1)^{0.05} \qquad t_2 \approx 8600r^2 (p_2/p_1)^{1/3} \tag{4.3}$$

where t_1 and t_2 are the cooling times of the bubble by 10 and 90%; r is in meters. Applying the said estimate to polyurethane foam and assuming $r \approx 10^{-3}$ m (typical value of the radius of pores), we find that in acoustic approximation $t_1 \approx 5 \cdot 10^{-5}$ sec and $t_2 \approx 10^{-2}$ sec. In case the propagation velocity of acoustic disturbances in the medium exceeds the value r/t_1, it is expedient to assume the absence of interphase heat transfer and the compression process of the gas in the pores of the material to be adiabatic. When $r \approx 10^{-3}$ m and $t_1 \approx 5 \cdot 10^{-5}$ sec the ratio $r/t_1 \approx 50$ m/sec. This value is noticeably less than the experimentally measured sound speed in polyurethane foam [136] which is equal to $\sim 2 \cdot 10^2$ m/sec. Consequently, one can disregard the interphase heat transfer in these cases and assume that in relation (4.2) $(a^0)^2 \simeq \gamma p^0/\rho^0$, where p^0 is the gas pressure in the pores of the medium; γ is the adiabatic index of the gas. An expression for the determination of the "frozen" sound velocity in a porous medium can hence be obtained

$$a^2 \approx (\gamma p^0)/(\alpha(1-\alpha)\rho^1) \tag{4.4}$$

The density of the homogeneous mixture is determined by expression

$$\rho = \alpha\rho^0 + (1-\alpha)\rho^1 \tag{4.5}$$

In the case when the gas pressure in the pores $p^0 \leqslant 1$ atm expression (4.4) is further simplified by accounting for (4.5):

$$a^2 \approx (\gamma p^0)/(\alpha\rho) \tag{4.6}$$

The use of the simplified relations (4.4) and (4.6) for the calculation of the frozen speed of sound in a porous medium is particularly convenient in the case when the medium has an open lattice structure, since here $\alpha = \alpha_0$ and is independent of the gas pressure. When $\alpha = 0$, it follows from (4.4) that the speed of sound tends to infinity as a result of neglecting the compressibility of the solid phase of the porous medium. When $\alpha = 0.5$, the sound speed tends to a minimum value of $a_1 = 2(\gamma\alpha p^0/\rho^1)^{1/2}$, which is equal to ~ 21.6 m/sec at $p^0 = 1$ atm and $\rho^1 = 1.2 \cdot 10^3$ kg/m^3; this is noticeably lower than the sound speed in air at normal conditions.

Let us now discuss the relations at the front of the steady shock wave propagating in a two-phase medium which is regarded as a continuous quasi-gas. Presented are calculations for the case of thermally insulated phases (heat transfer behind the shock wave is absent), and the calculations for the equilibrium case can be found in [140]. The assumption that elastic forces of the material skeleton are negligibly small compared with the pressure forces of the gas in the pores leads to a significant simplification of the system of balance equations of mass, momentum, and energy at the shock wave. Further simplification of the problem is contingent upon neglecting gas filtration through the permeable structure of the porous material, which is not self-evident and requires experimental verification. Analysis of the experimental results, obtained in [137], shows that for relatively strong waves in elastic small lattice polyurethane foam when the pressure gradient at the front $\Delta p \sim (10\text{–}15)$ atm and the initial pressure $p_0 = 1$ atm, the effect of gas filtration ahead of the wave front is insignificantly pronounced, which is an indication, in this case, of the admissibility of neglecting filtration. When porous plastic materials, having closed lattice structure, are used, filtration is automatically excluded.

Let us assume that interphase heat transfer at the shock wave can be neglected (an "adiabatic" flow model for two-phase medium is considered). The indicated simplification is based on the evaluation of the time for cooling the isolated gas bubble by 10% to be $t_1 \approx 5 \cdot 10^{-5}$ sec. It then follows that when the shock front lasts for $\tau \approx 10^{-4}$ sec [137], the gas temperature in the pores changes insignificantly, which allows one, henceforth, to ignore interphase heat exchange.

Let us go over to the system of coordinates of the shock. Let subscript "1" correspond to the parameters of the medium ahead of the shock wave, and subscript "2" behind the shock wave. Taking into account the assumptions made, the system of equations of balance of mass, momentum, and energy are identical in form to the corresponding equations for a perfect gas:

$$\begin{aligned} &\rho_1 u_1 = \rho_2 u_2 \\ &\rho_1 u_1^2 - \rho_2 u_2^2 = p_2 - p_1 \\ &p_1/\rho_1 + U_1 + u_1^2/2 = p_2/\rho_2 + U_2 + u_2^2/2 \end{aligned} \tag{4.7}$$

where ρ_i, u_i, p_i, and U_i ($i = 1, 2$) stand for density, mass velocity, pressure, and specific internal energy of the medium, respectively, in a specified region of flow. The simultaneous solution of equations (4.7) allows one to obtain an expression for the Hugoniot curve

$$U_2 - U_1 = ((p_2 + p_1)/2)\,(1/\rho_1 - 1/\rho_2) \tag{4.8}$$

The difference between the two-phase medium, represented as a quasi-gas, and the perfect gas is manifested first and foremost in the expression for internal energy U. The specific internal energy for the perfect gas is $U^0 = p^0/\rho^0\,(\gamma - 1)$ (from now on, the superscript "0" denotes the parameters of the gaseous phase). Since the elastic forces of the skeleton of the porous material are assumed to be negligibly small, then $p = p^0$; therefore, $U = p^0/\rho^0 \times (\gamma - 1)$. It is evident that in the absence of interphase heat transfer, when the internal energy of the medium is assumed to be equal to the internal energy of the gas in the free volume of the pores, the relation $U\rho = U^0\rho^0\alpha$ is true and it follows that

$$U = (1/(\gamma - 1))\,((p\alpha)/\rho) \tag{4.9}$$

We note that the values of the porosity coefficient α_1 and α_2, which are required to compute the values U_1, U_2 from Eq. (4.9), are not equal and are linked by a relation which can be readily obtained from the definition of the coefficient α in the two-phase medium:

$$\alpha = (v - v^1)/v \tag{4.10}$$

where v and v^1 are specific volumes of the medium and of the solid phase, respectively, trapped in volume v. Since the compressibility of the solid phase is ignored (v^1 = const), the value $v(1 - \alpha)$ = const, from which $\rho_2/\rho_1 = v_1/v_2 = (1 - \alpha_2)/(1 - \alpha_1)$.

From this expression the formal relation between the porosity coefficients on both sides of the discontinuity follows

$$\alpha_2 = 1 - (1 - \alpha_1)(\rho_2/\rho_1) \tag{4.11}$$

Substituting in (4.8) the values of U_1 and U_2, computed from Eq. (4.9), we obtain

$$\frac{p_2}{p_1}\left(\frac{2}{\gamma - 1}\frac{\alpha_2}{\rho_2} - \frac{1}{\rho_1} + \frac{1}{\rho_2}\right) = \frac{2}{\gamma - 1}\frac{\alpha_1}{\rho_1} - \frac{1}{\rho_2} + \frac{1}{\rho_1} \tag{4.12}$$

Multiplying both sides of (4.12) by $\rho_2(\gamma - 1)$ and substituting the right-hand side of expression (4.11) for α_2, we obtain the required relation linking pressure and density ratios at a shock wave

$$\frac{p_2}{p_1} = \frac{\rho_2/\rho_1(2\alpha_1 + \gamma - 1) - (\gamma - 1)}{\rho_2/\rho_1(2\alpha_1 - \gamma - 1) + (\gamma + 1)} \tag{4.13}$$

This expression is analogous to the Hugoniot adiabate for a two-phase medium and, as can be readily shown, when $\alpha_1 = 1$ (the case of pure gas) it acquires the classical form [148]:

$$\frac{p_2}{p_1} = \frac{\rho_2/\rho_1(\gamma + 1) - (\gamma - 1)}{\rho_2/\rho_1(1 - \gamma) + (\gamma + 1)}$$

This equation can be rewritten after solving for ρ_2/ρ_1 with respect to p_2/p_1:

$$\frac{\rho_2}{\rho_1} = \frac{p_2/p_1(\gamma + 1) + (\gamma - 1)}{p_2/p_1(\gamma + 1 - 2\alpha_1) + (2\alpha_1 + \gamma - 1)} \tag{4.14}$$

In a shock wave of high intensity we have $p_2/p_1 \to \infty$: hence, from (4.14) we obtain the value of the compression limit of the medium

$$\left(\frac{\rho_2}{\rho_1}\right)_\infty = \frac{\gamma + 1}{\gamma + 1 - 2\alpha_1}$$

For example, when $\gamma = 1.4$ and $\alpha_1 = 0.95$ the compression limit $(\rho_2/\rho_1)_\infty = 4.8$, i.e., noticeably less than in the case of a pure gas, when $(\rho_2/\rho_1)_\infty = 6$.

Using the shock adiabatic equation (4.13), it is possible to find, in explicit form, expressions for the velocity of the shock wave in material D_2 and mass velocity behind the front u_2. We use the known expressions [148]

$$D_1 = v_1 \sqrt{\frac{p_2 - p_1}{v_1 - v_2}} = \sqrt{\frac{p_1}{\rho_1}\left(\frac{p_2}{p_1} - 1\right)\frac{\rho_2}{(\rho_2 - \rho_1)}}$$

$$u_2 = \sqrt{(p_2 - p_1)(v_1 - v_2)} = \sqrt{\frac{p_1}{\rho_1}\left(\frac{p_2}{p_1} - 1\right)\frac{(\rho_2 - \rho_1)}{\rho_2}} \tag{4.15}$$

From Eq. (4.14) it follows that

$$\frac{(\rho_2 - \rho_1)}{\rho_2} = \frac{2\alpha_1 (p_2 - p_1)}{p_2(\gamma + 1) + p_1(\gamma - 1)} \tag{4.16}$$

Substituting this expression in (4.15), we obtain

$$D_2 = \sqrt{\frac{p_2(\gamma + 1) + p_1(\gamma - 1)}{2\alpha_1 \rho_1}}$$

$$u_2 = (p_2 - p_1)\sqrt{\frac{2\alpha_1}{\rho_1 (p_2(\gamma + 1) + p_1(\gamma - 1))}}$$

For low initial pressures (more precisely, when $p_2 >> p_1$), the expressions for the velocities are simplified to

$$D_2 \approx \sqrt{\frac{\gamma + 1}{2\alpha_1}\frac{p_2}{\rho_1}}, \quad u_2 = \sqrt{\frac{2\alpha_1}{\gamma + 1}\frac{p_2}{\rho_1}}$$

Using relation (4.16) and the expression for the frozen sound speed (4.6), the shock-wave Mach number can be expressed as a function of the gasdynamic parameters

$$M_2 = D_2/a \simeq \sqrt{(p_2(\gamma + 1) + p_1(\gamma - 1))/(2p_1\gamma)}$$

If we formally assume $\gamma \rightarrow 1$ (as noted in [140], this assumption corresponds to the case of interphase thermal equilibrium when the temperature change at the shock $\Delta T \rightarrow 0$), we obtain $M_{21} = \sqrt{p_2/p_1}$. An analytical expression for M_{21}, similar to the one presented, was obtained in [140] in connection with an isothermal liquid–gas bubble system.

Let us discuss, with the help of the presented relations, the process of reflection of an air shock wave from a wall covered by a layer of polyurethane foam. For the calculation of gasdynamic parameters of the shock wave developing in the porous material, it is convenient to use the method of shock polars in a gas [6]. This allows one to obtain, in addition to the magnitude of the pressure in the shock wave p_2, the value of the velocity u_2 of the interface between the gas and the porous material at the given intensity of air shock wave (Mach numbers M_0).

Some results of the calculation of pressure p_2 and the coefficient of pressure decrease at the interface $k(M_0) = P_2/P_{20}$ (where p_{20} is the pressure of normal reflection of the incident shock wave from a rigid wall) are given below for various values of M_0. The initial air pressure is assumed as $p_2 = 1$ atm, the density of the material $\rho = 23.5$ kg/m^3, and the coefficient $\alpha = 0.98$.

M_1	1.3	1.5	1.7	1.9	2.1	2.3	2.5
P_2	2.59	4.12	6.1	8.52	11.37	14.66	18.37
K	0.83	0.76	0.71	0.67	0.65	0.62	0.6

It is important to note that the experimentally measured values of $k(M_0)$ in the range of $M_0 = 1.6–2.1$ for polyurethane foam PPU-EM-1 with the indicated parameters agree with the computed data with an accuracy of ~5%.

When calculating the interaction of the shock wave in a porous compressible medium with a rigid wall, one must use the equation of balance of momentum for the incident and reflected shock waves which leads to the relation [148]

$$((p_3 - p_2)/\rho_2)(1 - \rho_2/\rho_3) = ((p_2 - p_1)/\rho_1)(1 - \rho_1/\rho_2)$$

where subscript "3" denotes the parameters of the reflected wave. Substituting in this expression ρ_2/ρ_1 and ρ_3/ρ_2 from (4.14) and accounting for the interdependence (4.11) of coefficients α_2 and α_1 allows one to obtain the relation between the reflection coefficient $R_0 = p_3/p_2$ and the pressure ratio p_2/p_1 at the shock wave front propagating in a porous medium:

$$R_0 = \frac{(3\gamma - 1)p_2/p_1 - (\gamma - 1)}{(\gamma - 1)p_2/p_1 + (\gamma + 1)} \tag{4.17}$$

This expression is identical to the classical expression [148], obtained for the case of a pure gas. The agreement of the results is easily explained, since the pressure on the wall p_3 is equal to the gas pressure in the pores of the material which is compressed by the reflected shock wave. This pressure is completely determined by the classical expression (4.17) for any intensity of the reflected shock wave. Assuming formally that $\gamma = 1$, we have $R_0 = p_2/p_1$, which agrees with the result obtained in [140]. We note that the interphase thermal equilibrium in the reflected shock wave from the wall is established somewhat more quickly than in the incident wave. This is so since the characteristic radius of the pores in the material decreases following the passage of the shock wave and the cooling times t_1 and t_2 decrease according to relation (4.3).

Denoting the initial air pressure by P_0, we write the expression for the coefficient of pressure increase $R_1 = p_3/p_{20}$:

$$R_1(M_0) = k(M_0)\frac{(3\gamma - 1)k(M_0)p_{20}(M_0)/p_0 - (\gamma - 1)}{(\gamma - 1)k(M_0)p_{20}(M_0)/p_0 + (\gamma + 1)} \tag{4.18}$$

For interphase thermal equilibrium, when the specific heat capacity of the solid phase can be assumed to be infinitely large compared with the gaseous-

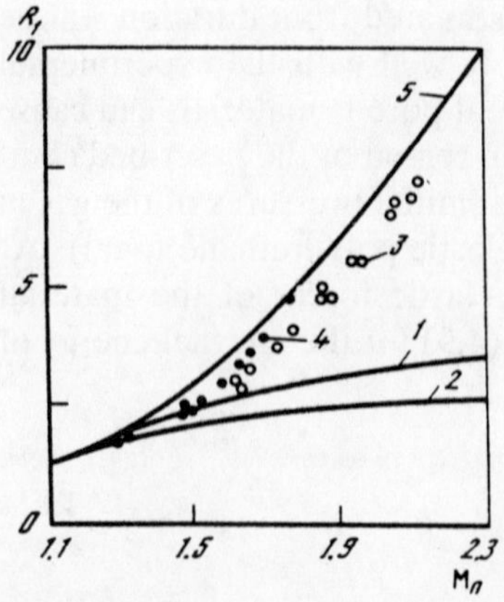

Figure 4.3. Dependence of the coefficient of pressure increase at the wall $R_1 = p_3/p_{20}$ on the Mach number of the air shock wave M_0.

phase specific heat capacity, with $\gamma = 1$, we have

$$R_1(M_0) = k^2(M_0)p_{20}(M_0)/p_0 \tag{4.19}$$

Figure 4.3 depicts the dependences of the coefficient $R_1(M_0)$ obtained from Eqs. (4.18) and (4.19) (curves *1* and *2*, respectively) for the initial pressure $p_0 = 10^5$ Pa, $\rho = 23.5$ kg/m^3, and $\alpha = 0.98$. Also shown in the figure are the results of the experimental measurement $R_1(M_0)$, obtained by using helium (*3*) and nitrogen (*4*) as driving gases. When helium was used the shock wave in the region of the test section was characterized by a falling pressure profile (pressure drop amounted to 10–15% in 1 m sec). This effect, which is not observed when nitrogen is employed, is due to the fact that the expansion wave reflected from the end of the high-pressure chamber catches up with the shock wave and forms unsteady pressure profile in the latter. The Mach number of the incident shock wave was varied in the range between 1.3–2.3. It should be noted that a satisfactory qualitative agreement between the calculated and experimental values of $R_1(M_0)$ takes place at $M_0 \leqslant 1.6$ only; when $M_0 > 1.6$ the experimental points lie above the calculated curve. Figure 4.3 also shows the calculated curve *5* plotted with the help of relation (4.15) for the coefficient $k(M_0)$, which is computed within the bounds of the adiabatic flow model ($\gamma = 1.4$). Curve *5* gives the maximum values of $R_1(M_0)$ at the given initial conditions. It is evident that when M_0 increases the experimental values of R_1 is described approximately by curve *5*, which is an indirect confirmation of the occurrence of interphase heat transfer in the shock wave in the porous material. Without going into detailed discussion of the processes leading to the development of the shock wave front in a porous compressible medium, we can note that the duration of the front must considerably exceed the values which are typical for shock waves in gases. In the general case, the thickness of the shock wave in a medium of fine structure corresponds to several characteristic dimensions of the medium d which, in our case, is the diameter of the pores. Assuming, for the sake of simplicity, that the front develops at a distance $l \approx 10d$, when $d \approx 10^{-3}$ m and the shock-wave velocity in the medium $D_2 \approx 2 \cdot 10^2$ m/sec, the estimated front duration will be $\tau \approx l/D_2 = 5 \cdot 10^{-5}$ sec, which agrees reasonably well with the experimental data [137]. The permeability of the skeleton of real porous materials can cause further increase of front duration. The application region of the described computational method is limited by the relatively high initial pressures of the gas in the compressible porous medium ($\gtrsim 10^4$ Pa for elastic polyurethane foam). At low pressures one must take into account the elastic forces of the material skeleton and appropriately transform expression (4.9) for the internal energy of the porous medium.

CHAPTER

FIVE

REFLECTION OF SHOCK WAVES BY A PLANE SURFACE

5.1 TYPES OF WAVE CONFIGURATIONS UPON REFLECTION OF SHOCK WAVES

The main results obtained during the investigation of shock-wave reflection are presented in the monographs and review papers [1, 6, 49, 148–162]. Distinction is made between steady, quasi-steady, and unsteady reflection. The steady reflection in gas flowing in a channel with Mach number M is depicted schematically in Fig. 5.1*a*, *b*. The shock wave *IA*, developing at the wedge in the upper part of the channel, is reflected from the lower wall of the channel.

Two types of reflections are observed: regular (Fig. 5.1*a*), when the point of reflection of the wave *A* lies on the surface; irregular (Mach) reflection (Fig. 5.1*b*), when the incident wave *IA* and the reflected wave *AR* intersect at a point no longer located on the wall, and when a third wave develops—Mach wave *AM* with a contact discontinuity *AE* (ω_1 and ω_2 are angles of incidence and reflection, respectively).

If, on the other hand, the shock wave travelling at a Mach number M_0 impinges on a wedge with a vertex angle α_0, then a quasi-steady reflection takes place (Fig. 5.1*c*, *d*). The reflection from the wedge is quasi-steady, since the wave travel takes place in a self-similar manner and, by going over to a system of coordinates moving with the intersection point *A*, one can obtain a steady wave pattern near this point. This pattern corresponds to the steady flow of the gas with Mach number $M = M_0/\sin \omega_1(\chi_1)$ is the angle made by the trajectory of the triple point.

Unlike the pattern of the steady reflection, the change in direction of the gas flow when it moves over a concave angle $O'OO''$ causes compression of the gas and formation of a detached NR (Fig. 5.1*c*) or an attached OR (Fig. 5.1*d*) waves which are conjugate with the reflected wave AR.

In the quasi-steady regime, reflection also takes place in a regular (Fig. 5.1*c*) or an irregular (Fig. 5.1*d*) manner. However, for this case, a large diversity is observed of the irregular reflection due to the different forms of the conjugacy.

Apart from the simple Mach reflection (Fig. 5.1*d*) shapes of a reflected wave with deformation have been observed: complex Mach reflection with a kink on the reflected wave (Fig. 5.2*a*) and double Mach reflection with a second triple point on the reflected wave having a positive (Fig. 5.2*b*) or a negative (Fig. 5.2*c*) value of the reflection angle ω_2. In the limiting case of the double Mach reflection the second triple point R moves along the wedge surface (Fig. 5.2*d*).

Simple Mach reflection takes place at small wedge angles and sufficiently weak shock waves. For strong shock waves and small wedge angles, reflection takes place in the forms of complex Mach and double Mach reflection. The boundaries for different forms of configuration for reflection in air presented in Fig. 5.3, where R is a regular reflection; DM is a double Mach reflection; CM is a complex Mach reflection; M is a simple Mach reflection; *1* is the calculated boundary between regular and irregular reflections; *2* is the calculated boundary between simple and complex Mach reflections; *3* is the experimentally determined boundary between complex and double Mach reflections; *4* are the experimental points; and *5* is the result of numerical calculations.

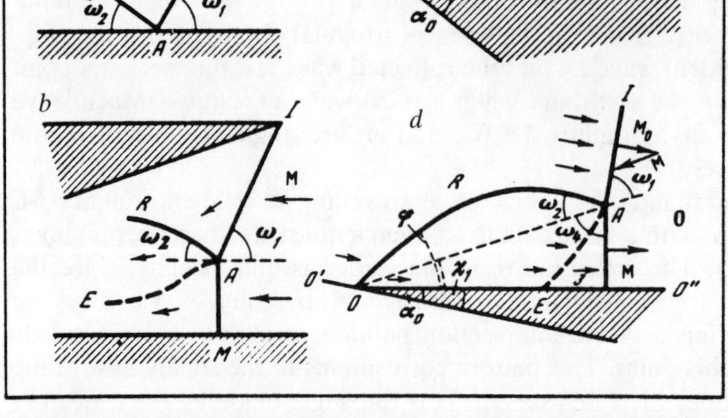

Figure 5.1. Steady (*a*, *b*) and quasi-steady (*c*, *d*) reflection of shock waves.

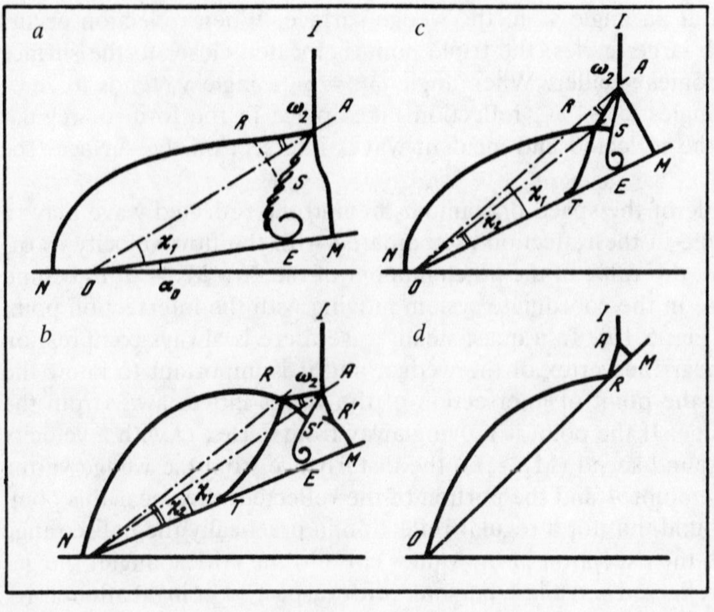

Figure 5.2. Shapes of quasi-steady reflection of shock waves.

Let us discuss the reflection of weak shock waves which was investigated in detail in [163–172]. Let us take a small wedge angle α_0 and weak wave with Mach number M_0. According to Fig. 5.3, reflection occurs in the shape of a simple Mach reflection (Fig. 5.1*d*).

The contact discontinuity in this case breaks up into a chain of vortexes and spreads over the surface. The triple point *A* moves along the straight line *OA*

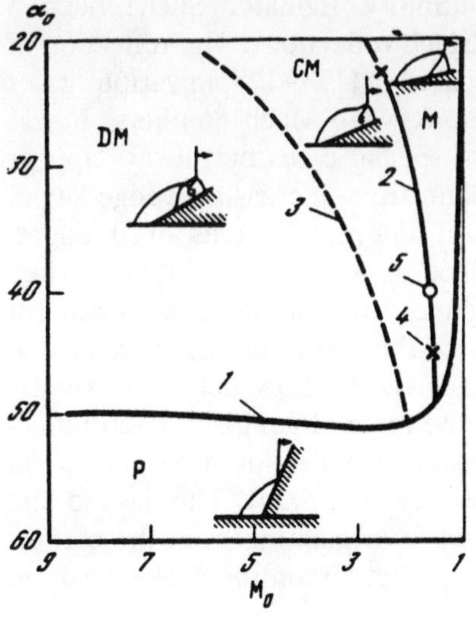

Figure 5.3. Domains of the different types of reflections.

which is inclined at an angle χ_1 to the wedge surface. When reflection occurs from wedges with large angles, the triple point is located closer to the surface and angle ω_1 becomes smaller. When angle $\omega_1 = \omega_{cr}$, angle χ_1 tends to zero, and for all the angles $\omega_1 < \omega_{cr}$ reflection takes place in the form of regular reflection when the reflected and incident waves intersect on the surface (see Fig. 5.1*c*).

The magnitude of the speed of sound a_2 behind the reflected wave plays a very important role in the reflection as compared with the flow velocity $\bar{u}_2$ or, equally important, the value of the Mach number of the flow $\overline{M}_2 = \bar{u}_2/a_2$ behind the reflected wave in the coordinate system moving with the intersection point of waves *A* (see Fig. 5.1*d*). In a quasi-steady case there is always compression of the gas flow near the vertex of the wedge, and it is important to know the velocity at which the point of intersection of the waves moves away from the vertex of the wedge. If the point is moving away from vertex *O* with a velocity higher than the sound speed ($\overline{M}_2 > 1$), the disturbance from the wedge vertex does not overtake point *A* and the portion of the reflected wave near this point is straight. It is found that for a regular reflection in practically the entire range of angles α_0 (with the exception of the values close to the critical angle) the gas flow behind the reflected wave is supersonic with respect to point *A*; therefore, the portion of the reflected wave *AR* near this point is straight (see Fig. 5.1*c*). In a simple Mach reflection, the Mach number of the gas $\overline{M}_2$ behind the reflected wave with respect to point *A* is less than unity, i.e., all the signals from the vertex of the corner reach this point, the reflected wave is bent from point *R* up to point *A* (see Fig. 5.1*d*).

Let us discuss the reflection of a strong shock wave, which was investigated in great detail in [173–200]. For very small wedge angles the reflection takes place in the form of a simple Mach reflection as shown in Fig. 5.1*d*. In [165] it is mentioned that when the intensity of the shock waves was increased a kink and, sometimes, a second triple point was observed on the reflected wave from which a shock wave and a second contact discontinuity originated. Such reflection forms were later called complex and double Mach reflections. The reflection of strong shock waves was systematically investigated in [173–175] in various gases for a wide range of reflection angles and shock-wave Mach numbers. It was established that such reflection types are not special cases but always appear with an increasing Mach number of the incident wave, increasing wedge angle, and most readily in gases with excitation of internal degrees of freedom. The creation of new configurations is explained as follows: they appear when the flow behind the reflected wave becomes supersonic and the compression of the gas at the wedge vertex ceases to influence the gas state in the vicinity of the triple point. When, for the given M_0 and ω_1, the flow behind wave *AR* becomes supersonic, the reflection occurs in the form of complex Mach reflection. Part of the reflected wave becomes straight from point *A* to point *R* in which a kink on the reflected wave develops (see Fig. 5.2*a*). The dashed line *RS* shows the boundary of the compression wave originating from the corner. Under the influence of this compression, the contact discontinuity *AE* is curled into a large-scale vortex. The pressure transducers on the wedge surface begin to show additional pressure increase on the surface after the passage of point E following the first pressure rise, which is obtained when the Mach wave AM passes near the transducer [175, 176, 179, 181]. When angle α_0 is increased, the compression of the gas increases and a second shock wave RS is formed (see Fig. 5.2*b*, *c*). Its intersection with the reflected wave AR leads to the appearance of a second triple point. Reflection takes place as a double Mach reflection with a second triple point R, from which a secondary contact discontinuity RT originates. In the coordinate system moving with point R, the gas stream flows into the incident wave, then into the reflected wave and runs over the wedge surface in the form of a jet transforming the first contact discontinuity AT into a vortex. Two consecutive pressure rises are also seen on pressure oscillograms during double reflection.

The complex Mach reflection is, in essence, a degenerated form of a double Mach reflection. The second triple point can move at an angle χ_2, which is less than χ_1; thereupon, the reflection angle becomes negative and the configuration acquires the same shape as in Fig. 5.2*c*. In double Mach configurations a kink or a new triple point can also appear at point R' (see Fig. 5.2*b*) on the Mach wave, as noted in [180, 186]. The height of the Mach wave decreases at large wedge angles α_0. In [187], a configuration resembling the one illustrated in Fig. 5.2*d* was obtained upon reflection from angles close to the critical when the second triple point practically moves over the surface. At wedge angles larger than α_{cr} reflection occurs in the form of a regular reflection.

Photographs of reflection in nitrogen showing the form of a complex Mach reflection and in carbon dioxide with a pronounced form of double Mach reflection are presented in Fig. 5.4. As was noted before, forms of simple Mach and regular reflections have been revealed so far during steady reflection. It is probable that for strong shock waves at a particular combination of parameters M and ω_1 (at which double Mach reflection with negative reflection angle ω_2 exists in the quasi-steady case) special forms of reflection are created in the steady state; the process in this case will most likely take place in vibrational form.

5.2 COMPARISON OF EXPERIMENTAL DATA WITH REFLECTION CALCULATIONS FOLLOWING THE DOUBLE- AND TRIPLE-SHOCK THEORIES

The flow pattern in a quasi-steady reflection of shock waves is a combination of reflection and compression of the gas near the vertex of the wedge. Let us go over to the coordinate system moving with a gas particle behind the incident shock wave (see Fig. 5.1*c*, *d*). With respect to this particle the shock wave moves away to the right with a subsonic velocity and is reflected from the surface. The wedge, on the other hand, moves from right to left relative to the gas particle. If the velocity of the gas flow and the wedge angle are such that the flow takes

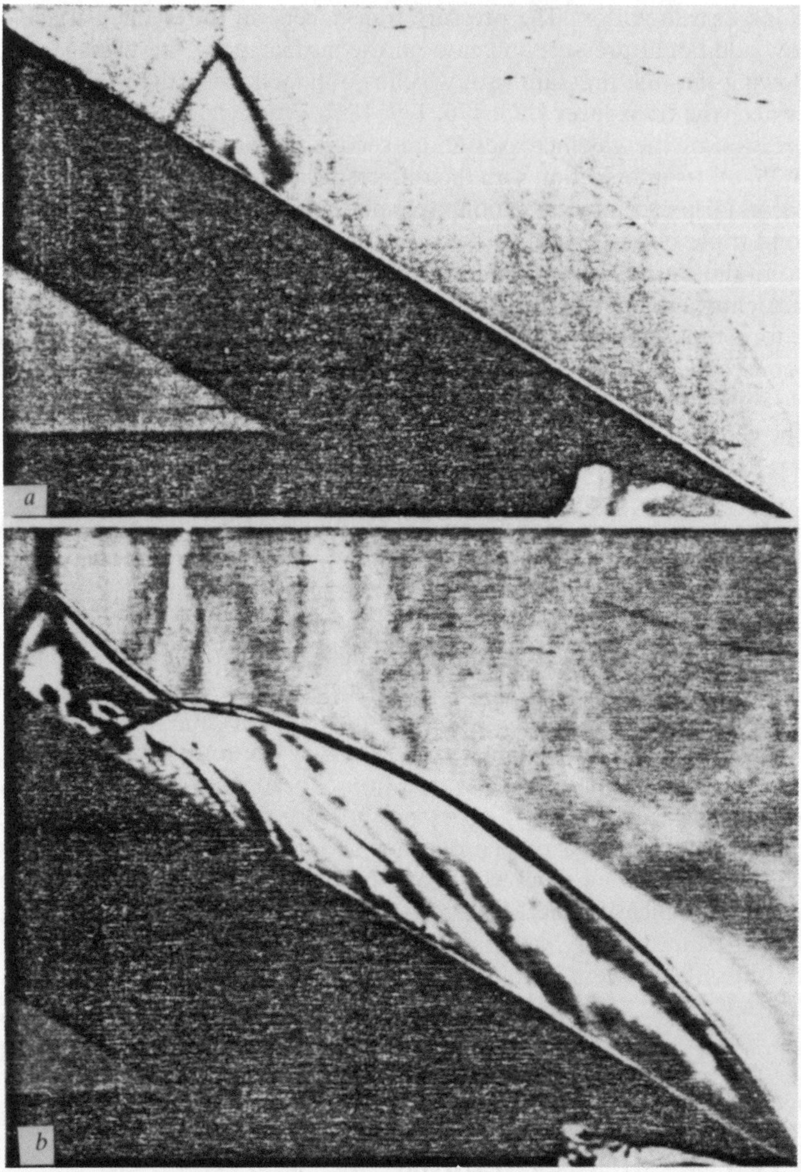

Figure 5.4. Complex Mach reflection in nitrogen, $M_0 = 3.92$, $p_0 = 5$ mm Hg, $\alpha_0 = 32°$ (*a*) and double Mach reflection in carbon dioxide, $\alpha_0 = 32°$, $p_0 = 20$ mm Hg, $M_0 = 5.18$ (*b*).

place with an attached wave, then the wave pattern near the wedge vertex corresponds to the stationary flow over the wedge (see Fig. 5.1*d*). This pattern is easily determined analytically. Otherwise, a detached wave is formed which, unlike the attached wave, does not correspond to the wave pattern of steady gas flow over the wedge. This wave also exists when the gas flow behind the incident wave is subsonic, since this is an unsteady wave which can be imagined as forming as a result of motion of an infinite inclined piston in the shock tube. The detached wave propagates slowly upstream.

The flow pattern for an ideal gas is self-similar. This follows from theoretical concepts and is verified experimentally in [174]. Therefore, we can go over to a system of coordinates moving with the point of intersection of the waves and examine its region [148, 155]. In this coordinate system, the gas flows into the wave at an angle ω_1, which is equal to the angle of incidence with a Mach number $M = M_0/\sin \omega_1$ and is deflected at the incident wave through an angle θ_1. When the reflection is regular the flow is deflected again at an angle θ_2 so that condition $|\theta_1| - |\theta_2| = 0$ is fulfilled, since the gas must flow parallel to the wall (Fig. 5.5*a*). Taking into account that the incident and reflected waves are straight, the laws of conservation can be written at these waves and a closed system of equations can be obtained. This system allows one to determine all the gas parameters, for the given M_0 and ω_1, in the vicinity of the intersection point, angles of deviation θ_1 and θ_2, and the reflection angle ω_2. For the solution of the system of equations the method of shock polars can be used [155].

The use of the polar method for the case of regular reflection is illustrated in Fig. 5.5*b*. Polar AI_1I_2GG' is constructed for the Mach number $M = M_0/\sin \omega_1$. The initial state of the gas behind the wave with Mach number M_0, which impinges on the wedge at an angle of incidence ω_1, is represented on the polar diagram by the point I_1. The polar diagram for the reflected wave is depicted by the curve $I_1R_1R_1G_1$. It is evident from the figure that condition $|\theta_1| = |\theta_2|$ is satisfied by two points on the polar diagram: R_1 and R_1'. Usually the regime corresponding to point R is realized (the weaker solution). It can be seen from the figure that the regular reflection can exist only when the polar diagram of the reflected wave intersects with the *y* axis. The boundary state is determined by the polar $I_2R_2F_2G_1$ when the maximum deviation of the flow θ_{cr} is achieved. The critical angle ω_1 corresponding to the boundary state is denoted by ω_e. The critical wedge angle is equal to $\alpha_{cr} = 90° - \omega_e$.

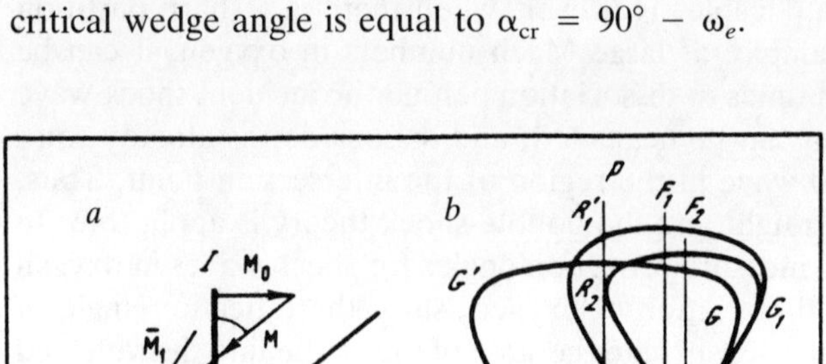

Figure 5.5. Regular reflection of shock waves *a*) schematic of wave configurations; *b*) shock polars in plane (p, θ) for regular reflection.

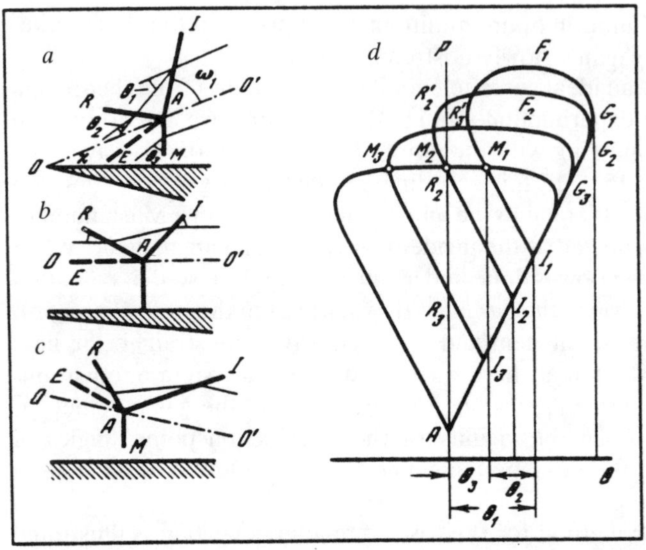

Figure 5.6. Irregular reflection of shock waves: Location of waves in normal (*a*), steady (*b*), and inversed (*c*) Mach configurations; shock polars for irregular reflection (*d*).

Experimental verification indicated that the double-shock theory describes the wave configurations well. Usually, the reflection angle ω_2 is measured as a function of the incident angle ω_1. For regular reflection of both weak and strong shock wave angles ω_2 correlate well with the experimental data.

If physicochemical transformations are taking place in the gas, then the self-similarity of the reflection process is violated and, consequently, the double-shock theory becomes inapplicable. However, when there is a sharp partition of relaxation time, for example, at large Mach numbers in oxygen, it can be assumed that the reflection times of dissociation behind the incident shock wave are so small that relaxation can be neglected, and dissociation is already completed behind the reflected wave in the region of the intersection point. Thus, both waves must remain straight and the double-shock theory is applicable. In [202–204] it is proposed to measure reflection angles for shock waves in oxygen to determine constants of dissociation of oxygen, since the reflection angle of the wave is a function which is sensitive to the state of the gas behind the reflected wave.

The critical angle ω_e of the existence of a regular reflection depends to a large extent on the degree of physicochemical transformations. This was shown in [205], where the boundaries of existence of the regular reflection were computed for various values of the effective adiabatic index γ from 1.01 to 1.6. The boundaries of a regular reflection for nitrogen and oxygen were computed in [191, 203] for the initial pressures from 1 to 1000 mm Hg accounting for excitation of molecular vibrations and their dissociation. The shift of the values of the boundary angle α_0 due to the influence of dissociation amounts to 10° towards increasing the existence region of regular reflection.

To solve the problem about irregular reflection in the triple-shock theory one must go over to the coordinate system moving with the triple point A. By analogy with the case of regular reflection we call the angle between the direction of travel OO' and the incident wave IA the angle of incidence ω_1 (Fig. 5.6*a*). The shock polars for irregular reflection are constructed in Fig. 5.6*d*. Analysis of the possible solutions for the intersection of shock waves in an ideal gas is conducted in [206]. The polar diagram corresponding to the reflected wave is the polar $I_1M_1F_1G_1I_1$. The polar diagram corresponding to the Mach wave is the same as that for the incident wave—polar $AI_1M_1M_3A$. The initial state is point I_1. The point of intersection of M_1 gives a solution corresponding to the configuration in Fig. 5.6*a*. Here, the gas is deviated by the reflected wave at an angle θ_2, which is less than angle θ_1, and the contact discontinuity AE along which the gas flows are directed towards the wall. When α_0 is increased, the trajectory of the triple point lies nearer to the wedge surface and angle ω_1 decreases. It is seen from the figure that similar Mach configurations can exist up to the moment when the trajectory of the triple point becomes parallel to the surface. In this case the magnitude of angle θ_2 is equal to the magnitude of angle θ_1; consequently, the configuration looks like the one in Fig. 5.6*b*. The critical angle ω_1, which corresponds to such a configuration, is designated by ω_N

This so-called steady Mach configuration is the limiting condition of these irregular configurations [150] in which the flow is deviated towards the wall behind the Mach wave.

In principle, it is possible for the point of intersection M_3 to exist when the deviation of flow behind the reflected wave is greater than that behind the incident wave (Fig. 5.6*c*). Such a configuration is not realized by quasi-steady reflection at the wedge, but its appearance is possible when the shock wave is reflected from a convex surface.

Checking the applicability of the triple-shock theory for the calculation of triple configurations was carried out in many works by measuring angle ω_2 as a function of ω_1 at a given M_0 for reflection from a wedge or values of M for steady reflection. Experiments showed that the triple theory does not give correct solutions for weak shock waves when simple Mach configuration is achieved, and is applicable for strong shock waves when the gas velocity behind the reflected wave becomes supersonic and reflection is realized in the form of a complex Mach or a double Mach reflection.

The results of experimental data on reflection of shock waves are given in Fig. 5.7. It is seen that the experimental data on regular reflection correlate well with the calculations in accordance with the double-shock theory, and the calculations in accordance with the triple-shock theory do not correlate with the experiments for weak shock waves (Fig. 5.7*a*). An explanation of the discrepancy was given in [207], where attention was paid to the Mach number of gas flows behind the reflected shock wave $\overline{M}_2$ in a system of coordinates moving with the

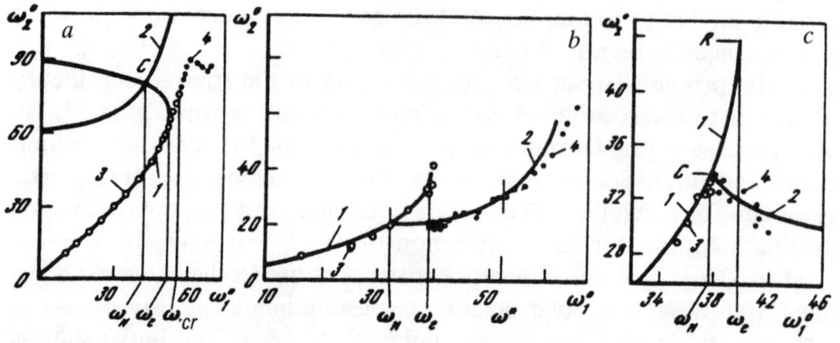

Figure 5.7. Dependence of the reflection angle on the incident angle: weak shock waves, $\xi = p_0/p_1 = 0.8$ [163] (*a*); strong shock waves, $\xi = 0.54$ [172] (*b*); steady reflection, Mach number of incoming gas flow $M = 2.95$ [190] (*c*). *1*) computation according to double-shock theory; *2*) computation according to triple-shock theory; *3*) regular reflection; *4*) irregular reflection.

triple point. In the vicinity of the triple point at $\overline{M}_2 \leq 1$, local subsonic and transonic regions are created. In this case, the flow near the triple point is determined by the entire interaction region. This was confirmed by the numerical calculations conducted in [208].

Fig. 5.7*b* shows the experimental results of the data of [172] on the reflection of strong shock waves, which indicate that the calculation for strong shock waves according to the triple-shock theory gives good correlation when reflection takes place in the form of a complex and a double Mach reflection. At constant Mach number M_0 angle α_0, and accordingly angle ω_1, was varied, i.e., in accordance

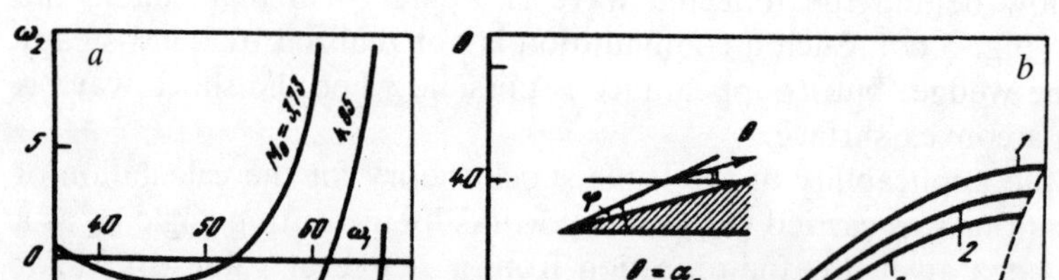

Figure 5.8. Dependence of the reflection angle ω_2 on the incidence angle ω_1 during a reflection in carbon dioxide, all modes of the molecular vibrations of the gas are excited (*a*) and effect of the degree of physicochemical transformations on the location of the attached wave in carbon dioxide (*b*).

with Fig. 5.3 the type of reflection must have been changed when ω_1 was increased from regular reflection to double Mach reflection at angle ω_e and to simple Mach reflection at angle ω^*. It is evident that the experimental points (Fig. 5.7*b*) correlate well for regular reflection and double Mach reflection, and that they begin to deviate from the theoretical curve for simple Mach reflection.

It is interesting to know the effect of the physicochemical transformations on the location of the waves in a triple configuration. Calculations showed that accounting for the real gas sharply changes the mutual location of the waves in the triple-shock configuration. Fig. 5.8*a* depicts the curves for the dependence of the reflection angle ω_2 on the incident angle ω_1 for carbon dioxide, taking into consideration vibrational excitation behind the waves. All similar curves for the ideal gas lie in the positive range, whereas part of the values for carbon dioxide at Mach number $M_0 = 3.79$ falls into the negative region, i.e., angle ω_2 becomes negative and the reflected wave lies beneath the line of motion of the triple point. When the Mach numbers M_0 are increased, the curves shift further to the negative region.

Calculations showed that accounting for the physicochemical transformations also leads to a change in the location of the attached waves, and namely, to a sharp decrease of the angle of the attached wave and acceleration of wave transition from detached to attached at the tip of the wedge when the wave velocity is increased. Fig. 5.8*b* depicts the dependence of the angle of the attached wave φ on the angle of flow deflection θ, which is equal to the wedge angle α_0, for a velocity of incident shock wave of 1.9 km/s in CO_2. Curve *4* represents calculations for an ideal gas, curve *1* shows calculations for completely dissociated gas, curves *2* and *3* represent no dissociation; with all vibrational modes excited on (*2*), and only asymmetric modes excited on (*3*). It follows from the figure that if carbon dioxide is regarded as an ideal gas, there is no attached wave when the wave impinges on a wedge with a vertex angle of more than 18°. As a result of excitation of the internal degrees of freedom and dissociation of the molecules, the wave which is formed at the inclined surface, becomes attached up to an angle of 42°. Such a dependence of angles ω_2 and φ on the degree of the physicochemical transformations causes the creation of double Mach configuration for the same wedge angles at lower Mach numbers in a gas where the physicochemical processes have developed further.

With the help of the triple-shock theory, the parameters in the vicinity of the triple point are calculated and angles ω_2 and ω_4 are determined for given M_0 and ω_1 (see Fig. 5.1*d*), i.e., if angle χ_1 is known for a given α_0, then angle ω_1 is found and the entire configuration is computed. If, on the other hand, angle χ_1 is unknown and only a wedge with vertex angle α_0 and a wave with Mach number M_0 are given, then the triple-shock theory can be supplemented by an additional assumption stating that the Mach wave is normal to the surface and that it is straight over the entire length up to the triple point. Thus, we obtain the so-called theory of elementary Mach reflection, which allows one to calculate the entire pattern of Mach reflection from M_0 and α_0.

Experiments show that the calculation of angle χ_1 according to the elementary theory correlates with the experiment only within a restricted range of wedge angles [49, 180, 182, 188, 191]. There is no correlation at small wedge angles and at angles close to the critical. Therefore, for the calculation of wave location in triple configurations it is reasonable to adopt the triple-shock theory with experimentally determined angles χ_1. The calculated values of χ_1, obtained by the equation presented in section 3.5, can also be used for small angles.

For determination of the pressure behind a Mach wave on the surface, both the triple-shock theory and the theory of elementary Mach reflection can be used. The triple-shock theory enables one to calculate the parameters near the triple point only. The Mach wave originating from the triple point bends and approaches the wall at a right angle; consequently, the pressure behind the wave, though almost independent of its curvature, can vary. The possible pressure values behind the Mach wave are represented by the points lying on the upper horizontal parts of the polar diagram. Therefore, the calculations conducted by the triple-shock theory and the theory of elementary Mach reflection correlate well. These theories are also applicable for the calculation of the second pressure rise in complex and double Mach configurations [179]. It is computed as the stagnation pressure of the gas in the region behind the reflected wave relative to the wall from equation

$$p_E/p_2 = (1 + (\gamma - 1)M_{2E}^2/2)^{\gamma/\gamma-1}$$

where M_{2E} is the Mach number of the gas in region *2* with respect to point *E* (see Fig. 5.1*d*).

Fig. 5.9 shows the results of the calculations and measurements of pressure from oscillograms in the first (*6*, *7*) and the second (*8*, *9*) rises for an angle $\alpha_0 = 24°$ during reflection in nitrogen and air. Nitrogen behaves as an ideal gas up to the Mach number $M_0 = 5.6$, and in air only molecular vibrations of oxygen

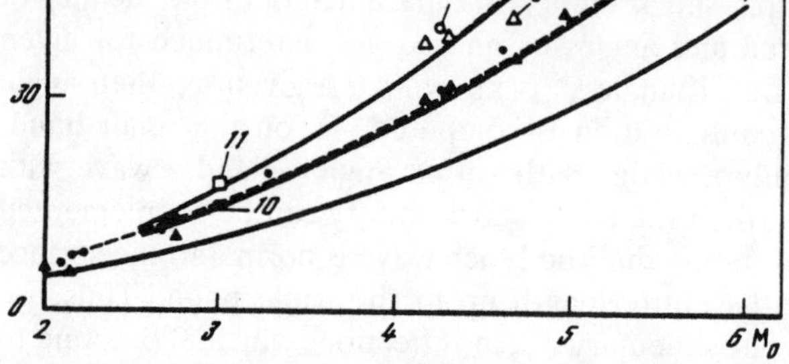

Figure 5.9. Comparison of experimental and calculated values in the first (on the Mach wave) and the second pressure rises.

are excited, contributing slightly to the pressure change. Computed curves for nitrogen and air differ by less than 5%; therefore, the experimental values for nitrogen and air are plotted in one graph ($\gamma = 1.4$). The calculated curves are presented also for a pressure drop in the incident wave p_1/p_0 (curve *1*), a pressure drop in the Mach wave p_3/p_0 (curves *2*, *3*), and the ratio of deceleration pressure p_E to pressure p_0 ahead of the incident wave (curves *4*, *5*).

The calculations based on the triple-shock theory with the experimental angle χ_1 (curves *3*, *5*) yield results which are close to the calculations according to theory of elementary Mach reflection (curves *2*, *4*). The graph shows that the pressure following the first rise corresponds to the pressure behind the Mach wave, and the pressure in the secondary rise correlates satisfactorily with the pressure of isentropic deceleration with respect to point *E* in region *2* (see Fig. 5.1*d*). The results of the numerical computations (see Chap. 8) are shown as curves *10* and *11*.

5.3 TRANSITION FROM ONE FORM OF REFLECTION TO ANOTHER

Transition from simple to complex and double Mach reflection. Transition from a simple Mach reflection to a complex reflection, as was mentioned in previous sections, takes place at the condition when $\overline{\mathrm{M}}_2 \geqslant 1$. Strictly speaking, it is more accurate to designate all regimes with $\overline{\mathrm{M}}_2 \geqslant 1$ as a double Mach reflection, since a complex Mach reflection is virtually a variation of the double Mach reflection when the reflected wave is straight and the compression is weak so that there is only a kink on the reflected wave. The condition for transition from a simple reflection to a complex reflection is formulated in [175, 179]. The appearance of the configuration with a kink can be experimentally located from the pressure oscillograms and Schlieren photographs by noting the occurrence of the kinks in the photos.

The crosses in Fig. 5.3 denote the experimental values of the transition angles which are obtained from the pressure oscillograms for angles 24° and 45°. The points were obtained from the intersection of the curves of the dependence on the Mach number in the first and second pressure rises. A point is given for angle 40°, which is determined from the pressure distribution on the wall obtained by a numerical method (chap. 8). All three points lie on the calculated curve of $\overline{\mathrm{M}}_2 = 1$. The determination of the transition boundary from the appearance of kink points on a reflected wave entails large errors. However, as experiments show, all regimes with kinks exist on the left from the calculated curve [185, 189, 191].

To determine the boundary of transition from a complex to a double Mach reflection one must know the location of the second triple point on the reflected wave. There are two hypotheses for the determination of the location of the kinks or the second triple point on the reflected wave. One hypothesis was stated in [178] in which the following procedure was proposed for finding the kink or the second triple point. The second triple point is the common point of the reflected wave at the triple configuration and the bow wave. Therefore, the lower boundary of these points can be defined as the intersection point of the boundaries of sonic disturbances in region *2* with the reflected wave. Another hypothesis is stated in [182]. According to this hypothesis, the kink is located at the intersection of the reflected wave with the line parallel to the incident wave and marks the location of the gas particle which was previously in the undisturbed gas in the cross section of the wedge tip. In [186], location of the triple point on the reflected wave was experimentally verified and it was revealed that, for small wedge angles, the scheme of [178] gives good correlation with the experiment. For large wedge angles, a better correlation is obtained with the hypothesis suggested in [182]. However, calculations conducted near the critical angle according to [182] also fail to show agreement with experiments.

The hypotheses about the criterion of transition to a double configuration with the second triple point are proposed in [179, 191]. In the first reference it is suggested that the configuration with a kink in the reflected wave exists when the triple-shock configuration at this point can exist only as a degenerate configuration. Another hypothesis, suggested in [191], deals with the transition from a complex Mach reflection to a double Mach reflection: transition takes place when the Mach number behind a part of the reflected wave, originating from point R, becomes greater than unity. It is likely that this criterion coincides with the criterion of [191]. In order to represent the transition criterion in the coordinate system (α_0, M_0), an additional hypothesis must be introduced to deal with the second triple point, which does not give complete correlation with experiments. Therefore, comparison of experiment with calculation cannot be regarded as an approbation of the criterion, although the calculation of the transition boundaries from this condition does not contradict the experimental data [179, 191, 198].

Transition from irregular to regular reflection. From the theory of regular reflection and triple-shock theory it follows that there exist two possible transition limits. One corresponds to the limit of existence of regular reflection and is determined by the condition $|\theta_1| = \theta_{cr}$; the other conforms to the existence of a steady Mach reflection and is determined by the condition $\theta_1 - \theta_2 = \theta_3 = 0$. This criterion is often referred to as the von Neumann condition. This same criterion was suggested by Henderson and Lozzi as a "criterion of mechanical equilibrium," since the pressure behind the reflected wave changes continuously during its transition through this limit as does the angle ω_2.

The angles of incidence ω_1, corresponding to these limits, are denoted by ω_e and ω_N. Fig. 5.10 illustrates the boundaries of the regions of existence of regular reflection ω_e (*1*) and boundary of Mach reflection ω_N (*2*). There is a region between the curves where both Mach and regular reflections are possible. Both curves intersect at point A. The Mach number M_0 at which this takes place is determined in [190, 191, 207]. For an ideal gas with adiabatic index $\gamma = 7/5$,

$M_0 = 1.45$. This Mach number is close to M_0 [207] at which the flow velocity behind the reflected wave is equal to the sonic velocity when angle $\omega_1 = \omega_e$.

Numerous experiments show that transition during reflection at a wedge occurs near the boundary ω_e which corresponds to the boundary of regular reflection [6, 163, 168, 186, 191]. The experimental value of ω_{cr}, at which transition takes place, is several degrees greater than ω_e (curve *3*, Fig. 5.10). Transition for steady reflection takes place at the boundary ω_N as was first shown in [209] and confirmed in [192, 211].

How transition takes place is well illustrated in Fig. 5.7, where the dependencies of the reflection angle ω_2 on the incident angle ω_1 are presented. At large wedge angles α_0, i.e., small angles ω_1, reflection takes place in a regular manner. Angle ω_2 corresponds to the calculated value by the double-shock theory. These values are illustrated in the figure by white circles. It is apparent that regular reflection exists up to the angle ω_e in quasi-steady reflection for strong shock waves (see Fig. 5.7*b*) and up to the angle $\omega_{cr} = (\omega_e + 1.5°)$ in reflection of weak waves (see Fig. 5.7*a*). In a steady reflection, transition to a regular reflection commences with an angle ω_N (see Fig. 5.7*c*).

The criteria of transition and their variation in steady and quasi-steady cases are considered in [194, 196, 211]. It is revealed that the reconstruction of the flow is linked with the possibility of transmitting information from downstream of the flow to the point of intersection of shock waves. Therefore, transition for the quasi-steady case takes place at the limit of regular reflection, since this limit practically coincides with the limit of the supersonic flow behind the reflected wave. In the steady case, when the disturbance signals in the flow could also be the expansion waves, transition takes place at the von Neumann limit.

By changing the conditions downstream of the flow, one can shift the transition limit. The reflection of a shock wave from two wedges was consistently investigated in [193]. Experiments were conducted in carbon dioxide for Mach number $M_0 = 2.9$. The angle between the reflecting surface of the second wedge and the incident wave was $\omega_1 = 40°$. For a wave with such a Mach number, angle ω_1 theoretically lies between angles ω_e and ω_N, i.e., two types of reflections are possible. If the wave impinges on a single second wedge at an angle ω_1 =

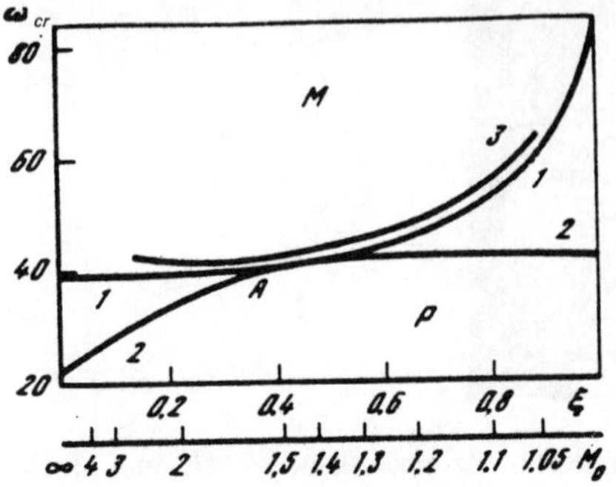

Figure 5.10. Boundaries of regions of regular reflection (*1*), steady reflection (*2*), experimental boundary (*3*).

40°, reflection occurs in the form of a regular reflection. The test was repeated on a double wedge. The wave is first reflected from the first wedge, so that the reflection took place in a Mach manner and the created Mach wave was reflected from the second wedge. In this case, reflection from the second wedge was a Mach reflection, since a supersonic region was initially created behind the reflected wave.

The shift of the experimentally determined angles ω_{cr} by several degrees relative to ω_e was investigated in [210] using large-size pictures. It was revealed that in the range of angles between ω_{cr} and ω_e, a regular configuration is seen with a distorted structure and with additional density gradient behind the reflected wave. An explanation of such shift of angle ω_{cr} is given in [168, 212], where this effect is related to the effect of the boundary layer behind the wave. Indeed, increasing the effect of viscosity by introducing artificial wall roughness causes a similar shift of ω_{cr} [212–215].

Fig. 5.7*b* shows that angle ω_2 changes discontinuously during the transition at the limit ω_e. Similarly, the transition at the limit ω_e for strong shock waves causes a pressure jump (Fig. 8.7, Chap. 8). It is rather odd that a small change in a wedge angle would cause a sharp change of pressure and angle of reflected wave. This, however, is explained by the fact that transition for strong shock waves takes place in the form of double Mach reflection where a region of elevated pressure, close to the pressure of regular reflection, is formed in the wake of the Mach wave.

It was revealed [188] that the double Mach configuration exists up to the moment of transition to regular reflection; however, it undergoes certain changes: upon approaching the boundary regime the first and second triple points and the root of the Mach stem draw together and, at the moment of transition to regular reflection, all three points merge into one point of regular reflection on the wedge surface.

Measurements by piezoelectric transducers showed that with a double Mach reflection from the wedge (near the critical), the following pressure events on the reflecting surface take place. First, the pressure rises behind the Mach wave;

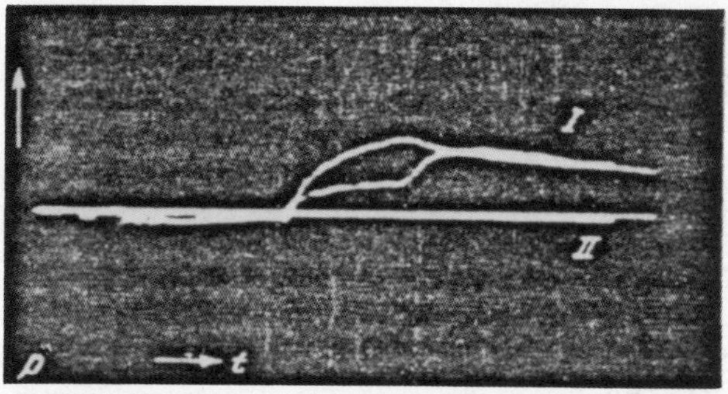

Figure 5.11. Superimposed oscillograms for regular reflection at angle $\alpha_0 = 47°$ (*I*) and double Mach reflection at angle $\alpha_0 = 43°$ (*II*) (carbon dioxide, $M_0 = 3.68$, $p_0 = 12.7$ mm Hg).

then in the second rise, linked with the presence and passage of a second triple configuration near the pressure gauge, the pressure rises up to the pressure of regular reflection. Fig. 5.11 shows two superimposed pressure oscillograms obtained during different tests in carbon dioxide. At angle $\alpha_0 = 47°$, regular reflection is observed with one pressure rise. At $\alpha_0 = 43°$, double Mach reflection is noticed with two pressure rises, with the rise in the second instance reaching the pressure of regular reflection. As angle α_0 is increased, the moment of the second rise approaches the moment of passage of the Mach stem near the gauge. The moment of transition to regular reflection cannot be recorded with a piezoelectric transducer, since, as was shown earlier, points *A* and *M* in this case are so close that the space-time resolution of the gauge becomes inadequate.

Thus, upon transition to Mach reflection, the pressure behind the Mach wave drops; however, the flow parameters change sharply near the root of the Mach wave. In the wake of the supersonic region, a region of flow stagnation sets in, where the parameters are close to the parameters of regular reflection.

CHAPTER

SIX

DIFFRACTION OF A SHOCK WAVE AT A CONVEX CORNER

6.1 SPECIFIC FEATURES OF THE WAVE PICTURE UPON DIFFRACTION OF SHOCK WAVES

When a shock wave moves along the sides of a convex corner *jom* (Fig. 6.1), a diffraction occurs, i.e., turning of the wave around the vertex *o*. Fig. 6.1*a* shows a schematic of diffraction when the gas flow *1* behind the wave SW_s is subsonic, $u_1 < a_1$; Fig. 6.1*b*, $u_1 = a_1$, i.e., $M_1 = 1$; in Fig. 6.1*c*, *d* the flow is supersonic. In all cases, an expansion wave propagates in the flow region *1* with the speed of sound a_1. When $u_1 \leqslant a_1$, the boundaries of the disturbed region are the sonic circles illustrated in Fig. 6.1*a*, *b* by the dashed lines. The center of these circles, according to [73], lies on the extension of line *jo*—the x–axis at a distance $u_0\tau$ from the corner vertex; the radius $R = a_1\tau$; τ is the time interval from the moment the wave passes the corner vertex. When $u_1 > a_1$, an expansion fan appears centered at point *o* bounded by the characteristics RW_1 and RW_2 in Fig. 6.1*c*, *d* (Prandtl-Meyer solution). The Mach number of the incoming flow M_1 can be determined from the angle of inclination of the characteristic RW_1. The disturbed flow of the region *1*, when $u_1 > a_1$ is bounded by the Mach line RW_1, which adjoins the sonic circle. It follows from this argument that the construction of the sonic circle in the case of diffraction does not differ from the case $\alpha_0 = 0$, which was discussed in Chap. 3 (see Fig. 3.6).

Inspection of triangles OZO″ and O′ ZO″ shows that the angle χ between the trajectory of point z and axis x is determined from equation

$$\operatorname{tg}\chi = \sqrt{a_1^2 - (u_0 - u_1)^2}/u_0$$

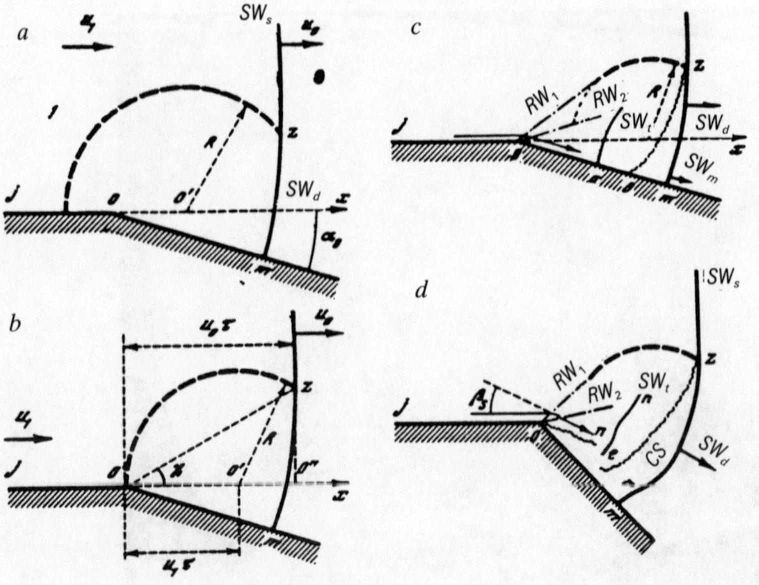

Figure 6.1. Schematic of the diffraction of shock waves near a corner *jom*. SW_s of the incident wave, SW_d is the diffracted part of the wave; SW_t is the stagnation wave; *a*) gas flow behind SW_s is subsonic; *b*) sonic; *c*) supersonic, diffraction without separation; *d*) supersonic, diffraction with separation; u_0 is the shock-wave velocity; u_1 is the gas velocity behind the shock wave.

where u_0 and u_1 are the propagation velocity of the shock wave SW_s and the flow behind it, respectively; a_0 and a_1 are the sound speeds ahead of and behind shock wave SW_s, respectively.

The diffracting part of the shock wave SW_d is weakened in comparison with that of SW_s and when $M_0 = u_0/a_0$ is small, the front of the SW_d is close to a sonic circle.

Diffraction without separation occurs when the inclination angles of the wall are small $\alpha \lesssim 20°$—stream *1* after turning at the corner vertex flows along the wall *om* (Fig. 6.1*c*). When $20° \lesssim \alpha_0 \lesssim 40°$, flow with partial separation is observed, and at $\alpha_0 \gtrsim 40°$, the separation line *oe* starts right from the vertex of the corner *o*, along which the stream flows after it turns at the tip by an angle β_s (Fig. 6.1*d*).

The problems of diffraction at small angles α_0 were discussed in [73]. By linearizing the differential equations of gasdynamics, the process of diffraction of the shock wave was analyzed at angles 0–7°. It is revealed that when the effects of viscosity are neglected, the diffraction process is self-similar. Theoretical curves of the pressure distribution along the surface of the corner were found. The pressure at each point of the surface must change as a result of flow adherence from the pressure behind the shock wave slipping along the wall SW_m up to the pressure at the end of the Prandtl-Meyer fan.

Shock-wave diffraction at small angles was investigated in [216]. It was revealed that in a real diffraction situation even at small angles α_0 a stagnation

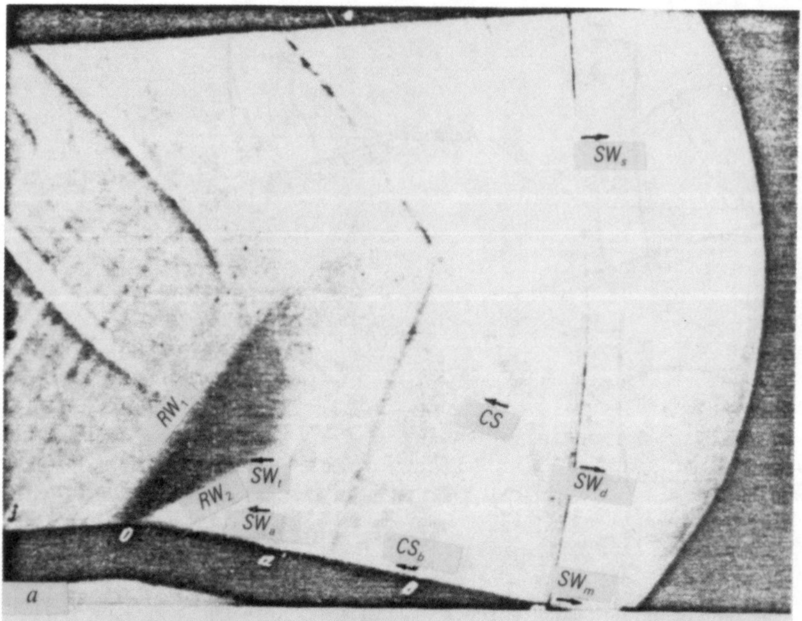

Figure 6.2. Schlieren photographs of diffraction *a*) M_0 = 2.3, α = 15°, oxygen; *b*) M_0 = 2.77, α_0 = 30°, air; *c, d*) M_0 = 4.71, α_0 = 65°, air; *e*) M_0 = 3.64, α_0 = 65°, freon-13; *f*) M_0 = 5, α_0 = 90°, CO_2.

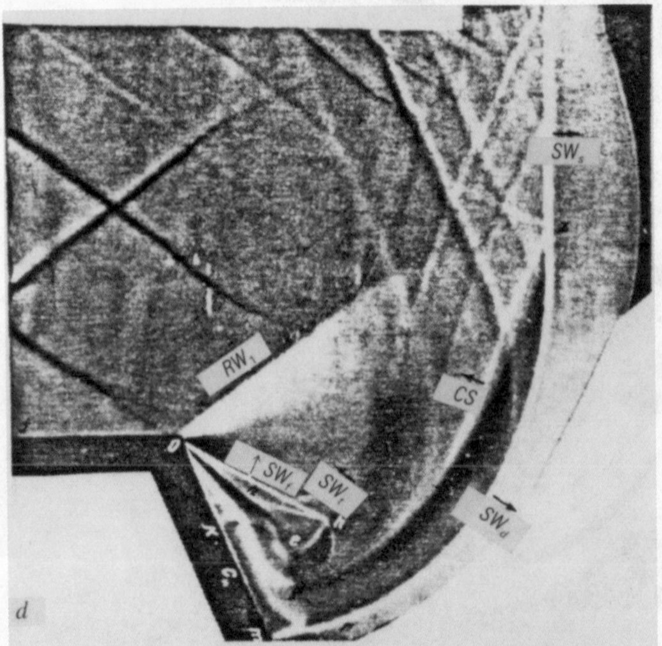

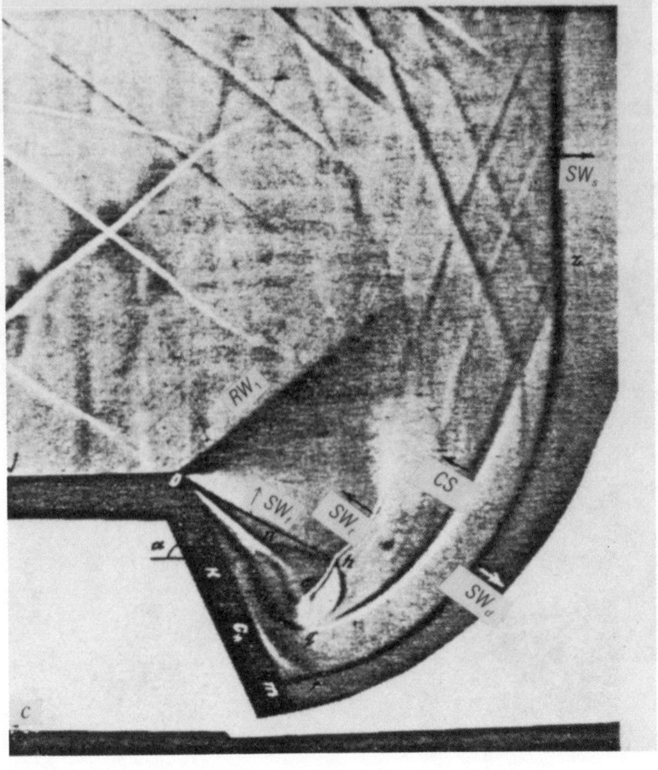

Figure 6.2. (*Continued*)

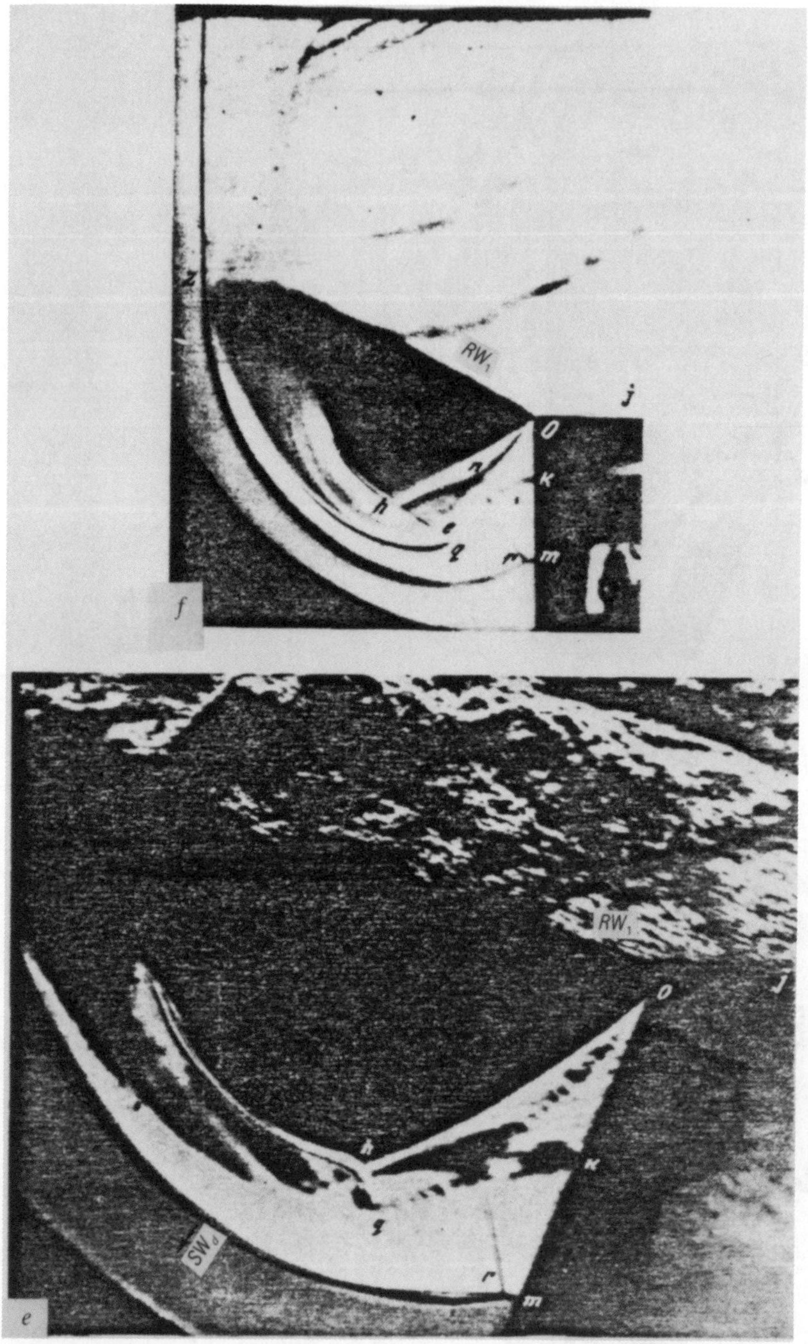

Figure 6.2. (*Continued*)

Figure 6.3. Diffraction interferogram, $M_0 = 2.1$, $\alpha = 90$, air.

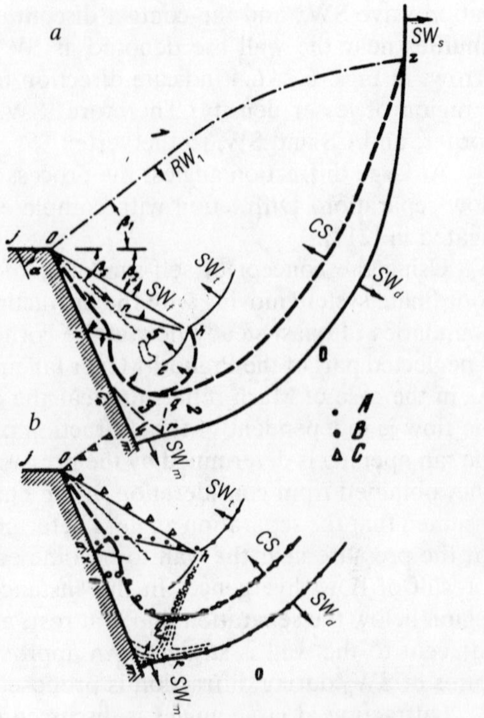

Figure 6.4. Schematic of diffraction corresponding to the photographs in Fig. 6.2*c*, *d* with the main symbols (*a*), and comparison with numerical calculations (section 8.3) (*b*) *A*, *B*, *C* are locations of discontinuities from the results of the numerical calculations.

wave SW_t, which is not taken into account by the linear approximation, appears together with the contact surface *zb* (Fig. 6.1*c*). Gas flow behind the diffracting part of the wave SW_d is slowed in comparison with the flow behind the undisturbed part SW_s. Simultaneously, the gas flow passing through the expansion wave is accelerated in comparison with the flow in region *1*. For the purpose of matching these currents, the stagnation wave SW_t opposing the flow is formed, in which the accelerated gas in the expansion wave is slowed down to the velocity behind SW_d.

Fig. 6.2*a–f* shows Schlieren photographs of the diffraction process of the shock wave at a convex two-sided corner; Fig. 6.3 shows an interferogram [233], Fig. 6.4*a*, *b* show schematic diagrams with symbols of the shock wave configurations corresponding to Fig. 6.22*c*, *d*.

The photograph in Fig. 6.2*a* is related to the case of diffraction at small angles ($M_0 \approx 2.3$; $\alpha_0 = 15°$ and $\gamma = 1.4$), at which oxygen flows over the vertex of the corner without flow separation from the wall. The structure of the gas-dynamic discontinuities here is the simplest: the expansion fan $RW_1 - RW_2$ originating from the vertex of the corner *o*, the diffracting wave SW_d, the stagnation wave SW_t, and the contact discontinuity CS. The parts of these discontinuities near the wall are denoted as SW_m, SW_a, and CS, respectively. The arrows in Figs. 6.2–6.4 indicate direction from the region of greater density to a region of lesser density. Therefore, SW_s is directed from the vertex of the corner, and CS and SW_t to the vertex.

At large diffraction angles, the process takes place with partial or complete flow separation. Diffraction with complete separation was theoretically investigated in [217].

Using the concept of self-similarity of the process and going over to the coordinate system moving with the diffracting wave, the authors determined the boundaries of quasi-steady flow at the corner. It was shown that when viscosity is neglected part of the Prandtl-Meyer fan near the vertex of the corner is steady. As in the case of Mach reflection near the corner, one can find a region where the flow is independent of the diffraction phenomenon. However, the angle of the fan opening is determined by the pressure in the region under the separation line, obtained from consideration of the entire diffraction pattern. The authors assumed that the separation angle is determined by the flow divergence required for the pressure near the wall to become equal to pressure p_0 ahead of SW_d as a result of flow divergence. In this instance, it is assumed that the gas in the region below the separation line is at rest, and part of the diffracting wave SW_m adjacent to the wall is straight. An approximate method of calculation of the forms of SW_d during diffraction is proposed in [218].

Diffraction at large angles is discussed experimentally in [6, 219–228].

It is known theoretically that the diffraction process must progress self-similarly [73]. This assumption was confirmed in [221] for air at various diffraction angles. In [228], it is experimentally shown that the vortex motion during diffraction is also self-similar.

The self-similarity of the process was experimentally checked [6] by comparing single pictures taken at the same flow Mach number and at different moments in time for CO_2, N_2, and air in the range of Mach numbers 2–10 and angles 15–120°. In addition, self-similarity was tested using the streak-record technique. Streak-records of propagation of undisturbed and disturbed parts of the diffracting wave were carried out by the CFR camera operating in the waiting mode [49]. The slits were located along the angle generators; therefore, all traces of the disturbances in the wall region could be seen in the record. These traces are straight lines. The diffraction process was found to be self-similar with an accuracy of $\pm 3\%$. The profiles of the discontinuities do not change with motion in the case of self-similarity. As a result, the velocity of each section of the front of the diffracting wave can be computed from a single picture.

Experiments revealed the role of the stagnation wave SW_t in creating the separation. The appearance of the stagnation wave and its interaction with the boundary layer on the wall leads to a separation flow near the wall [6]. The stagnation wave interacts with the unsteady boundary layer on the wall *om*, formed behind the wave SW_m, and causes its separation. The process is similar to bifurcation upon reflection from the tube end, with the difference that the stagnation wave in the laboratory system of coordinates moves downstream of the flow. For small angles α_0 the stagnation wave is weak and its interaction with the boundary layer does not lead to separation. When angle α_0 is increased from 0 to 50°, the intensity of the stagnation wave increases. At a particular intensity separation takes place.

At diffraction angles 20–30° the process occurs with partial flow separation. Separation commences at some point *i* (see Fig. 6.2*b* below the tip of the corner and takes place with the formation of a λ-configuration. Fig. 6.2*b* depicts a regime with $M_0 = 2.77$, $\alpha_0 = 30°$, $\gamma = 1.4$ (air) when the flow separates from the wall under the influence of SW_t and λ-configuration forms. This configuration is made up of three shock waves: the upper part of SW_t, its lower part *he*, and SW_f. The arc of the contact discontinuity *hg* is barely visible in the photograph, though clearly seen in Fig. 6.2*c*. The thickness of the separation zone decreases gradually towards the vertex of the corner, forming a distinctive wedge which, when a supersonic flow passes over it, creates an oblique shock wave. Such a shock—SW_f—is visible and begins at a point *l*. Apparently, the action of the separation zone is manifested at this point, and on section *il* this zone is simply unnoticed. Thus, we assume that the separation zone begins at point *l* and extends as a thin layer to point *i*, where it thickens noticeably. From this place additional shock waves, similar to SW_f, can also be created. A thin system of disturbances, which are clearly shown by the pressure transducer, is formed under the separation line. The gas flow is caught in the region under the separation line and revolves around the point *e*—the center of the degenerating vortex. The separation point moves downstream of the flow at a constant velocity; after a period of time a pattern of nonseparating flow is established near the vertex of the corner, which corresponds to the steady supersonic flow over the corner. When diffraction takes place at an angle of 60° or more, the separation line *on* and SW_f emanate from the tip of the corner. The process takes place with complete flow separation. Separation angles in this instance lie within the range 40–70° as opposed to the steady flow when the Prandtl-Meyer solution in the limiting case allows the flow of the ideal gas to turn at angle $\alpha \approx 130°$ for $\gamma = 1.4$. The gas in this case would have expanded to vacuum. The separation line in carbon dioxide and air remains stable (see Fig. 6.2*f*, *d*), and in freon (see Fig. 6.2*e*) it is evident that the line is made up of a whole system of vortexes.

Fig. 6.2*c*, *d* shows the diffraction process of SW_s at $M_0 = 4.71$, $\alpha = 65°$, $\alpha = 1.4$; the Schlieren photographs are taken with different locations of the knife edge (Fig. 6.2*c*—the edge is vertical on the right, Fig. 6.2*d*—the knife edge is vertical on the left) so that the gasdynamic discontinuities are revealed more clearly. In Fig. 6.2*e*, *f* the knife edge is horizontal. It is seen that the structure of the discontinuities are much more complex, although the upper part of the flow pattern remained the same as in Fig. 6.2*a*—SW_d, CS, and SW_t. From the vertex of the corner *o* below the expansion fan three separations emanate now: SW_f, slip surface *one*, and the contact discontinuity *oqz*, which turned under the separation line as a result of vorticity at point *e*. The presence of vorticity is clearly seen in the interferogram of the diffraction process (see

Fig. 6.3). The upward bending of the slip surface *one* at point *n* causes the formation of an oblique shock wave *nh*, which is directed against the flow (see arrow in Fig. 6.4*a*).

The part of the diffracted shock wave SW_d near the wall has a kink at the point *r* (see Fig. 6.2*c*). Analysis of photographs of various regimes shows that two lines *rq* and rc_* originate at this triple point. Line rc_* in Fig. 6.2*c, d* characterizes the surface of weak discontinuity, as RW_1, where $\rho_4 > \rho_3$ (see Fig. 6.4). For the traditional Mach reflection for a similar triple-shock configuration an inverse relation is fulfilled: $\rho_4 < \rho_3$. Line *rq* can be identified with the surface of a weak discontinuity or with the shock wave. The distribution of the isobars, obtained in numerical calculations (see section 8.3), is such (see Fig. 6.4*b*, dashed line) that $P_2 > P_3$. This agrees with the distribution of density on the interferogram of Fig. 6.3. Therefore, the configuration in the vicinity of the point *r* in the photographs of Fig. 6.2*c, d* differs from the traditional one which takes place in a Mach reflection of a shock wave from a wedge. The distortion of SW_d at point *r* was predicted by the Lighthill theory for small angles α [73]. When a shock wave is reflected from a wedge, the kink under consideration is formed under conditions of excess pressure in the disturbed region and the Mach wave "protrudes" forward, and when diffracted at a convex corner under expansion conditions wave SW_m near the wall is "drawn in" toward the corner *o*. In photographs of other regimes (see Fig. 6.2*e, f*) line *rq* is similar to a reflected shock wave as in ordinary Mach triple-shock configuration, i.e., $\rho_4 < \rho_3$.

In the photographs of Fig. 6.2*c–f* and Fig. 6.3, a shock wave originating from a point *k* is seen, which is directed from the vertex of the corner. It is possible that this wave was formed due to reexpansion of the flow in region *mk*, as it happens in the so-called *N*-wave. It is evident that shock wave SW_k is reflected from the surface, and a configuration similar to the one appearing in regular reflection can sometimes be seen. It is seen that SW_k is almost parallel to the side *oj* at various α_0.

The solid lines in Fig. 6.4*a, b* correspond to the locations of the shock waves depicted in Fig. 6.2*c, d*; the dashed lines denote the contact discontinuities; curved dashed lines represent the slip surface (arrows near this surface characterize qualitatively the values of the velocity vectors of the flows). The dash-dot line represents the boundary characteristic of the fan of expansion wave RW_1, and the other boundary characteristic RW_2 transforms into SW_f. Curve 5 in Fig. 6.4*a* denotes the location of the local extreme P_{min} (see Fig. 6.3). Noted also are *A*, *B*, *C* locations of gasdynamic discontinuities according to the results of numerical calculations, which will be covered in more detail in section 8.3. A recurrent flow appears under the separation line *on*. The contact surface CS turns under the separation line and ends at the wall near the vertex of the corner. This surface separates gas 0, compressed by the diffracting wave SW_d, from the flow issuing out of the channel. Such a location means that the outflow of the gas behind the incident wave does not penetrate the region near the wall. As a result of the contact surface moving to the vertex of corner, the gas velocity in sector *qok* (see Fig. 6.4*a*) near the tip *o* is almost equal to zero. Meanwhile, the gas along the wall behind SW_m moves away from the vertex of the corner. Consequently, the gas on sector *om* is accelerated from rest near the vertex of the corner to the flow velocity behind the shock wave SW_m. This process is recorded by the pressure transducers.

6.2 SHAPE OF A SHOCK WAVE DURING DIFFRACTION

Experimental results. Let us consider the shape of the shock-wave front. When the wave turns around the convex corner the speed of the part of the wave near the wall decreases as a result of larger volume of gas being entrapped. The change in velocity involves the part of the shock-wave front to which the interaction spreads with a centered expansion wave originating at the vertex of the corner.

The shape of the diffracting part of the shock wave is, on the whole, a convex line with respect to the undisturbed gas. In the limiting case of $M_0 = 1$ this is the boundary of the sonic circle. The shape of the diffracting part of the wave is smooth at low Mach numbers and small angles (see Fig. 6.1*a*). When the Mach number is increased, a kink or even a triple point appears from which a reflected wave and tangential surface originate at large Mach numbers and angles α_0 of the order 60–90° (see Fig. 6.2*e*). Upon diffraction at angles 120° and 170°, the triple point on the diffracted wave is not observed.

Fig. 6.5 [6] shows experimentally obtained profiles of the shock wave during diffraction. Profiles *3* are computed from the Whitham theory for large Mach numbers of the flow [218]. According to Whitham, the profile is a curved line (main curve) up to the angle (shown in dashed line) beyond which the flow parameters become the same as those at the wall; therefore, part of the wave behind the point *B*—intersection with the dashed line—is straight and perpendicular to the wall.

It is evident that the experimental profiles differ from the theoretical both quantitatively and qualitatively. At low Mach numbers ($M_0 < 3$) the wave is more gently sloping and the wave velocity at the wall is greater than calculated;

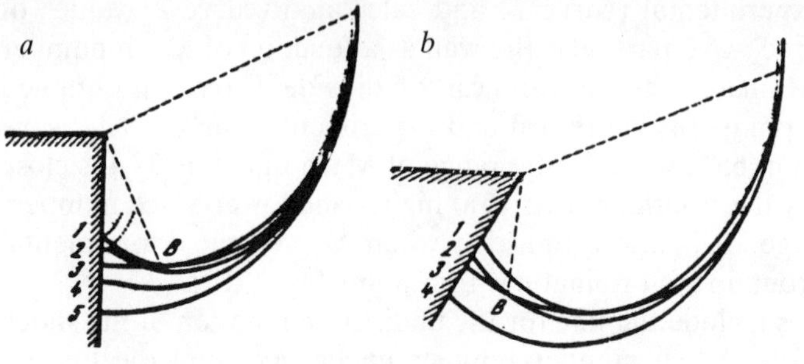

Figure 6.5. Profiles of shock waves. *a*—$\alpha_0 = 90°$: *1*) $M_0 = 7.85$; *2*) $M_0 = 5.73$; *3*) calculated; *4*) $M_0 = 3.34$; *5*) $M_0 = 2.06$; *b*—$\alpha_0 = 120°$: *1*) $M_0 = 6.75$; *2*) $M_0 = 4.44$; *3*) calculated; *4*) $M_0 = 2.04$.

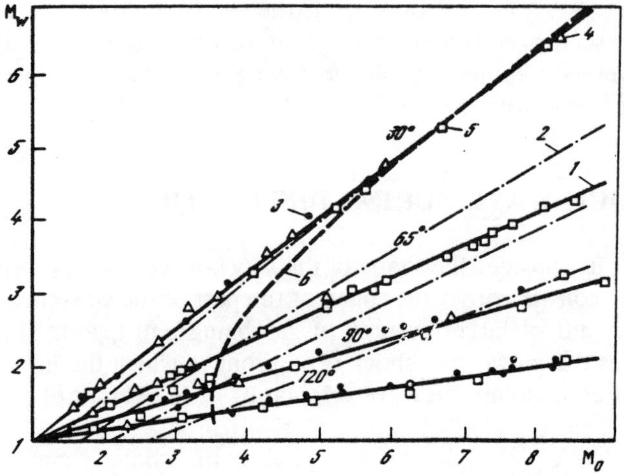

Figure 6.6. Dependence of the Mach number of the shock wave M_w at the wall on the Mach number of the incident wave M_0 *1*) experiment; *2*) theory; *3*) nitrogen; *4*) air; *5*)carbon dioxide.

at large Mach numbers ($M_0 > 3$) the velocity at the wall is less than calculated. In the average range of angles, a triple point is formed on the wave, and at small and large angles the front line has a kink.

The appearance of an inflection point on a diffracted wave at an angle of flow inclination $\alpha_0 = 5°$ and Mach number of the incident wave $M_0 = 2.53$ was predicted in [73] for a gas with an adiabatic index $\gamma = 1.4$. In [220], the formation of the inflection point is observed and the quantitative deviation from calculations by Whitman is indicated. However, since experiments in this work were conducted with small Mach numbers M_0, only an increase in velocity compared with the calculation was observed. Experiments in [219] were also conducted at small Mach numbers; therefore, the suggested method to modify the Whitham theory by the authors to decrease velocity is not, evidently, applicable for larger M_0.

Fig. 6.6 shows experimental (curve *1*) and calculated (curve *2*) values of Mach number M_w of the wave part near the wall as a function of Mach number of the incident wave M_0 in air, nitrogen, and carbon dioxide. Curve *6* is obtained from the intersection points of theoretical and experimental curves. The wave form for the angles from 65° to 170° in the range of Mach numbers 3–4 is close to the one predicted by the Whitham theory, at higher and lower Mach numbers M_0 these values diverge. It is interesting that when $M_0 > 2$ the experimental points are grouped around rays originating from point $M_0 = 1$, $M_w = 1$.

The boundary lines include the line for the undisturbed motion of the shock wave ($\alpha_0 = 0$) when the Mach number remains unchanged, and the line on which the Mach number of the diffracted wave is equal to unity at such large turning angles, because of which the diffracted wave is so weakened that it transforms into a sonic wave. Approximating formulas, based on experiments, have been found for the shock-wave front in the polar coordinates r and α:

$$r = 1 + 2^{\chi_0 + \chi_1 + \alpha} \tag{6.1}$$

where the value of the characteristic angle χ_0 is determined from the equation $M_0/\cos\chi_1 = 1 + 2^{\chi_0}$.

Angle χ_1 is determined from the location of point A at which the front begins to distort. The approximating formula (6.1) satisfactorily describes the wave form within the diffraction angles $0° < \alpha_0 < 170°$ and Mach numbers of the incident wave $2 < M_0 < 6$. Evidently, the formula does not describe the behavior of the wave near the wall. Another formula has been obtained [226] for the velocity of the wave motion at the wall.

The experimental dependence is approximated by the relation

$$M_w = (1 - \sin(\alpha_0/2))(M_0 - 1) + 1 \tag{6.2}$$

The distance of various points of the front from the vertex of the corner is proportional to the velocity of these points.

Apparently, the obtained dependence of shock-wave attenuation at a thin wedge has a more general character, since it was revealed that this dependence is true in the three-dimensional case [222]. The attenuation of the shock wave at low Mach numbers ($M_0 < 2$) was investigated in [229–232]. It is shown in these experiments that when the shock wave exits from the tube in a practically unbounded space, the pressure ratio at the diffracted wave is linked with the increase of cross sectional area by the relations obtained in [234].

Modification of the Whitham theory for the calculation of the shape of a diffracting wave. Experiments showed that the profile of the shock wave for the same Mach number and different angles lie on a single common curve [220, 226]. A situation corresponding to the Mach reflection arises near the wall. The boundary condition in this case is not violated, the gas behind the Mach wave flows along the wall; however, the entire wave form is more convex than predicted by the Whitham theory [218]. Therefore, we assume that the change in the front shape takes place right at the wall and not before reaching the point B (Fig. 6.7*a*), as was suggested in the Whitham theory. The diffracted wave travels along the main shock curve. The emergence of the triple point is linked to the fact that wave reflection occurs in accordance with the Mach number M_ϕ of the diffracting wave at the wall and with the angle between the tangent to the main shock curve and the wall ϕ_0. M_ϕ can be computed as follows (Fig. 6.7*a*) [226]. We construct the main shock curve up to the intersection with the wall (point P). The Mach number is determined graphically at this point by drawing a perpendicular line from the vertex of the corner to the tangent to the curve at point P. We find the intersection angle of these lines—angle ϕ_0. By considering the reflection of the curvilinear shock as a straight line reflection, as a first approximation, we find an angle χ_w at which the triple point will move during Mach reflection. The location of point R is determined. From this point we draw a wave perpendicular to the wall and obtain point *M*. Behind the Mach wave the gas flows parallel to the wall so that the condition on the surface is satisfied. The Mach number at the wall M_w is determined by the distance *OM*.

Such modification of the Whitham theory gives better qualitative correlation with the experiment, since it predicts the appearance of triple points. However, it is apparent that there are no kinks on the shock waves at small and large angles. The same result is obtained in the modified theory in the following manner. When the angle of travel of the triple point is close to the angle at which point *B* moves, the Mach line in the triple configuration becomes tangent to the main curve and the solution agrees with the ordinary solution. According to Whitham there will not be a triple point. Additionally, a triple point cannot exist when, instead of this point, there are no conditions for branching of the shock (analogous to the conditions in the second triple point during Mach reflection).

Fig. 6.7*b* presents the results of comparison of calculation and experiment. Given are the experimentally obtained profiles of the diffracting waves for Mach numbers $M_0 = 2.4$ and 4.7 and diffraction angles $\alpha_0 = 90°$ and 170°. The profiles for identical initial intensity agree along the entire length. Divergence is observed near the wall. The dashed line in Fig. 6.7*b* represents the calculation according to the Whitham technique. The modification of the technique gives very small correction at angle 90° for $M_0 = 2.4$ and a significant change for $M_0 = 4.7$. However, the quantitative discrepancies remain noticeable. It is possible that this is a result of the crude simplifications which are introduced during the calculations in replacing the curvilinear shock by a rectilinear shock.

6.3 FLOW AROUND A CORNER IN AN EXPANSION FAN DURING DIFFRACTION OF SHOCK WAVES

Expansion of the gas, in which the internal degrees of freedom are not excited and the effect of viscous forces can be neglected, takes place during diffraction of the shock wave in the centered Prandtl-Meyer fan.

With the help of photographs in [225], the angle corresponding to the first characteristic of the Prandtl-Meyer fan, β_1, the angle β_s determining the location of the discontinuity line, and the angle β_2 determining the direction of the last characteristic in the Prandtl-Meyer fan were measured (see Fig. 6.4). From the first characteristic the Mach number of the flow M_1 behind the shock wave in the undisturbed region can be determined. Comparison of the measured Mach number with the calculated value allows one to come to a conclusion about the excitation level of the internal degrees of freedom in the gas behind the incident wave and to find the value of γ. Analysis of these measurements indicated that, under the conditions of the experiment, dissociation of the gas did not take place during the observation time. Vibrations of the nitrogen molecules were not excited up to $M_0 = 5.5$ in nitrogen and air. When the Mach number of the shock wave was increased to 7, the vibrations of nitrogen were excited in the

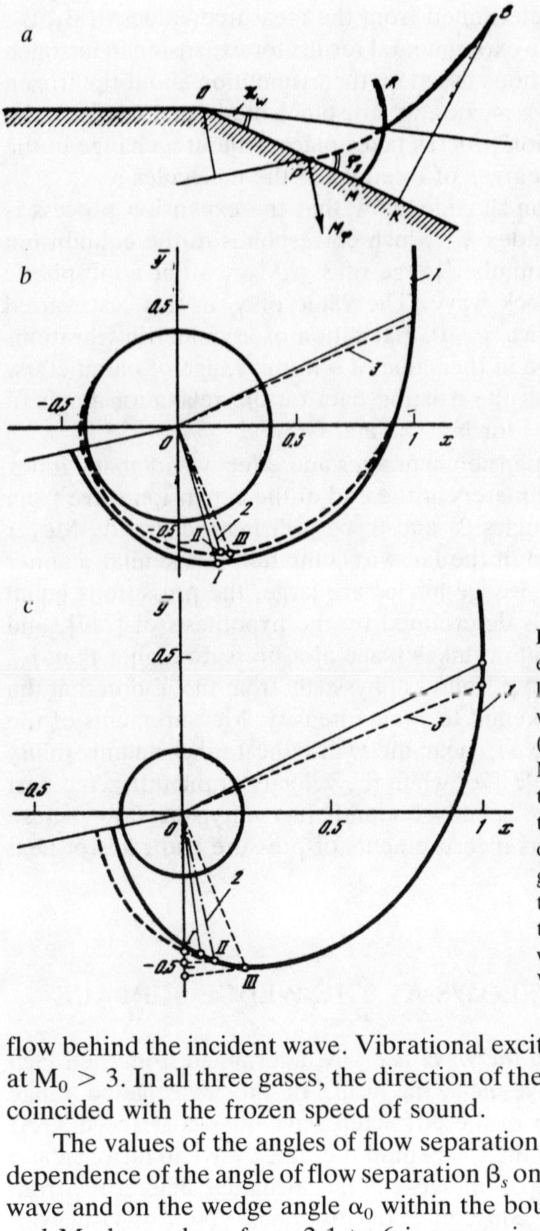

Figure 6.7. Comparison of calculated and experimental shock-wave profiles: calculation schematic (*a*); diffraction of shock waves at angles 90° and 170°, $M_0 = 2.4$ (*b*); diffraction at angles 90° and 170°, $M_0 = 4.7$ (*c*). *1*) experiment; *2*) calculation according to [226]; *3*) calculation by the Whitham method; *I, II*) break points on a shock wave during diffraction at angle 90° obtained experimentally and by the modified Whitham theory, respectively; *III*) contact point of the main curve with the perpendicular to the wall in the Whitham theory.

flow behind the incident wave. Vibrational excitation in carbon dioxide occurred at $M_0 > 3$. In all three gases, the direction of the first expansion fan characteristic coincided with the frozen speed of sound.

The values of the angles of flow separation in nitrogen are given in [6]. The dependence of the angle of flow separation β_s on the Mach number of the incident wave and on the wedge angle α_0 within the bounds of angle α_0 from 20° to 170° and Mach numbers from 2.1 to 6 is approximated, with an accuracy of up to 2%, by the following relation (both β_s and α_0 are expressed in degrees):

$$\beta_s = 30\left(2 - \frac{1}{M_0 - 1}\right)\left(1 - \frac{6}{\alpha_0 - 30}\right)$$

If β_s is known, the pressure at the end of the expansion fan can be determined from the Prandtl-Meyer formula. The gas parameters during expansion and the effective adiabatic index were determined from the measured values of β_1, β_2, and β_s [6]. It was revealed that the experimental results for expansion in nitrogen and air correlate with the calculations based on the assumption about the frozen value of the adiabatic index $\gamma = 1.4$, i.e., at a distance of about 2 cm from the vertex of the corner, the expansion process takes place without a change in the excitation level of the internal degrees of freedom of the molecules.

The results of tests in carbon dioxide show that the expansion process is described well by the effective index γ_e, which corresponds to the equilibrium excited vibrations in the Mach number range of $3 < M_0 < 10$ at atmospheric pressure behind the incident shock wave. The value of γ_e in this case varied from 1.21 at $M_0 = 3$ to 1.15 at $M_0 = 10$. Excitation of asymmetric vibrations introduces an insignificant change in the value of γ in this range of parameters.

These conclusions agree with the existing data on the relaxation times of excitation of molecular vibrations for N_2, O_2, and CO_2.

The determination of the expansion character and effective adiabatic index γ_e allows one to find the gas parameters at the end of the expansion wave from the Mach number M_0 and the angles β_2 and β_s by utilizing the Prandtl-Meyer function. The pressure at the end of the fan was computed in a similar manner and it was shown that when the wedge angles are large, the pressure is equal to p_0, i.e., the flow separation is determined by the hypothesis of [217], and when the angles are small separation takes place at a pressure higher than p_0. The assumption that p_0 is equal to p_2 follows physically from the notion that the shock wave near the wall is weakened to sonic intensity. Measurements of the Mach number of the shock wave M_w near the wall indicate the inadmissibility of such an approach even for angle 170° when the velocity of the diffracted part of the shock wave is close to the speed of sound. In reality, the flow picture near the wall is more complex, as measurements of pressure and rates of heat transfer indicate.

6.4 PRESSURE AND HEAT FLOWS AT THE WEDGE SURFACE

The pressure immediately behind the wave is easily determined from the Mach number of the part of the wave near the wall. In the investigated range $(2 < M_0 < 6)$, the Mach number of the diffracted wave does not exceed 5. At Mach numbers of this magnitude the gas behind the shock wave in nitrogen and air can be regarded ideal and its parameters can be calculated analytically from the laws of conservation with a constant adiabatic index $\gamma = 1.4$. Using approximation (6.2), we write an empirical equation for the pressure behind the diffracted shock wave as a function of the value of wedge angle and Mach number of the incident shock wave [6]:

$$\frac{p_w}{p_0} = \frac{2\gamma}{\gamma + 1} M_w^2 - \frac{\gamma - 1}{\gamma + 1} \tag{6.3}$$

The pressure at the wall was measured by a miniature piezoelectric transducer. Fig. 6.8 shows a comparison of the pressure calculated by equation (6.3) with the measured values of pressure amplitude at the front of the shock wave at the wall of a wedge with different angles and at different Mach numbers of the incident shock wave. The approximating equations (6.3) correlate well with the experiment.

Inspection of the Prandtl-Meyer fan allowed the computation of the pressure at the end of the fan. This pressure in the vicinity of the vertex of the corner must be equal to the pressure on the surface, since during flow separation this region is detached from the wall by two tangential discontinuities *on* and *oq* across which the pressure stays unchanged (see Fig. 6.4). This pressure is taken as a reference point. Thus, the possible limits of pressure change at the wedge wall were determined, namely, behind the part of the wave near the wall and at the end of the expansion fan. The pressure at the wall must, therefore, decrease along the surface. However, it is evident from the photographs that kink and Mach configurations appear on the diffracted wave, and waves are seen near the wall. This creates a more complex picture of pressure distribution.

Pressure and wall temperature measurements were carried out in [223] using a pressure transducer and a thin film sensor mounted in one plane at a distance of 15–30 mm from the wedge boundary.

Pressure recordings for an angle of 15°, where the flow is attached, indicate the presence of a system of diffracted wave and stagnation wave. The stagnation wave is weak and the pressure drop is insignificant. Upon diffraction at corners of 30° and 40°, the transducer registers the pressure rise in the diffracted part of the wave and a smoother drop at the reversed stagnation wave which interacts with the boundary layer. When diffraction occurs at a corner of 60°, a triple point appears on the diffracted wave and, consequently, the pressure behind

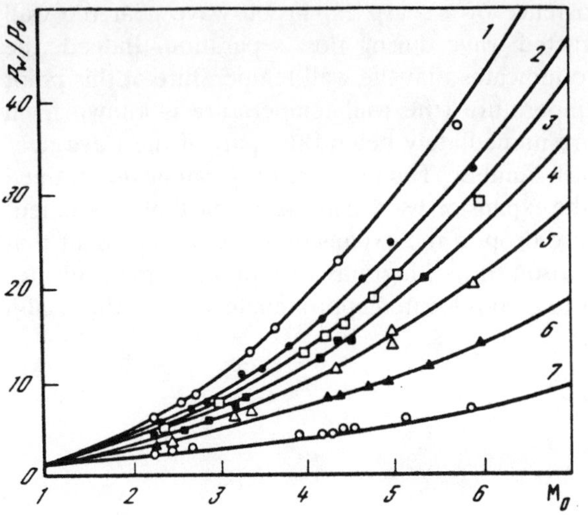

Figure 6.8. Experimental points and approximated curves for the pressure on the surface during diffraction of shock waves *1*) $\alpha_0 = 0°$; *2*) 15; *3*) 30; *4*) 30; *5*) 40; *6*) 60; *7*) 90°.

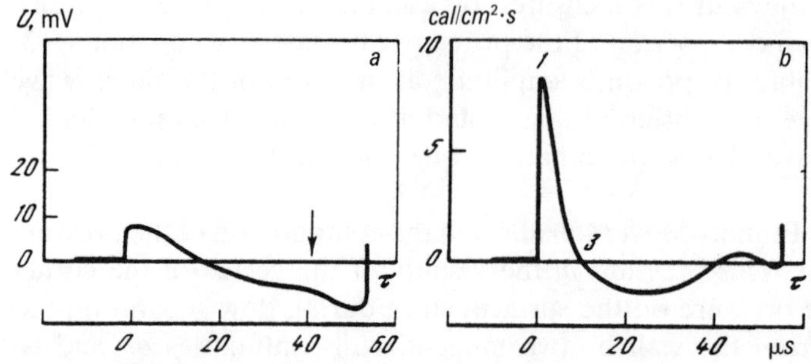

Figure 6.9. Pressure oscillograms (*a*) and heat flow (*b*) on the surface of a corner 90° with diffraction in nitrogen, $M_0 = 4.92$.

the diffracted wave drops in the expansion wave. The curve for the 90° corner indicates a pressure drop below the initial pressure (Fig. 6.9*a*).

Measurements of the surface temperature were conducted simultaneously with the pressure measurements during diffraction, as practiced in investigation of other processes. Comparison of oscillograms with Schlieren photographs of the diffraction process allowed one to observe the character of temperature change at the moment when various flow discontinuities pass near the sensor. Based on the readings of thin film sensor, the heat flows in the wall of the wedge were determined. For $40° < \alpha_0 < 90°$, an interesting fact is noted: the intensive expansion of the gas in the expansion wave leads to its sharp cooling and negative values of heat flow in this region (Fig. 6.9*b*), i.e., upon diffraction at angles 40–90° an unsteady region of flow separation exists with heat flow from the wall to the gas. It is interesting that the wall temperature at this moment falls below the initial temperature.

These data confirm the results obtained from the analysis of pressure curves for the corner of 90°: the existence of a sharp expansion wave near the wall immediately behind the diffracted wave during flow separation. Indeed, the tendency for heat flow to zero indicates that the wall temperature at this point becomes equal to the gas temperature (the wall temperature is known from oscillograms). The temperature immediately behind the part of the wave near the wall is known from the Mach number. The pressure drop can be determined from the temperature drop in the expansion wave if an isentropic flow is assumed. From the evaluation of pressure drop in the expansion wave, it is evident that it must drop below p_0. Comparison of oscillograms with photographs indicates that the point at which q tends to zero is located approximately below the region of the vortex *e* (see Fig. 6.2).

CHAPTER
SEVEN

UNSTEADY INTERACTION OF SHOCK WAVES WITH CURVILINEAR SURFACES

7.1 REFLECTION OF SHOCK WAVES FROM CURVILINEAR SURFACES

The process of shock-wave reflection from curvilinear surfaces is not self-similar. When a plane shock wave, propagating along surface P (Fig. 7.1*a*), is reflected from a concave round cylindrical surface with sufficiently small angle α_0, Mach reflection takes place first, then regular reflection follows starting at some transition angle ω_{cr}. On the other hand, when the wave is reflected from a convex surface with sufficiently large angle α_0 (Fig. 7.1*b*), regular reflection takes place first followed by Mach reflection. Here, α_0 is the angle between the plane and the cylindrical surfaces (angle between plane P and the plane F tangent to the cylindrical surface at the line of intersection with the plane); T stands for trajectories of the triple points for three values of the angle α_0; ω is the angle of incidence of the shock wave (angle between the wave front and the plane tangent to the cylindrical surface at the intersection line with the shock wave); ω_1, ω_2 are the angles of incidence and reflection (angles between the front of the incident and reflected waves, respectively, and the plane tangent to the trajectory of the triple point at point A).

The reflection of a shock wave from curvilinear surfaces were investigated experimentally in [6, 190, 192, 193, 215, 235–252, 260–262] and analytically in [253–258]. Great attention was given in all these investigations to the problem of transition between Mach and regular reflections. The theoretical transition angles ω_e and ω_N, which are determined by the theory of regular reflection and

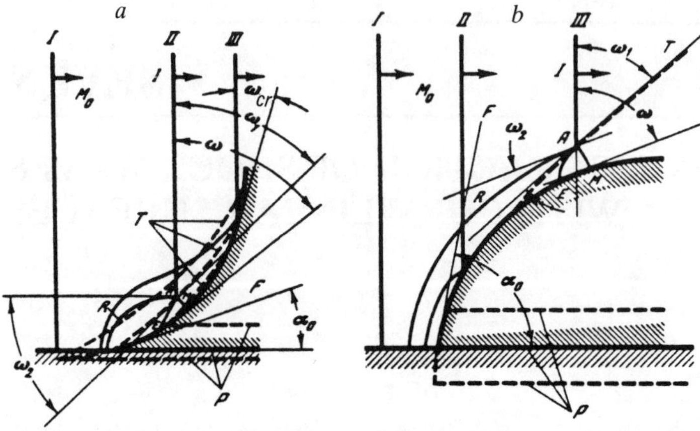

Figure 7.1. Reflection of the shock wave from a concave (*a*) and convex (*b*) round cylindrical surfaces. *I*, *II*, *III*) undisturbed motion, regular and Mach (*a*), or Mach and regular (*b*) reflections of the shock wave, respectively; *IA*, *AR*, *AM*) incident, reflected, and Mach shock waves; *AE* is the contact surface; *A* is the triple point.

by the criterion of formation of a steady Mach configuration [6], respectively, are not unique. There is a region of values ω_{cr} between curves *1* and *2* (Fig. 7.2) which permits both regular and Mach reflections. This non-uniqueness of the theoretical determination of ω_{cr} gave impetus to a large number of experimental investigations [6]. However, the more the question about the criterion and tran-

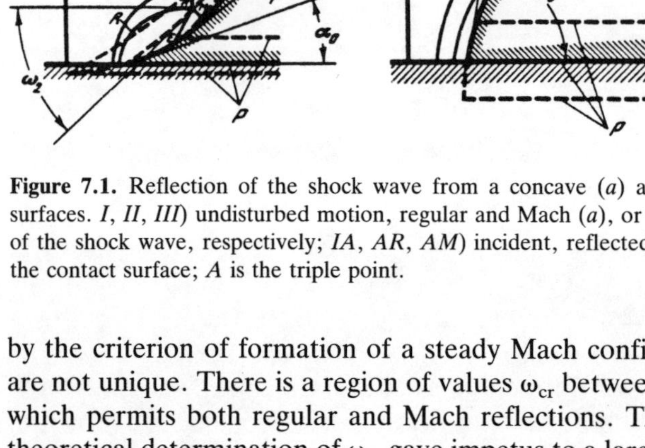

Figure 7.2. Experimental and theoretical data of the dependence of transition angle ω_{cr} on the intensity of the incident shock wave $\xi = p_0/p_1$. Curves *1*, *2*) transition angles according to the theory of regular reflection and for a steady Mach configuration; curves *3*, *4*) plotted according to the theory of [243, 244] for convex *3* and concave *4* cylindrical surfaces; experimental data for convex cylindrical surfaces; *5*) region constructed from data of works [235, 244, 245]; *6*, *7*) [239]; *8*) [242]; *9–17*) experimental data for convex cylindrical surfaces; *9–11*) [239]; *12*) [238]; *13*) [243]; shaded region *14*) [244, 245]; *15–17*) [242]; experimental data for a wedge: curve *18*) [6].

sition angle is studied, the more complex it seems. It is revealed that the transition angle is dependent on whether the reflection is steady or unsteady, self-similar or not [6, 192, 235, 238, 240–244]. In addition, it was found that the transition angle is influenced by the properties of the real gases: by viscosity which leads to the creation of the boundary layer in the gas behind the shock waves [212], by physicochemical transformations in the gases behind the shock waves [194, 196, 259], and also by the roughness of the reflecting surface [215]. It was also revealed [6, 187] that there exists not only one type of reflection but no less than five types; this fact also complicates the discussion of the transition problem. It was also found that the transition from a Mach reflection to a regular one and that from a regular reflection to a Mach type does not take place at one and the same angle ω_{cr} [235, 238].

The transition angles are determined experimentally from the pictures of configurations of discontinuities obtained during shock-wave reflection. In this case, the experimental angles ω_{cr} are obtained either by constructing experimentally the dependences of the incident angles ω_2 on reflection angles ω_1 or from the disappearance (or manifestation) of Mach wave or contact surface. The accuracy of determining the transition angle by optical methods is limited by the optical resolution; therefore, the optical techniques cannot give exact values of ω_{cr}. Thus, for example, the configurations from which the experimental dependences ω_2 (ω_1) had been plotted for regular reflection in [190] were later (in [192]) ascribed to the double Mach reflection in which the first reflected wave was not resolved optically and, therefore, angle ω_2 of the wave, originating from the second triple point, was measured in the pictures.

The magnitude of the Mach wave can also be recorded by optical methods only up to a certain limit.

The experimental data on the dependence of the transition angle at concave and convex cylindrical surfaces on the shock-wave intensity are presented in Fig. 7.2. Analysis of the data shows that the transition angles for concave surfaces lie below and for convex surfaces above the experimental curve for the wedge.

The transition angle ω_{cr} in [240–242] was determined by the intersection of the trajectory of the triple point with the surface. Such a method does not give the exact values of the transition angle ω_{cr}, because the distance between the triple point and the surface becomes so small with decreasing Mach wave as to render it impossible to determine by optical techniques. As a result, the boundary curve was determined in the coordinate plane (ω_{cr}, α_0). On one side of this plane the real transition angles were present, and on the other side there could be no angles, since the triple configuration was clearly visible here in the entire region of angles. This boundary was determined with an accuracy of $\pm 0.5°$. Examples of interferograms of Mach reflection on a concave and convex surface are depicted in Fig. 7.3*a, b,* and the obtained transition angles are plotted in Fig. 7.2.

The dependence of the transition angle ω_{cr} on the history of the reflection process has been determined by the geometry of the reflection surface. The concave or convex round cylindrical surface is completely characterized by its radius r_0 and angle α_0 between plane P, intersecting the cylinder parallel to its

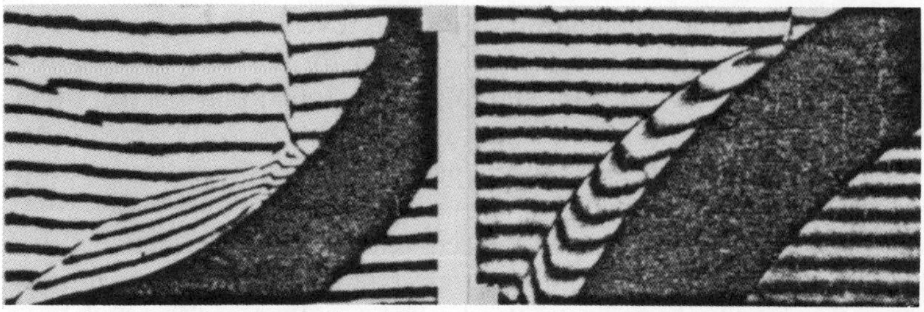

a *b*

Figure 7.3. Interferograms of Mach reflection of a shock wave in nitrogen (p_0 = 10 mm Hg, M_0 = 6.2) from a concave (*a*) and convex (*b*) surfaces.

axis, and plane *F* tangent to the cylinder at the intersection line of the surfaces (see Fig. 7.1*a, b*). The incident wave is directed relative to the reflecting surface so that its front is always parallel to the cylinder axis and perpendicular to the intersecting plane *P*.

The experimental transition angles ω_{cr} are represented on the coordinate plane (ω_{cr}, α_0) in Fig. 7.4 by points 1′, which are characterized by the following values (M_0; α_0, in degrees; r_0, in meters): (5.6–6; 0; 0.072), (6.1–6.7; 6; 0.1), (2.4–5.4; 28; 0.1) for concave surfaces and (4.6–6.6; 71; 0.123) for a convex surface (point 2′). Plotted in the same plane are the experimental points 5′ ($M_0 \approx 3$; $r_0 = \infty$) from [6] and point 6′ ($M_0 \approx 6$; $\alpha_0 = 90°$; r_0 = 0.012; 0.075, and 0.1 m) from [235] for a convex surface.

It was found that curve *1*, separating the coordinate plane (ω_{cr}, α_0) into two regions, can be drawn through these points (Fig. 7.4). Above this curve lies the region of Mach reflection in which the triple configuration of the waves is clearly distinguishable by the experimental (optical) methods. Below this curve is the region of regular reflection, the region of real transition angles, and the region of the probable Mach reflection which are not recorded by optical means. The indicated curve of the experimental values of the transition angles is obtained for strong shock waves (M_0 = 2.4–6.7) and reveals the increasing dependence of ω_{cr} on the angle α_0. On this curve lie points for convex surfaces with r_0 = 0.012; 0.075; 0.1, and 0.123 m, for plane surfaces ($r_0 = \infty$), and for concave surfaces with r_0 = 0.072 and 0.1 m.

By analyzing the experimental data of [243] on the plane (ω_{cr}, α_0) in Fig. 7.4, the plots of the dependence ω_{cr} (α_0) with parameter ξ are constructed. The obtained curves confirm the upward trend of the dependence ω_{cr} (α_0).

Since various researchers have conducted their experiments for the determination of ω_{cr} on cylindrical surfaces characterized by various α_0 and r_0, the observed difference in the results at identical gasdynamic conditions can be interpreted as dependence ω_{cr} (r_0) if dependence ω_{cr} (α_0) is ignored. Thus, in order to demonstrate that the Mach reflection near the transition point can be taken as regular reflection, two shadow photographs are presented in [192]. One of the photographs illustrates regular reflection from a cylindrical surface of one

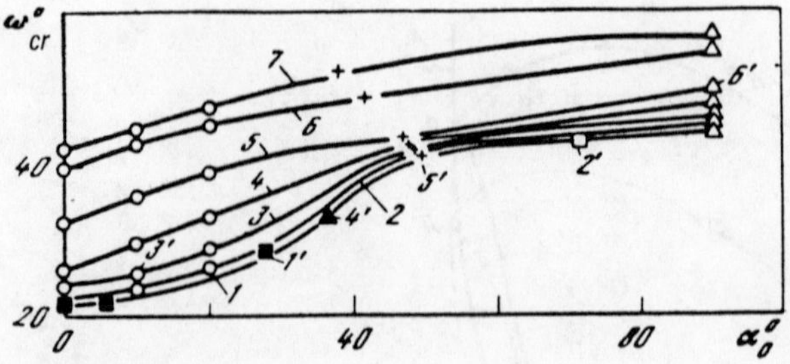

Figure 7.4. Dependence of the critical angle of incidence on angle α_0. Curves *1–7* correspond to the values of ξ = 0.05; 0.22; 0.27; 0.35; 0.47; 0.67; 0.73. *1′*, *2′*) data taken from work [242]; *3′–6′*) data from works [243, 192, 6, 236], respectively.

radius, and the other—a Mach reflection from a surface of different radius, the local angle of incidence being the same. A conclusion can be made that it only appears that a regular reflection is seen in the picture, and in reality this is a Mach reflection which is not recorded in the experiment because of the small dimensions of the Mach wave. Transition from one type of reflection to another, in the opinion of the authors of [192], is completely determined by the intensity of the incident shock wave. The regular reflection evident in the picture is apparent up to an angle of incidence, which is determined from the curve of steady Mach reflection (curve *2* in Fig. 7.2). However, it is shown in [240–242] that the transition angle is independent of r_0 in an ideal gas approximation; therefore, the presented pictures can be commented upon as follows. Surfaces with various angles α_0 are illustrated in the pictures, and where angle α_0 is small, a Mach configuration is observed, and on a surface with large angle α_0, a regular configuration is observed.

We note that when a shock wave is reflected from a concave cylindrical surface, the Mach wave is seen in the region of incidence angle ω, where it is not supposed to be according to the theoretical and experimental data for the case of reflection from plane surfaces. In other words, the shift of the boundary ω_{cr} downward (see Fig. 7.2) can be regarded to be experimentally proven. It thus follows that when the shock wave is reflected from the concave surface, it is possible for the reversed Mach configuration to be formed (Chap. 5, Fig. 5.7*a*, case 3) if the incidence angle ω_1 becomes less than the angle of formation of steady Mach configuration ω_N. When the shock wave is reflected from a convex cylindrical surface, the transition is also delayed in comparison with the transition on a plane surface. However, we have a case of a transition from regular to Mach reflection and it is hard to assert that there is no Mach wave even though it has not been recorded by an optical device. However, numerous experimental data (see Fig. 7.2) confirm the idea of boundary shift.

7.2 DIFFRACTION OF SHOCK WAVES AT ROUNDED CORNERS

If the angle ω between the shock-wave front and the interaction surface becomes greater than 90°, the process of shock-wave diffraction takes place, which is characterized by the formation of expansion waves in the gas flow behind the

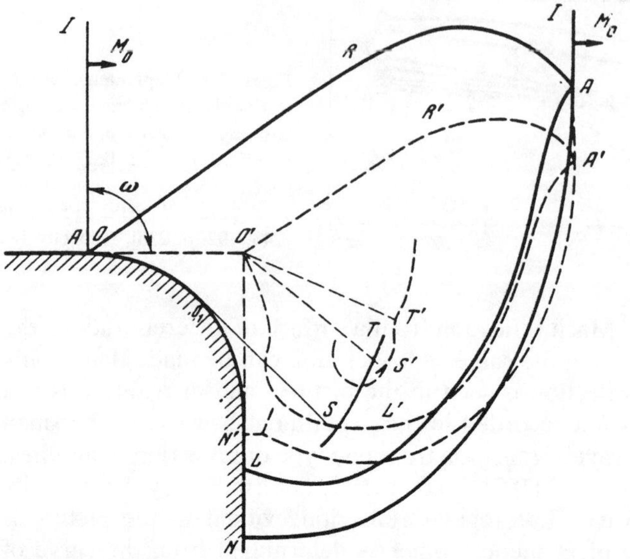

Figure 7.5. Schematic of shock-wave diffraction at a corner with a sharp edge (dashed line, notations with a prime) and at a rounded corner (solid line). *IA, AN* are the incident and diffracted shock waves; *OR* and *OT* are the first and the last characteristics of the expansion fan; *AL* is the contact surface; *AR* is the front of a sound wave; *TS* is the deceleration wave; *OS* is the line of flow separation.

shock wave (Figs. 7.5, 7.6). The unsteady process of shock-wave diffraction at rounded angles has been investigated experimentally in [241, 263–265]. Attempts were made in [263] to prevent flow separation by forming the surface adjoining the plane of a two-sided corner to an appropriate shape. This goal was not reached. Investigations of shock-wave diffraction at rounded angles were carried out in a shock tube with square cross section (72×72 mm) [241, 264, 265]. Experiments were conducted in nitrogen and air at pressures $p_0 = 10$–60 mm Hg and Mach numbers $M_0 = 2$–6 with models having the following diffraction angles α_0 (in degrees) and radius r_θ (in meters): (40; 0.005), (40; 0.04), (40; 0.095), (90; 0), (90; 0.005), (90; 0.032). As a result of the experiments, obtained were shadowgraphs, Schlieren photographs, interferograms, and holograms of images of gas discontinuities at different times of the process of shock-wave interaction with the models.

Fig. 7.6, *a, b* shows the interferograms of shock-wave diffraction at the same Mach number M_0 ($M_0 = 4.8$) at a corner with a sharp edge and at a rounded corner with $\alpha_0 = 90°$ and $r_0 = 0.032$ m, respectively. The incident shock wave was located at equal distances from the vertical walls. Comparison of the photographs in Fig. 7.6 shows that the structure of the shock, formed upon diffraction of the shock wave, at the corner with a sharp edge and at a rounded corner are different. The diffracted shock wave in Fig. 7.6*b* has smoother profile and does not exhibit a break point as opposed to Fig. 7.6*a*. The contact surface during

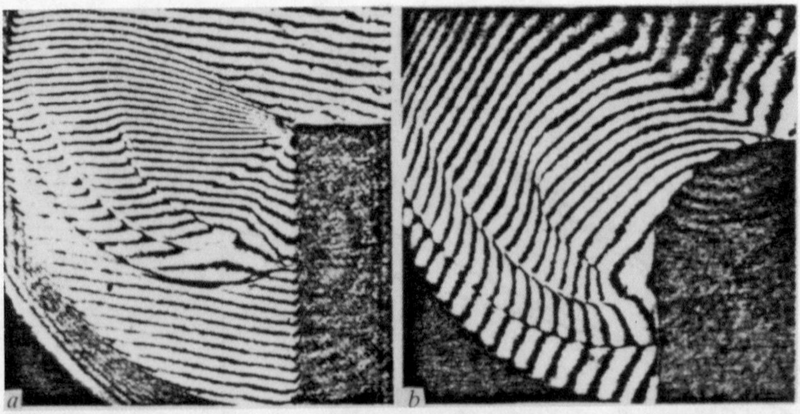

Figure 7.6. Interferograms of shock-wave diffraction in air (p_0 = 30 mm Hg; M_0 = 4.8) at a sharp-edged corner (*a*) and rounded corner (*b*) at identical distances from the wave front to the vertical wave.

diffraction at a rounded angle ends at the vertical surface, whereas, in the case of the corner with a sharp edge, it bends around the stagnation wave and ends at the vertex of the corner. The intersection point of the last characteristic of the expansion fan with the wall, and also the point of flow separation, are shifted downwards along the rounded surface unlike the diffraction at the sharp-edged corner.

The velocity of the part of the diffracted shock wave near the wall on the rectilinear section following the rounded corner u_w was determined with the help of a series of pictures of successive locations of the shock wave obtained in a single experiment, and also a series of pictures obtained in various experiments but at equal M_0. The distance S between two successive locations of the part of diffracted shock wave near the wall and the distance between the corresponding locations of the initial shock wave were measured in the photographs. From the known experimental value of the velocity of the initial shock wave u_0, the time corresponding to the displacement S_0 of the initial shock wave was obtained. Furthermore, from the known distance S, a velocity u_w was determined by equation $u_w = u_0 (S/S_0)$.

The distance S varied at a constant M_0 and α_0, but the velocity u_w remained unchanged. The variation in rounding-off radius had also no effect on the value of this velocity. Fig. 7.7 shows the experimental dependences of the Mach number M_w of the part of the diffracted shock wave near the wall on the rectilinear section following the rounded section on the Mach number M_0 for the following values of diffraction angles α_0 (in degrees) and rounding-off radii r_0 (in meters): (40; 0.005), (40; 0.04), (40; 0.095), (90; 0.005), (90; 0.032) marked with points *1′–5′*, respectively. Lines *1* and *2* in this figure represent the dependence M_w (M_0), for diffraction at the corresponding corners 40° and 90° without rounding-off, plotted from the data of [6, 220]. Curves *3, 4* represent calculations according to the Whitham theory for plane corners. As Fig. 7.7 shows, the obtained results

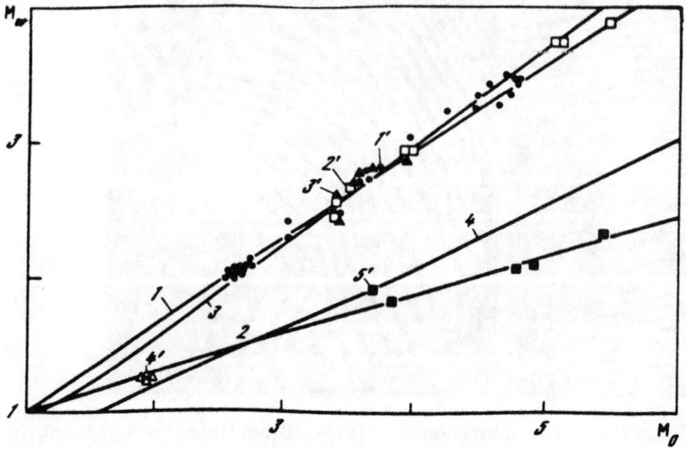

Figure 7.7. Experimental M_0 dependence of Mach number M_w on M_0. M_w is the part of diffracted shock wave near the wall in nitrogen at the plane section following the rounding-off.

in the present work correlate well with the corresponding graphs for diffraction at a sharp-edged corner within the bounds of the experimental error which amounts to 5%.

The process of bending of the shock wave around the cornered part of the wall is not self-similar. The velocity u_w' of the diffracted shock wave near the wall is decreased from the velocity of the incident wave u_0 down to the velocity

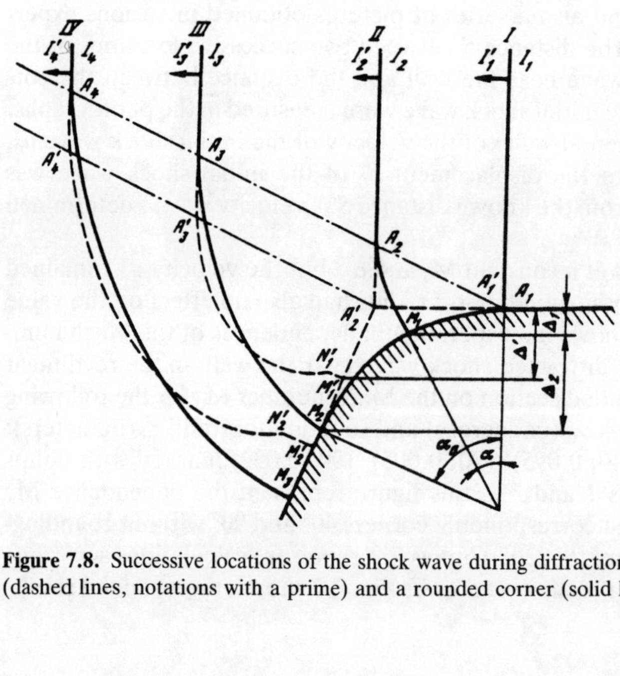

Figure 7.8. Successive locations of the shock wave during diffraction at a corner with a sharp edge (dashed lines, notations with a prime) and a rounded corner (solid lines).

u_w at the rectilinear section of the wall following the corner. Let us determine the dependence of velocity on time in the form

$$u'_w = r\frac{d\alpha}{dt} \tag{7.1}$$

where α is the angular coordinate of the shock wave on the wall of Fig. 7.8. Integrating Eq. (7.1) at $t = 0$ and $\alpha = 0$ at the point corresponding to the beginning of the diffraction process, we obtain

$$r\int_0^\alpha \frac{d\alpha}{u'_w(\alpha)} = t \tag{7.2}$$

In order to determine α (*t*) one must know the form of the function u'_w (α). If it is assumed that velocity u'_w reaches the value u_w on the line adjoining the rounded and plane sections of the wall (location *III* in Fig. 7.8), then at each point on the rounded section of the wall the velocity of the part of the diffracted shock wave near the wall can be taken as equal to its velocity along the plant tangent to the rounded section at this point. In this case, instead of u'_w (α), we can adopt the experimental dependence proposed in [6]:

$$u_w(\alpha) = u_0 - (u_0 - a)\sin(\alpha/2) \tag{7.3}$$

where *a* is the speed of sound. Substituting (7.3) in (7.2) and integrating, we obtain

$$\alpha = 4\,\mathrm{arctg}\left\{\frac{1}{\mu}\left[1 + \sqrt{\mu^2 - 1}\,\mathrm{tg}\,\nu(t)\right]\right\}$$

$$\mu = \frac{u_0}{u_0 - a}, \quad \nu(t) = \frac{u_0 - a}{4r_0}\sqrt{\mu^2 - 1}\cdot t - \mathrm{arctg}\frac{1}{\mu^2 - 1} \tag{7.4}$$

where r_0 and u_0 enter into Eq. (7.4) as parameters.

Substituting (7.4) and (7.1), we obtain the time dependence of the part of the diffracted shock wave near the wall at the rounded section of the wall:

$$\frac{M'_w}{M_0} = \frac{(\mu^2 - 1)\dfrac{1}{\mu^2}\sec^2\nu(t)}{1 + \dfrac{1}{\mu}\left[1 + \sqrt{\mu^2 - 1}\,\mathrm{tg}\,\nu(t)\right]} \tag{7.5}$$

For determination of the Mach number M'_w at reasonably large diffraction angles $\alpha_0(\alpha_0 < \pi/2)$ we assume the dependence of u'_w (α) to be linear (Fig. 6 [221]). Then, by executing similar manipulations we obtain the following simple equations for the determination of α (*t*) and M'_w (*t*):

$$\alpha(t) = \frac{u_0\alpha_0}{u_0 - u_w(\alpha_0)}\left(1 - \exp\left(-\frac{u_0 - u_w(\alpha_0)}{r_0\alpha_0}t\right)\right) \tag{7.6}$$

$$\frac{M'_w}{M_0} = \exp\left(-\frac{u_0 - u_w(\alpha_0)}{\alpha_0}t\right) \tag{7.7}$$

where u_w (α_0) is determined from Eq. (7.3).

All the errors inherent in the determination of Mach number M'_w from Eqs. (7.5) or (7.7) become apparent when determining the distance S travelled by the part of the diffracted shock wave near the wall along the rounded section of the wall. Comparison of the distances S, determined from photographs, and computed from equation $S = \alpha r_0$, where α is determined from (7.4) or 7.6), showed that the computed values of S correlate well with the experimental data within the bounds of the experimental error which reaches 3–5%.

Figure 7.8 shows schematically the successive locations of the shock wave during diffraction at a rounded corner (solid lines) and at a sharp-edged corner (dashed lines, notations with a prime). After passing the start of wall inclination (location *I*), the shock wave *IA* travels along the rounded section. At location *II*, the diffracting shock wave at the rounded corner drops by a vertical distance Δ_1 relative to the wave diffracting on the flat surface. Furthermore, the diffraction process starts at the sharp-edged corner. The diffracted shock wave in the region near the wall starts to shift downwards vertically at a constant velocity $u_w \sin \alpha_0$. The shock wave which diffracts at the rounded corner continues to decelerate along the rounded section with a linear velocity greater than u_w as was indicated above. Thus, upon reaching location *III*, corresponding to the adjoining point of the rounded and rectilinear sections of the wall, component $(\Delta_2 - \Delta')$ is added to the travel difference Δ_1 in the vertical direction between the parts of the shock wave near the wall diffracting at the rounded corner and the sharp-edged corner. This component reaches the value $\Delta \sin \alpha_0 = \Delta_1 + \Delta_2 - \Delta'$.

From Fig. 7.8 and Eq. (7.4) we determine:

$$\frac{\Delta}{r_0} = \left(1 + \frac{M_w}{M_0}\right)\operatorname{tg}\frac{\alpha_0}{2}$$
$$- \frac{4M_w}{M_0 - 1}\sqrt{\left(\frac{M_0}{M_0 - 1}\right)^2 - 1}\ \operatorname{arctg}\frac{\sqrt{2M_0 - (1/M_0^2)}\ \operatorname{tg}(\alpha_0/4)}{1 - (2M_0 - 1)/(M_0(M_0 - 1))\operatorname{tg}(\alpha_0/4)}$$

From this equation, and also from Fig. 7.8, it is seen that Δ can become an arbitrary large value when $\alpha_0 \to \pi$ and $r_0 \to \infty$.

At fixed α_0, r_0, and M_0, the difference in travel Δ is independent of time, i.e., it can be concluded that the rounded corner affects the profile of the shock wave so that it “remembers” about this influence during the entire subsequent propagation process. This conclusion has been experimentally substantiated within the bounds of the test section, namely the ratio l/r_0, where l, the length of the plane section of the wall following the rounded corner for points *1′–5′* in Fig. 7.7, was taken to be equal to 17.6; 1.9; 0.6; 13.2; 1.2, respectively.

CHAPTER

EIGHT

NUMERICAL SIMULATION. INTERACTION OF SHOCK WAVES WITH A WEDGE AND DIFFRACTION AT A CONVEX CORNER

8.1 THE SHOCK-CAPTURING METHOD

The theoretical study of the flow pattern, which includes nonlinear processes of shock-wave interaction, is generally possible only with the use of numerical methods. This approach makes it possible to determine the flow parameters at the interior points of the flow where experimental measurements entail great difficulties. Despite the fact that numerical methods are powerful tools in the investigations, they nevertheless cannot replace analytical solutions. The latter are indispensable in regards to obtaining results in the vicinity of special points (for example, on edge of the sharp wedge) or (limiting) cases of values determining parameters: $\to 0$ or $\to \infty$.

The choice of a numerical method is mainly dictated by the character of flow and the required accuracy. The specific feature of gasdynamic problems is the possibility of spontaneous emergence of discontinuities of flow parameters even with smooth initial and boundary conditions. The release of these discontinuities leads to the violation of the uniformity of the numerical algorithm, a property which is considered as an attractive quality of the methods of differences when solved on computers. Efforts to find the means for the solution of unsteady gasdynamic problems without the formation of discontinuities led to the development of the so-called shock-capturing method [266, 267].

In this method the utilization of gasdynamic equations with "artificial viscosity" or "numerical"* viscosity leads to the smearing of discontinuities, and the flow formally becomes smooth without discontinuities. The shock waves and contact surfaces appear as narrow regions with large gradients.

In the shock-capturing method we use explicit finite-difference schemes, approximating differential equations with first, second, third, and so forth, order. The advantage of the first order scheme is the monotonous behavior of the solution near strong discontinuities. One of the disadvantages is the relatively large width of the transition zone characterizing the shock wave, which increases with decreasing shock-wave amplitude. This is explained by the presence of a large uncontrolled numerical viscosity in the corresponding difference equations. Schemes of higher order allow one to obtain narrow transition zones. At the same time, schemes of an even order, in particular the second, are characterized by the presence of oscillations of the solutions in the vicinity of discontinuities. Schemes of the third order possess the best properties of the first and second order schemes; however, the realization of boundary conditions for them is more complex. The reason for this being the need for a larger number of points at the layer t for the calculation of one point at the layer $t + \Delta t$. Experience with calculations indicates that the optimum in many cases is the use of schemes of the second order of accuracy.

Gasdynamic equations. In the problems under consideration, the shock generator gas flow is sufficiently well-described by a system of ideal gas (inviscid, non-heat-conducting) equations, which can be written in traditional notations and expressed in the following form

$$\frac{\partial \rho}{\partial t} + \operatorname{div}(\rho \mathbf{W}) = 0$$

$$\frac{\partial \mathbf{W}}{\partial t} + \mathbf{grad}\,\frac{W^2}{2} - \mathbf{W} \times \mathbf{rot}\,\mathbf{W} + \frac{1}{\rho}\,\mathbf{grad}\,p = 0 \qquad (8.1)$$

$$\frac{\partial e}{\partial t} + \operatorname{div}((e + p)\mathbf{W}) = 0$$

Here, p and ρ stand for pressure and density; $\mathbf{W}$ is the velocity vector; $e = \rho(\varepsilon + 0.5\, W^2)$ represents the complete energy of a unit volume of gas; ε is the internal energy of a unit mass. For a wide range of problems the system of equations can be completed by the equation of state of a calorically perfect gas: $p = \rho\varepsilon(\gamma - 1)$, $\gamma = c_p/c_\upsilon$, $\varepsilon = c_\upsilon T$, where γ is the specific heat ratio. Equation (8.1) can be written in any convenient coordinate system.

In the general case, for the right-handed triple orthogonal vectors $x_1x_2x_3$ (Fig. 8.1a), the equation for unsteady motion of an ideal gas for plane flows or

*Numerical viscosity (diffusion)—due to the finite-difference approximation.

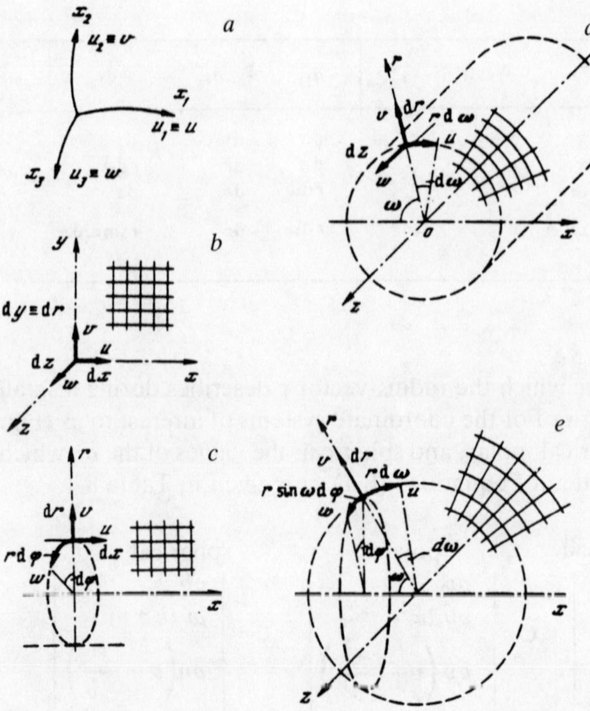

Figure 8.1. Orthogonal systems of coordinates. *a*) arbitrary; *b*) Cartesian; *c*) cylindrical (*x* is the axis of symmetry); *d*) polar (*zox* is the plane of symmetry); *e*) spherical (*x* is the axis of symmetry).

with axial symmetry ($\delta/\delta x_3 = 0,\ u_3 = 0$) can be written as follows

$$\frac{\partial \mathbf{f}}{\partial t} + \frac{1}{h_1}\frac{\partial \mathbf{A}}{\partial x_1} + \frac{1}{h_2}\frac{\partial \mathbf{B}}{\partial x_2} + \frac{\mathbf{C}}{h_1} = 0 \tag{8.2}$$

$$\mathbf{f} = \left|\begin{matrix}\rho \\ \rho u \\ \rho v \\ e\end{matrix}\right| \quad \mathbf{A} = \left|\begin{matrix}\rho u \\ \rho u^2 + p \\ \rho u v \\ (e+p)u\end{matrix}\right| \quad \mathbf{B} = \left|\begin{matrix}\rho v \\ \rho u v \\ \rho v^2 + p \\ (e+p)v\end{matrix}\right| \quad \mathbf{C} = \left|\begin{matrix}\rho b \\ \rho u\left(b + v\dfrac{\partial h_1}{\partial x_2}\right) \\ \rho v\left(b - \dfrac{u^2}{v}\dfrac{\partial h_1}{\partial x_2}\right) \\ (e+p)b\end{matrix}\right|$$

$$b = \frac{1}{h_3}\left(u\frac{\partial h_3}{\partial x_1} + v\frac{\partial h_1 h_3}{\partial x_2}\right).$$

Here u and v are the projections of the velocity vectors (Fig. 8.1). Lame's metric coefficients h_1, h_2, h_3 depend on the selection of the coordinate system and are determined as follows [272]:

$$h_i = \frac{dS_i}{dx_i} = \left|\frac{d\vec{\mathbf{r}}}{dx_i}\right|,\ i = 1, 2, 3 \tag{8.3}$$

Table 8.1

System of coordinates	x_1	x_2	x_3	ds_1	ds_2	ds_3
Cartesian	x	y	z	dx	dy	dz
Cylindrical	x	r	φ	dx	dr	$r d\varphi$
Polar	ω	r	z	$r d\omega$	dr	dz
Spherical	ω	r	φ	$r d\omega$	dr	$r \sin\omega d\varphi$

where ds_i is the length of arc which the radius-vector **r** describes during a small change of the corresponding x_i. For the coordinate systems of interest to us (Fig. 8.1*b*–*e*)—Cartesian, cylindrical, polar, and spherical, the values of the h_i, which are needed to write the system of equations (8.1), are given in Table 8.1.

Cartesian: cylindrical: polar: spherical:

$$\mathbf{C} = \begin{vmatrix} 0 \\ 0 \\ 0 \\ 0 \end{vmatrix} \quad \mathbf{C} = \frac{v}{r}\begin{vmatrix} \rho \\ \rho u \\ \rho v \\ e + p \end{vmatrix} \quad \mathbf{C} = \begin{vmatrix} \rho v \\ \rho u\, 2v \\ \rho v\left(v - \dfrac{u^2}{v}\right) \\ (e+p)v \end{vmatrix} \quad \mathbf{C} = \begin{vmatrix} \rho b \\ \rho u(b+v) \\ \rho v\left(b - \dfrac{u^2}{v}\right) \\ (e+p)b \end{vmatrix}$$

It is convenient to combine the writing of the systems of equations (8.2) in various coordinate systems in the numerical algorithm

Cartesian cylindrical

$$\frac{\partial \mathbf{f}}{\partial t} + \frac{\partial \mathbf{A}}{\partial x} + \frac{\partial \mathbf{B}}{\partial y} + \nu \frac{v}{y} \mathbf{C}_0 = 0 \tag{8.4}$$

$$\mathbf{C}_0 = \begin{vmatrix} \rho \\ \rho u \\ \rho v \\ e + p \end{vmatrix}$$

polar spherical

$$\frac{\partial \mathbf{f}}{\partial t} + \frac{1}{r}\frac{\partial \mathbf{A}}{\partial \omega} + \frac{\partial \mathbf{B}}{\partial r} + \frac{\mathbf{C}_1}{r} = 0 \tag{8.5}$$

$$\mathbf{C}_1 = \begin{vmatrix} \rho b \\ \rho u(b+v) \\ \rho v(b - u^2/v) \\ (e+p)b \end{vmatrix} \qquad \begin{aligned} b &= h + v \\ h &= \nu(v + u \operatorname{ctg}\omega) \end{aligned}$$

$$\nu = \begin{cases} 0 - \text{horizontal flow} \\ 1 - \text{axisymmetric flow} \end{cases}$$

h_1	h_2	h_3	b	$\frac{\partial h_1}{\partial x_2}$	Type of flow
1	1	1	0	0	Planar
1	1	r	v/r	0	Axisymmetrical
r	1	1	v	1	Planar
r	1	$r\sin\omega$	$\frac{u}{\operatorname{tg}\omega} + 2v$	1	Axisymmetrical

Vectors **f**, **A**, and **B** have the same values as in (8.2). The system of equations of one-dimensional flow is written from (8.4) and (8.5) as

with the calculation $\frac{\partial}{\partial x} = \frac{\partial}{\partial \omega} = 0$ and $u = 0$

$$\frac{\partial \mathbf{f}}{\partial t} + \frac{\partial \mathbf{B}}{\partial r} + \nu \frac{v}{r} \mathbf{C}_2 = 0, \quad \mathbf{C}_2 = \begin{vmatrix} \rho \\ \rho v \\ e + p \end{vmatrix} \tag{8.6}$$

$$\nu = \begin{cases} 0 - \text{horizontal flow} \\ 1 - \text{cylindrical symmetry} \\ 2 - \text{spherical symmetry} \end{cases}$$

Method of finite differences. Let us discuss the numerical method with the help of which the results of interaction of the shock wave with various bodies have been obtained (sections 8.2, 8.3, Chap. 9). As an example, we use the divergent form of the system of gasdynamic equations for one- and two-dimensional flows analogous to (8.4) and (8.5):

$$\frac{\partial \mathbf{f}}{\partial t} + \frac{\partial \mathbf{A}}{\partial x} + \frac{\partial \mathbf{B}}{\partial y} + \mathbf{C} = 0 \tag{8.7}$$

$$\frac{\partial \mathbf{f}}{\partial t} + \frac{\partial \mathbf{B}}{\partial r} + \mathbf{C} = 0 \tag{8.8}$$

The numerical simulation of unsteady shock-wave flows implies the statement of the Couchy problem with initial and boundary conditions. The value of the function at some moment of time $\mathbf{f}_{k,l}^{n+1}(t + \Delta t)$ is computed using the known values $\mathbf{f}^n(t)$ at several nodes of the computational mesh (Figs. 8.2, 8.3). The superscript n here is the value counter along the time variable $t = n\Delta t$, and the subscripts k and l are ascribed to the space variables: $x = k\Delta x$, $y = l\Delta y$. The aggregate of the nodes in the time layer t, which is needed for the calculation of one value in the layer $t + \Delta t$, is called the pattern.

Let us consider two double-step schemes of the second order of approximation which are based on a pattern from three points in the one-dimensional

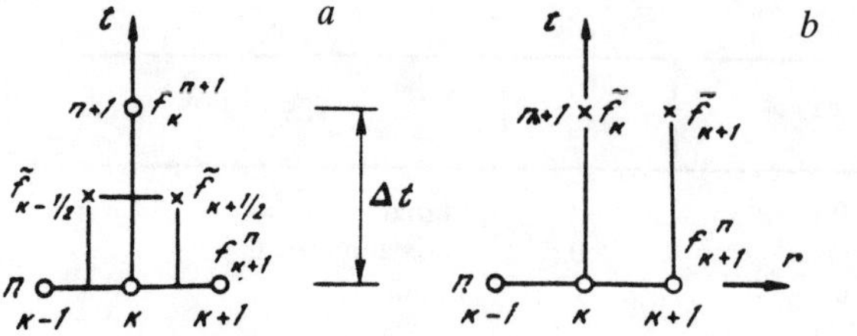

Figure 8.2. Patterns from three points in a one-dimensional case *a*) Lax-Wendroff (Richtmyer variant) scheme, *b*) MacCormack scheme.

case (Fig. 8.2), and from nine or seven points in the two-dimensional case (Fig. 8.3).

Richtmyer scheme (Lax-Wendroff type) [266] (Fig. 8.2*a*). The one dimensional equation (8.8) is first approximated with the first order of approximation—Lax scheme (Fig. 8.2*a*):

$$\frac{\tilde{\mathbf{f}}_{k+1/2} - 0.5(\mathbf{f}_k^n + \mathbf{f}_{k+1}^n)}{0.5\Delta t} + \frac{\mathbf{B}_{k+1}^n - \mathbf{B}_k^n}{\Delta r} + 0.5(\mathbf{C}_k^n + \mathbf{C}_{k+1}^n) = 0$$
$$\Delta t = t^{n+1} - t^n, \quad \Delta r = r_{k+1} - r_k \tag{8.9}$$

From this difference expression we determine the value of $\tilde{\mathbf{f}}_{k+1/2}$ (predictor step):

$$\tilde{\mathbf{f}}_{k+1/2} = 0.5(\mathbf{f}_k^n + \mathbf{f}_{k+1}^n) - \frac{\Delta t}{2\Delta r}(\mathbf{B}_{k+1}^n - \mathbf{B}_k^n) - \frac{\Delta t}{4}(\mathbf{C}_k^n + \mathbf{C}_{k+1}^n) \tag{8.10}$$

The value $\tilde{\mathbf{f}}_{k-1/2}$ is sought for in a similar way. Before the following step (corrector), which ensures the second order of approximation, $\tilde{B}_{k+1/2}$ and $\tilde{C}_{k+1/2}$ are calculated with the help of $\tilde{\mathbf{f}}_{k+1/2}$ and $\tilde{\mathbf{f}}_{k-1/2}$. Equation (8.8) is approximated on the corrector step by the central scheme "cross," which ensures the second

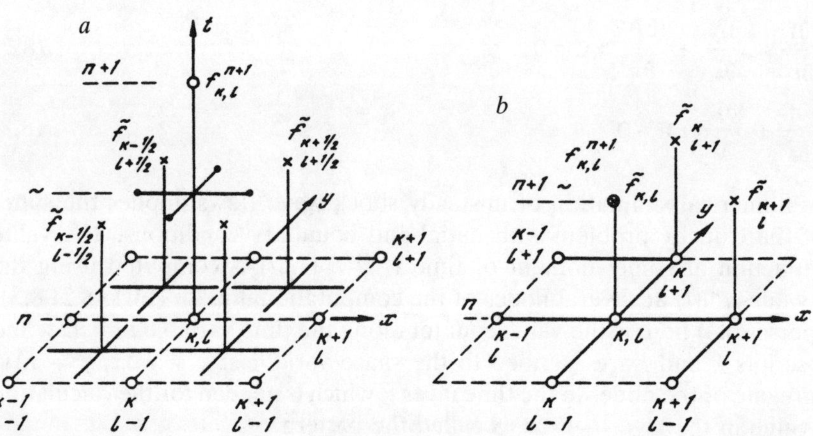

Figure 8.3. Patterns for a two-dimensional case *a*) Nine-point Chudov-Roslyakov scheme; *b*) MacCormack scheme—seven points.

order, from which the final formula for the calculation of $\mathbf{f}_k^{n+1}$ is obtained

$$\mathbf{f}_k^{n+1} = \mathbf{f}_k^n - \frac{\Delta t}{\Delta r}(\tilde{\mathbf{B}}_{k+1/2} - \tilde{\mathbf{B}}_{k-1/2}) - \frac{\Delta t}{2}(\tilde{\mathbf{C}}_{k+1/2} + \tilde{\mathbf{C}}_{k-1/2}) \tag{8.11}$$

The two computation steps, the predictor (8.10), and corrector (8.11), make up an explicit Richtmyer scheme (of the Lax-Wendroff type) of the second order of approximation.

For the simplification of the structure of difference schemes, the averaging of operators and difference taking are introduced in the literature:

$$\mu_k \mathbf{f}_k = 0.5(\mathbf{f}_{k+1/2} + \mathbf{f}_{k-1/2}); \qquad \mu_k \mathbf{f}_{k\pm 1/2} = 0.5(\mathbf{f}_k + \mathbf{f}_{k\pm 1}) \tag{8.12}$$

the central difference: $\delta_k \mathbf{f}_k = \mathbf{f}_{k+1/2} - \mathbf{f}_{k-1/2}$.

the difference forwards: $\delta_k^+ \mathbf{f}_k = \mathbf{f}_{k+1} - \mathbf{f}_k$.

the difference backwards: $\delta_k^- \mathbf{f}_k = \mathbf{f}_k - \mathbf{f}_{k-1}$

The difference equations (8.10, 8.11) can thus be reduced to:

$$\begin{aligned} \tilde{\mathbf{f}}_{k+1/2} &= \mu_k \mathbf{f}_{k+1/2}^n - \frac{\Delta t}{2\Delta r}\delta_k \mathbf{B}_{k+1/2}^n - \frac{\Delta t}{2}\mu_k \mathbf{C}_{k+1/2}^n \\ \mathbf{f}_k^{n+1} &= \mathbf{f}_k^n - \frac{\Delta t}{\Delta r}\delta_k \tilde{\mathbf{B}}_k - \Delta t \mu_k \tilde{\mathbf{C}}_k \end{aligned} \tag{8.13}$$

In the last decade or so, the explicit two-step scheme of MacCormack [267, 270] with noncentral differences has become popular (Fig. 8.2, *b*). It allows one to obtain the same results as scheme (8.10), (8.11) with a lesser number of arithmetic operations in the two-dimensional case. It is of paramount importance that the MacCormack scheme utilizes the values of the quantities only at the nodes of the mesh (i.e., there is no averaging). With the accounting for the notation in (8.12) of the MacCormack schemes, the predictor and the corrector can be written in the following manner

$$\begin{aligned} \tilde{\mathbf{f}}_k &= \mathbf{f}_k^n - \frac{\Delta t}{\Delta r}\delta_k^- \mathbf{B}_k^n - \Delta t \mathbf{C}_k^n \\ \mathbf{f}_k^{n+1} &= 0.5(\mathbf{f}_k^n + \tilde{\mathbf{f}}_k - \frac{\Delta t}{\Delta r}\delta_k^+ \tilde{\mathbf{B}}_k - \Delta t \tilde{\mathbf{C}}_k) \end{aligned} \tag{8.14}$$

or

$$\begin{aligned} \tilde{\mathbf{f}}_k &= \mathbf{f}_k^n - \frac{\Delta t}{\Delta r}\delta_k^+ \mathbf{B}_k^n - \Delta t \mathbf{C}_k^n \\ \mathbf{f}_k^{n+1} &= 0.5(\mathbf{f}_k^n + \tilde{\mathbf{f}}_k - \frac{\Delta t}{\Delta r}\delta_k^- \tilde{\mathbf{B}}_k - \Delta t \tilde{\mathbf{C}}_k) \end{aligned} \tag{8.15}$$

Schemes (8.14) and (8.15) are used in practical computations depending on the direction of flow velocity or shock-wave motion [270, 273]. MacCormack used both versions (8.14) and (8.15), changing them at each step with respect to t. For this purpose parameter $\lambda = 0$ or 1, which alternates at each Δt, was introduced into the algorithm

$$\begin{aligned} \tilde{\mathbf{f}}_k &= \mathbf{f}_k^n - \frac{\Delta t}{\Delta r}(1 - \lambda\delta_k^-)\delta_k^+\mathbf{B}_k^n - \Delta t\mathbf{C}_k^n \\ \mathbf{f}_k^{n+1} &= 0.5\,[\,\mathbf{f}_k^n + \tilde{\mathbf{f}}_k - \frac{\Delta i}{\Delta r}(1 + \lambda\delta_k^+)\delta_k^-\tilde{\mathbf{B}}_k - \Delta t\tilde{\mathbf{C}}_k] \end{aligned} \tag{8.16}$$

The two-dimensional scheme, proposed by L. A. Chudov and G. S. Roslyakov in 1962 [268] (a modification of the Lax-Wendroff scheme), has the following form (Fig. 8.3*a*)

$$\begin{aligned} \tilde{\mathbf{f}}_{k+1/2,\, l+1/2} &= \mu_l\mu_k\tilde{\mathbf{f}}_{k+1/2,\, l+1/2} - \frac{\Delta t}{2\Delta x}\delta_k\mu_l\mathbf{A}^n_{k+1/2,\, l+1/2} - \frac{\Delta t}{2\Delta y}\delta_l\mu_k\mathbf{B}_{k+1/2,\, l+1/2} - \frac{\Delta t}{2}\mu_l\mu_k\mathbf{C}^n_{k+1/2,\, l+1/2} \\ f^{n+1}_{k,l} &= f^n_{k,l} - \frac{\Delta t}{\Delta x}\delta_k\mu_l\tilde{\mathbf{A}}_{k,l} - \frac{\Delta t}{\Delta x}\delta_l\mu_k\tilde{\mathbf{B}}_{k,l} - \Delta t\mu_l\mu_k\tilde{\mathbf{C}}_{k,l} \end{aligned} \tag{8.17}$$

Here, the operators of averaging and difference taking are used in the direction of change of the subscripts k and l, respectively; for example

$$\delta_k f_{k,l} = f_{k+1/2,l} - f_{k-1/2,l}, \qquad \mu_l f_{k,l} = (f_{k,l+1/2} + f_{k,l-1/2})/2$$

One of the versions of the two-dimensional MacCormack scheme is expressed as follows [273] (Fig. 8.3*b*):

$$\begin{aligned} \tilde{\mathbf{f}}_{k,l} &= \mathbf{f}^n_{k,l} - \frac{\Delta t}{\Delta x}\delta_k^+\mathbf{A}^n_{k,l} - \frac{\Delta t}{\Delta y}\delta_l^+\mathbf{B}_{k,l} - \Delta t\mathbf{C}^n_{k,l} \\ \mathbf{f}^{n+1}_{k,l} &= 0{,}5\,[\,\mathbf{f}^n_{k,l} + \tilde{\mathbf{f}}_{k,l} - \frac{\Delta t}{\Delta x}\delta_k^-\tilde{\mathbf{A}}_{k,l} - \frac{\Delta t}{\Delta y}\delta_l^-\tilde{\mathbf{B}}_{k,l} - \Delta t\tilde{\mathbf{C}}_{k,l}] \end{aligned} \tag{8.18}$$

In [270], a detailed description is given of the various MacCormack schemes of the second and third approximation orders, constructed according to the Runge-Kutta principle.

The smoothing operator and selection of Δt. Unfortunately, the use of the schemes described earlier entails significant oscillations in the vicinity of shock waves or contact surfaces. For the elimination of oscillations and increase of stability of the computations, the solution is smoothed out or a term of artificial viscosity is added to it:

$$\mathbf{f}_*^{n+1} = f_k^{n+1} + \Omega\Phi_k^j, \qquad j = \begin{Bmatrix} n+1 \\ n \end{Bmatrix} \tag{8.19}$$

Here $\Omega \approx 2$ is a constant coefficient, $\mathbf{f}_*$ is an improved (as regards oscillations) solution, Φ_k^j is an aggregate of dissipative terms.

Case $j = n$ corresponds to explicit implementation of "artificial viscosity"; when $j = 1$, the procedure, defined by expression (8.19), is called smoothing.

In [269], the following form of "artificial viscosity" ($j = n$) is recommended:

$$\Phi_k^n = \frac{\Delta t}{\Delta x}\left(\delta_k f_{k+1/2}^n \delta_k(|u|+a)_{k+1/2}^n - \delta_k f_{k-1/2}^n \delta_k(|u|+a)_{k-1/2}^n\right)$$
$$\sim O(\Delta x^3) \tag{8.20}$$

This "artificial viscosity" does not upset the second order approximation and ensures the effective dissipative process. The authors of [269] used an expression similar to (8.20), along the two coordinate directions, in two-dimensional problems. It is shown in [274] that when solving unsteady two-dimensional problems, it is expedient to use structure (8.20) in the form or a smoothing operator ($j = n + 1$) and also add diagonal differences of the type (8.20) when shock waves in the flow field travel diagonally and an inclined reflection of the shock waves takes place from a solid surface.

The authors of [273] employ a term of "artificial viscosity" ($j = n$) of the order $\sim\Delta x^4$

$$\Phi_k^n = -\Omega[\mathbf{f}_{k+2}^n + \mathbf{f}_{k-2}^n - 4(\mathbf{f}_{k+1}^n + \mathbf{f}_{k-1}^n) + 6\mathbf{f}_k^n]$$

Analysis of the problems of stability, dissipative and dispersive properties for the above presented schemes has been conducted in [269, 270, 275].

Since for the explicit schemes the time step is limited, it is being selected from the corresponding stability condition. In [269], a simplified Courant-Friedrichs-Lewy condition was used for scheme (8.17)

$$\Delta t = c_r \min\left(\frac{\Delta x}{|u|+a}, \frac{\Delta y}{|v|+a}\right), \quad a = \sqrt{\gamma\frac{p}{\rho}}$$

The constant coefficients $c_r \approx 1$ and Ω are determined experimentally according to two conditions: controlling the stability of the computations (elimination of overflow of the exponent of the numbers when calculations are conducted on a computer) and obtaining parameters of exact solutions at some control points or cases.

For example, at the frontal point of a body in a flow the parameters of the stagnation flow are known, or when a shock wave is reflected from the frontal point the value of the pressure of the direct reflection is known.

8.2 REFLECTION OF A SHOCK WAVE FROM A WEDGE

The difficulty of the theoretical study of the processes of shock wave interaction with various surfaces is caused by the significant nonlinearity of equations (8.2), which describe the unsteady motion of the gas.

Analytical solutions are possible only in limiting cases ($\to \infty$ or $\to 0$) involving, mainly, the flow of an ideal gas. Sommerfeld [276] was the first to use a method consisting of linearizing the principal equations of motion with the assumption that the incident discontinuity is weak. The indicated method was used for a problem of weak-wave diffraction at a semi-infinite plate. Later, the interaction of a weak wave with a wedge of finite angle was discussed in [277]. In [278], the linear problem of Mach reflection and shock-wave diffraction was solved. The angle between the direction of wave motion and the solid surface was assumed small, and the reflected wave regarded as a sonic disturbance. Currently, the reflection of a weak wave from a wall within the bounds of acoustic approximation is studied in sufficient detail [279].

Along with the acoustic reflection model, one must examine the flow with small but sharp changes of gasdynamic parameters occurring in a narrow region near the reflection point of the weak shock wave at an angle of incidence close to the critical. The acoustic approximation in this instance is qualitatively incorrect. As was shown in [280, 281], the theory of small perturbutions leads, in such cases, to nonlinear equations of "short waves" with nonlinear boundary conditions. The application of this theory can be seen in [282, 283]; and in [284] a comparison is conducted with the results of numerical calculations and a law of similarity formulated. Regular and irregular reflections and interaction of weak shock waves are discussed in [285–287].

Another possibility of solution of the reflection and diffraction problems is based on the fact that linearization of the main equations of motion for an arbitrary intensity of the incident shock wave is carried out on the assumption that an obstacle causes only minor disturbances in a continuous flow behind the shock wave front. This method was developed by Lighthill for a small wedge angle [73] and for a shock-wave reflection close to the normal [289], and also in [201, 209] for a small obstacle of an arbitrary shape. This method was later developed further by Arutyunyan [159, 291].

Mach reflection from a wedge with a small opening angle has been discussed in linear and nonlinear arrangements in [292, 293], where the terms of the second or higher order of infinitesimal are neglected and the remaining terms take into account the linearity near the diffracted shock wave. The solution for the pressure at the wall, obtained at some distance from the wave, correlates with the linear solution of Lighthill. The reflection of a shock wave from a thin wedge with small density ratio at the front was investigated in work [294]. Regular reflection taking into account vibrational relaxation was computed in work [295]. Problems about shock wave flows of a gas in a nonlinear arrangement was examined in works [296, 297].

Advances in computer technology and numerical techniques have led to the appearance of works in which the results of integration of a complete system of nonlinear equations of gasdynamics is presented. Initially, shock-wave reflection from a wedge or a wall in an ideal gas was discussed numerically by the shock-capturing method in works [298, 299], and the effect of viscosity was taken into account in work [208].

When unsteady problems are solved by the shock-capturing method, gasdynamic discontinuities are not excluded but are smeared over several intervals of the calculation mesh. In conditions of limited computer memory, situations may arise when the shock wave passes the entire computation region and the difference solution has not succeeded to form completely, since the dimensions of the transition zones characterizing the discontinuities are only slightly less than the computation region. This circumstance diminishes the accuracy of the results [298–301]. In this case an improvement in accuracy can be achieved by changing over to self-similar variables and obtaining the solution by the time asymptotic method by iteration.

In [302] formulation of the problem is given for a complete system of gas-dynamic equations in self-similar variables and a method of solution is proposed, but no results are given. Difficulties are cited with the instability of calculations of the slip surface, which have apparently been overcome in [303]. The formulation of the problem to be presented later is analogous to the one proposed in [302]. However, for the calculation of discontinuous flows, the shock-capturing method without exclusion of gasdynamic discontinuities in the flow field has been chosen [304, 305]. The shock wave and, in this case also, the slip surface are regions with sharply varying gasdynamic parameters. Calculations were conducted for a wide range of wedge angles, shock-wave intensities, and also values of the adiabatic index [306]. Generalization of these results is given in [307]. The interaction of shock waves with a wedge in supersonic flow is discussed in [308] and a comparison is made with the analytical solution for weak shock waves [309], which indicates good correlation of the results.

The accuracy of the results is improved when: one makes the transition from first to second order approximation of the difference scheme; one uses the self-similar variables and the time-asymptotic method; calculations are made with the exclusion of the principal gasdynamic discontinuities. These three conditions are met in [303, 310] for regular and Mach reflections. We note that the procedure of excluding the slip surface in [303] is not entirely successful. As a result, an inaccuracy results in the calculation of parameters at the wall in the vicinity of the slip surface (see discussion later, Fig. 8.8).

The steady Mach reflection was simulated by the Monte-Carlo method in [311] using the Boltzmann equations.

Effective shock-capturing methods are currently being developed with minimum smearing of gasdynamic discontinuities [312–315]. Comparison of the results obtained with the help of such methods is conducted in [315], and with experiments in [200].

Formulation of the problem. It is expedient to utilize in the solution the self-similar variables $\xi = x/(u_0 t)$, $\eta = y/(u_0 t)$, where x and y are the Cartesian coordinates with the center at the vertex of the wedge; u_0 is the velocity of the incident wave [305]. Instead of the dimensionless time coordinate t we introduce $\tau = \ln t$.

The finite-difference analog of the system of gasdynamic equations (8.4) in the dimensionless form is

$$\frac{\Delta \mathbf{f}}{\Delta \tau} + \frac{\Delta}{\Delta \xi}(\mathbf{A} - \xi \mathbf{f}) + \frac{\Delta}{\Delta \eta}(\mathbf{B} - \eta \mathbf{f}) + 2\mathbf{f} = 0 \tag{8.21}$$

$$\mathbf{f} = \begin{vmatrix} \rho \\ \rho u \\ \rho v \\ e \end{vmatrix} \quad \mathbf{A} = \begin{vmatrix} \rho u \\ \rho u^2 + p \\ \rho u v \\ (e+p)u \end{vmatrix} \quad \mathbf{B} = \begin{vmatrix} \rho v \\ \rho u v \\ \rho v^2 + p \\ (e+p)v \end{vmatrix} \quad e = \rho\left(\epsilon + \frac{u^2 + v^2}{2}\right)$$

Here, ρ is the density; p is the pressure with respect to ρ_0 and $\rho_0 u_0^2$, respectively. Subscript 0 is ascribed to the parameters of the undisturbed flow ahead of the incident wave *md* (Fig. 8.4), and subscript 1 to be used later—behind the wave; u and v are the projections of the velocity vector on axes x and y with respect to u_0; e is the total energy of a unit volume; $\varepsilon = c_v T$ is the internal energy of a unit mass. The equation of state of a calorically perfect gas $p = \rho\varepsilon(\gamma - 1)$ closes the system of equations (8.21), $\gamma = \text{const}$.

A schematic of the gasdynamic discontinuities evolving during shock-wave reflection from a wedge is illustrated in Fig. 8.4. The flow in the self-similar variables ξ, η is steady. The parameters of the undisturbed flow ($\mathbf{f} = \mathbf{f}_0$) are given on contour *dec*, and the parameters behind the incident wave *md* ($\mathbf{f} = \mathbf{f}_1$) on contour *agd*. In the presence of the condition of symmetry for line *ab* and absence of a normal flow on surface *bc*, the boundary conditions are completely determined.

The initial condition is to take $\mathbf{f} = \mathbf{f}_0$ to the right of line *md* (continued up to the intersection with the surface of the wedge), and $\mathbf{f} = \mathbf{f}_1$ to the left. Letting the derivatives with respect to τ stay in the system of equations (8.21) ensures that the equations are hyperbolic in the entire region under consideration, which allows one to use the time asymptotic method along this coordinate.

Integration of the system of equations (8.21) is carried out with the help of a nine-point difference scheme of the second order of accuracy (8.17).

Results of numerical calculations. When solving the problem, attention is mainly given to the Mach reflection and the regimes of regular interaction at angles of incidence close to the critical value. The results of the computations are compared with the theory of regular (theory *V*) and elementary theory of Mach

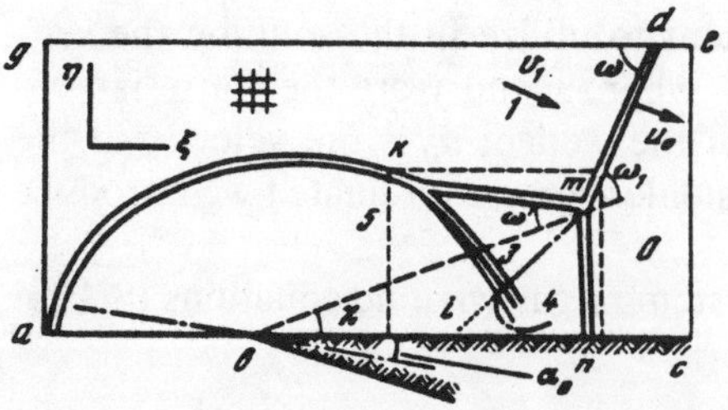

Figure 8.4. Schematic of the calculation field for double Mach reflection of the shock wave from a wedge with angle α_0. *md* is the incident shock wave; *mn* is the Mach wave; *m* and *k* are the first and second triple points; *mi* is the contact surface.

(theory *Y*) reflection [159, 155], in which the shock wave is assumed plane and the Mach wave perpendicular to the wedge surface.

The results of the numerical solution for a wave with Mach number $M_0 = 4$, wedge angle $\alpha_0 = 90° - \omega = 40°$, and $\gamma = 1.4$ are given in Fig. 8.5. The shock waves and contact surfaces that are generated in the flow are determined from the characteristic closeness of the isobars and isochors. The values of pressure and density depending on the curve numbers are given in table 8.2.

In the vicinity of the first triple point *m* good correlation is observed of the results of the numerical solution and theory *Y* with respect to gasdynamic parameters and geometry of discontinuities (discrepancy less that 2% with respect to reflection angle ω' and angle of trajectory of the triple point χ).

We note that the second configuration with an incoming reflected wave, which is feasible within the frame work of theory *Y* (terminology according to L. D. Landau [1]), is not realized when the boundary value problem is solved. This correlates with the experimental data [180].

Figure 8.5 shows the second triple point *k* (Fig. 8.4), which is formed by the intersection of the reflected shock wave with the internal compression wave. Analogous phenomenon is observed during experiments for gases whose adiabatic index γ changes sharply during passage through the shock [179], and for gases with almost constant index [180]. The present calculations were conducted with a constant $\gamma = 1.4$. The second triple point *k* is also observed when the reflected wave is located above the trajectory of the first triple point in accordance with theory *Y*.

The appearance of the compression wave *ki* (Figs. 8.4, 8.5) is more convenient to explain by considering the flow in a coordinate system linked with the first triple point *m*. In this case, the streams travel at velocities u_3 and u_4 from point *m* along the slip surface *mi*, and point *i* of the intersection of this surface with the wedge surface moves away from point *m* at a velocity $u_4 < u_3$. Thus, the gas of region *3*, which has a velocity $u = u_3 - u_4$ relative to point *i*,

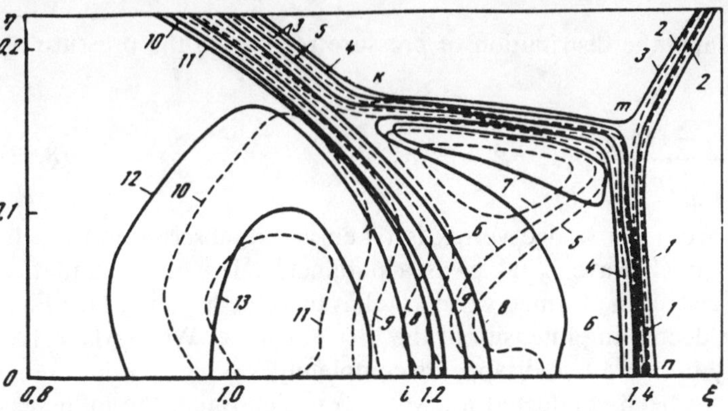

Figure 8.5. Lines of constant pressure (solid) and density (dashed) inside the contour bounded by the dashed lines in Fig. 8.4.

Table 8.2

N°	1	2	3	4	5	6	7	8	9	10	11	12	13
$\frac{p}{p_0}$	4.1	8.8	18.9	27.7	31.2	37.1	40.3	37.1	40.3	49.7	52.8	56.0	59.1
$\frac{\rho}{\rho_0}$	1.53	2.6	4.47	4.74	5.27	6.87	7.14	7.14	7.67	9.81	10.3	–	–

is decelerated toward point *i* and turns inside the angle *mib*. As a result of this compression, waves appear at specific flow regimes. These waves can then form a shock wave, which moves to the second triple point *k*. When the ratio u/a_3 (a_3—speed of sound in region *3*) exceeds unit significantly, the shock might be attached to the turning point *i*. The photographs in [176] show a configuration in which the shock almost originates from the point *i*. When the value u/a_3 is decreased, an unsteady wave moving away from point *i* is created. Finally, decreasing u/a_3 further leads to the establishment of weak compression waves which cause only a bending of the reflected SW or have no influence on the form of the latter.

The configuration near the second triple point is apparently closer to the type discussed in [316] when two incident SW intersect. In this case, contact discontinuity and compression (or expansion) wave must develop in region 5. The intensities of the latter are very low; therefore, they are difficult to detect in the numerical shock-capturing calculation.

The coordinate of the triple point *k* is determined fairly accurately on the basis of the scheme proposed in [179]. According to this scheme, point *k* is the point of intersection of the reflected SW with the circle of the sonic disturbances of radius a_3. The source of these disturbances is carried away by stream *3* with a velocity u_3 along the contact discontinuity *mi* from the first triple point. Such as assessment agrees well with the experiments [179] and with the present calculations.

Fig. 8.6*a* illustrates the distribution of pressure relative to the pressure of direct reflection p_n

$$p_n = \frac{1-\gamma+(3\gamma-1)p_{10}}{1+\gamma+(\gamma-1)p_{10}}\,p_1, \quad p_{10}=p_1/p_0 \tag{8.22}$$

The distance ξ from the wedge vertex is taken as the abscissa and Mach number M_0 of the shock wave is taken as a parameter. It is evident that a secondary pressure rise in the formed wave *ki* takes place near point *i* (see Fig. 8.4), dropping with decreasing intensity of the incident wave. When $M_0 \lesssim 1.9$ [307], the secondary pressure rise disappears completely.

Calculations have been conducted for $M_0 = 3$ to determine the influence of the angle of incidence $\omega = \pi/2 - \alpha_0$ on the structure of interaction. Figure 8.6*b* shows the curves of pressure distribution over the wedge surface for $\gamma = 1.4$

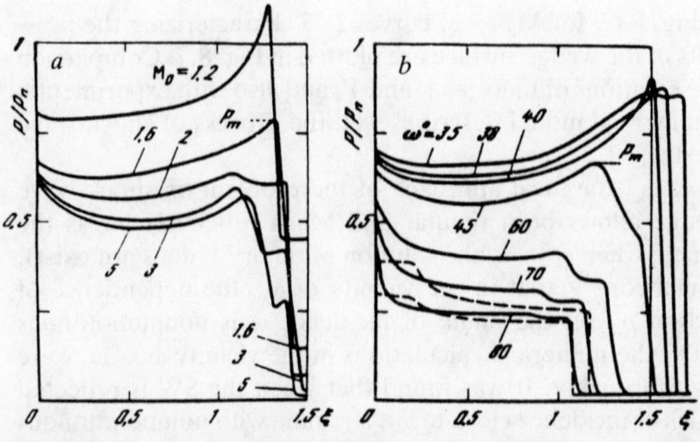

Figure 8.6. Pressure distribution over the wedge surface for $\gamma = 1.4$. *a)* for various Mach numbers at $\alpha_0 = 40°$; *b)* for various angles of incidence ω at $M_0 = 3$.

and different values of ω. The dashed lines show the solution of [73] which was obtained for small wedge angles. The difference in the character of the curves near $\xi = 0$ is explained by the fact that scaling by linear dependence is conducted for the pressure distribution curve over the surface for $p_{10} = 10$ ($M_0 \approx 2.952$) and $\alpha_0 = 0.1$ rad (5.7°). Such a regime of gas flow over a wedge behind an incident wave takes place with an attached shock wave, and a region of constant parameters develops near $\xi = 0$. This difference is insignificant when the flow is analyzed away from the wedge. A discrepancy of up to 6%, resulting from the nonlinearity of the reflection process, is observed in the region of Mach wave *mn* for $\omega = 80°$. Using a linear solution for lower angles of incidence ($\omega = 70°$) leads to an error of ~15–20%, which is caused by additional compression.

The pressure in the region of the wedge vertex can be estimated from the total stagnation pressure of flow stagnation *1* behind the incident wave if we ignore the fact that the reflected shock is unsteady, since the velocity at which it moves away from the wedge along the plane of symmetry is not high for a weak incident shock waves.

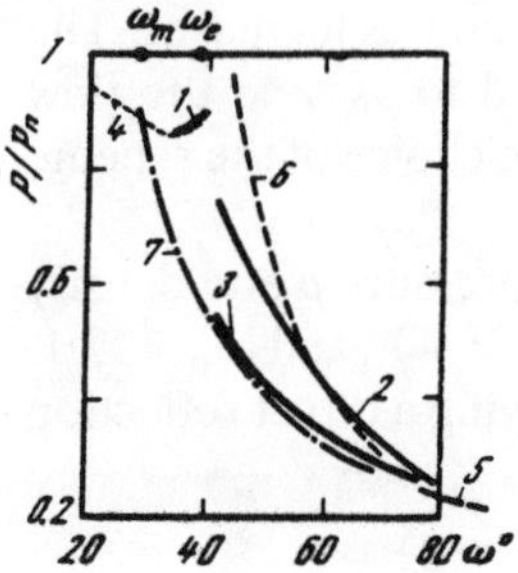

Figure 8.7. Pressure values at characteristic points of the wedge surface. *1)* local maximum pressure p_m in the region of reflection point (regular configuration); *2)* local maximum pressure p_m near the contact surface (Mach configuration); *3)* pressure on the wedge surface behind Mach wave; *4)* solution of the theory of regular reflection; *5)* linear solution [159]; *6)* estimate from the stagnation pressure (isentropic or Pitot for $u/a_3 < 1$ or $u/a_3 > 1$); *7)* solution of theory Y behind front of Mach wave.

From the data of Fig. 8.6*b* for $M_0 = 3$, curves *1–3* characterizing the pressure at individual points of the wedge surface are plotted in Fig. 8.7. Comparison of these curves with the solutions of theories *V* and *Y*, and also with experimental data, shows that the numerical model describes well the process of shock-wave reflection from the wedge.

In some region $\omega < \omega_e$, the local approach of the problem on shock wave interaction with a wedge allows both regular and Mach reflection (ω_e is the critical angle of incidence; when $\omega > \omega_e$ the solution of theory *V* does not exist). Hence, it follows from theory *V* that in the vicinity of ω_e, the dependence of the pressure of reflection p_m on the angle of incidence ω is nonmonotonous (curve 4). Consequently, the numerical calculations in the vicinity $\omega \sim \omega_e$ were conducted particularly thoroughly. It was found that when the SW is reflected from the wedge at angles of incidence close to ω_e, a regime with nonmonotonous dependence $p_m(\omega)$ is established and such flow is steady. The values of the local maximum pressures p_m (curve *1*) correlate well with the data of theory *V* (curve *4*). When $\omega \to \omega_e$, the domain of constant parameters on the surface of the wedge decreases (Fig. 8.6*b*) and the flow behind the reflected wave relative to the interaction point becomes subsonic.

However, when a SW is reflected from, for example, a sphere or cylinder, the nonmonotonous character of the dependence of the maximum pressure on the angle of incidence ω is not observed in the numerical solution. This is confirmed by the experimental data of [237] and the results of the numerical investigation of section 9.1.

The discussed method can, naturally, be applied to the case when the shock wave reflection occurs at the wedge, over which a supersonic stream flows [308].

Comparison of the shock-capturing method with the shock-fitting method. At the present time, the question of choice of a method for the numerical simulation of shock wave gas flows is considered to be of vital importance. There are two principal approaches: the shock-capturing method without exclusion of gas-dynamic discontinuities, and calculations with identification of the main discontinuities. The shock-capturing computer code is simpler to implement; however, the shock wave fronts and slip or contact surfaces are spread over several intervals of the computational mesh making their localization difficult. This situation is alleviated when the number of the calculated points is increased. The shock-fitting methods with discontinuities treated separately have more complex algorithms, but then the accuracy of computation of discontinuities increases. The main disadvantage of the shock-fitting method is the need to assume the flow scheme *a priori*, which could lead to inaccurate results if the choice of the scheme was made incorrectly.

Figure 8.8 shows a comparison of the distributions of pressure p and density ρ, similar to the one illustrated in Fig. 8.6, *a*, for $\gamma = 1.4$, $\alpha_0 = 40°$, and $M_0 = 2.1$. The value of the density of ρ_n of the gas near the wall following a direct reflection is given by

$$\rho_n = \rho_1 \times ((\gamma + 1)p_n/p_1 + \gamma - 1)/((\gamma - 1)p_n/p_1 + \gamma + 1)$$

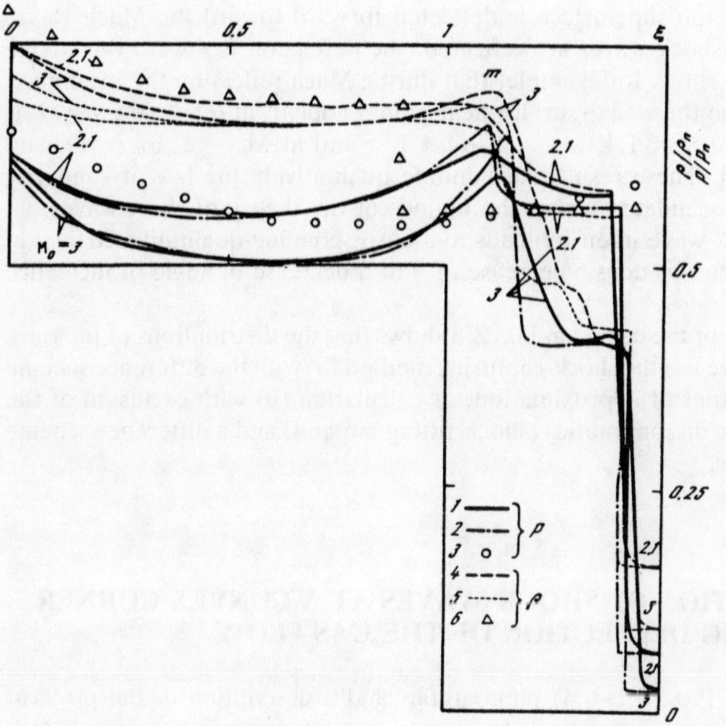

Figure 8.8. Pressure and density distribution over the wedge surface.

The data presented in Fig. 8.8 are obtained using three methods: (a) the shock-capturing method, which was discussed earlier (*p*—curve *1*, ρ—curve *4*); (b) with exclusion of SW and the entire slip surface in the self-similar variables and with the utilization of the MacCormack difference scheme of the second order of approximation [303] (*p*—symbol *3*, ρ—symbol *6*); (c) with exclusion of SW and part of the slip surface and retaining a small portion of the slip surface (up to 10%) for the shock-capturing method near the wedge surface [317] (*p*—curve *2*, ρ—curve *5*)—the Godunov method with difference scheme of the first order of approximation was used. Similar curves were plotted using methods (a) and (b) for the case when $\gamma = 1.4$, $M_0 = 3$, and $\alpha_0 = 40°$ in which the local maximum pressure is distinctively evidence in the vicinity of the point *i* (see Fig. 8.4).

From the comparison of the results for $M_0 = 2.1$ it follows that when using the shock-capturing method (a) and method (c), a characteristic maximum pressure is observed near $\xi = 1.1$, which is absent in the results obtained by method (b) [303]. The indicated rise for similar regimes was noted in the experiments of [200, 165, 176]. This indicates that the procedure employed in [303] for the calculation of the parameters near the slip surface needs to be revised, since the shape of this surface can vary. Physical [180, 6] and numerical experiments [304, 305] show that when sufficiently strong shock waves ($M_0 \geqslant$ are reflected, the lower section of the slip surface is deflected forward toward the Mach wave, and when weak shock waves are reflected, the deflection is toward the wedge tip. Calculations show, for example, that during Mach reflection the local maximum pressure on the wedge surface near point i appears at $\alpha_0 = 40°$, $\gamma = 1.4$, and $M_0 \geqslant 1.9$; at $\gamma = 1.4$, $M_0 = 3$, $\alpha_0 \geqslant 10°$; and at $M_0 = 3$, $\alpha_0 = 40$, and $\gamma \leqslant 2$ [304–307]. These results substantiate qualitatively the law of similarity [284], which is formulated within the bounds of the theory of short waves: an increase in shock wave intensity leads to a corresponding qualitative change in reflection character as does a decrease of γ or a decrease of angle of incidence ω from 90° to $\omega \sim \omega_e$.

Comparison of the curves in Fig. 8.8 shows that the distributions of pressure correlate when we use the shock-capturing method (a) with the difference scheme of the second order of approximation, or calculation (b) with exclusion of the main gasdynamic discontinuities (shock-fitting method) and a difference scheme of the first order.

8.3 DIFFRACTION OF SHOCK WAVES AT A CONVEX CORNER UPON STRONG DEFLECTION OF THE GAS FLOW

In Chap. 6 (see Figs. 6.1–6.4) photographs and a description of the physical pattern of shock wave diffraction at a convex two-sided corner were presented. Also presented were schemes with the principal notations which are used later when analyzing and describing the results of the numerical calculations. Discussed below are the regimes of flows in which physical and numerical experiments indicate discontinuous and reversed flows caused by the action of the physical or artificial viscosity at large angles of flow deflection. The results of systematic calculations are given in a wide range of determining parameters.

Literature review. The first and sufficiently detailed investigation of shock wave diffraction at a convex corner $\alpha = 90°$ was conducted by the authors of [318, 319] using the shock-capturing method. The method of large particles of the first order of approximation was employed. The cases discussed were $M_0 = 3.95$; $\gamma = 1.4$ and $M_0 = 10$; $\gamma = 5/3$ [318] and $M_0 = 15$ in oxygen with equation of state for a real gas [319]. Presented were the lines of equal level of p and ρ and velocity field from which it is clear that the numerical flow stops "following" the Prandtl-Meyer solution, and a separation zone, and even a reversal flow, is formed.

A similar zone of separation is obtained numerically in [320], where a scheme of second order of approximation is used on a weak irregular mesh. The regimes considered were $\gamma = 1.4$; $\alpha = 90°$; $M_0 = 3.95$ and $\alpha = 45°$; $M_0 = 4.01$, and also the flow in an L-shaped channel.

A scheme of third order of approximation was used in [321] for the calculation of shock wave diffraction at $\alpha = 90°$ and in a T-shaped channel.

An approach differing from the enumerated works, where the shock-capturing method was used, has been applied in [317, 322]. The distinctive feature of this approach is the initial layout of the computation domain on the basis of information obtained *a priori* from experience or previous computations. The mobile mesh, which is adapted to the results of the computations, allows an increase in the accuracy of the calculations significantly. Of particular importance is the formation of separation surfaces without which, as the authors note, the obtained results are unsatisfactory. The Godunov method of the first order of accuracy is used in the integration. Reference [317] gives an example of the calculation of shock wave integration with axially symmetric cylinder of a finite length in a tube. The final stage of the interaction process is shock-wave diffraction at angle $\alpha = 90°$ in axially symmetric case. The velocities on both sides of the formed separation surface differed by a factor of ten. In [322], the unsteady flow of a supersonic jet from a channel of circular cross section was discussed. The Mach number of the SW ahead of the outlet of the tube was equal to 2.14–3.49. In [317, 322], the fields of equal-level lines of p and ρ were presented and the locations of SW_d and SW_t and the separation lines were shown.

The method of discrete vortexes for the investigation of diffraction of weak SW was used in [323]. The medium is assumed incompressible and the flow is potential. The fields of the isobars and location of the SW ($p_{10} \approx 2$) are compared with the experiment.

The exclusion of the main discontinuities increases the accuracy of the calculation, although this occurs at the expense of increased complexity of the algorithm. Another method for increasing the accuracy of the results is based on the transition to self-similar variables [305] and obtaining a steady solution. This approach has been chosen in [324], and by the authors of the present work. The question of diffraction in [324] has been considered as a more general formulation where SW_s, whose front is perpendicular to the left side of *jo* ($\omega = 90°$), moves along this side (Fig. 8.9), or the angle of front inclination is such that a *V*-shaped configuration appears (regular reflection). The pressure distributions on the surface of the body are given for the case $\omega = 90°$, $M_0 = 10$, $\gamma = 1.4$,

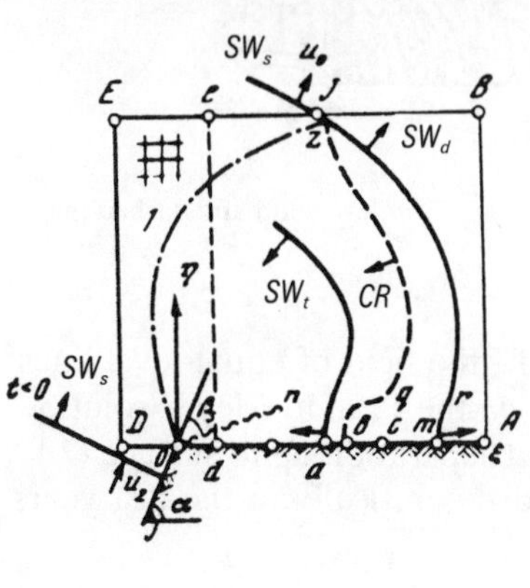

Figure 8.9. Schematic of SW diffraction at a convex two-sided corner *jOA*.

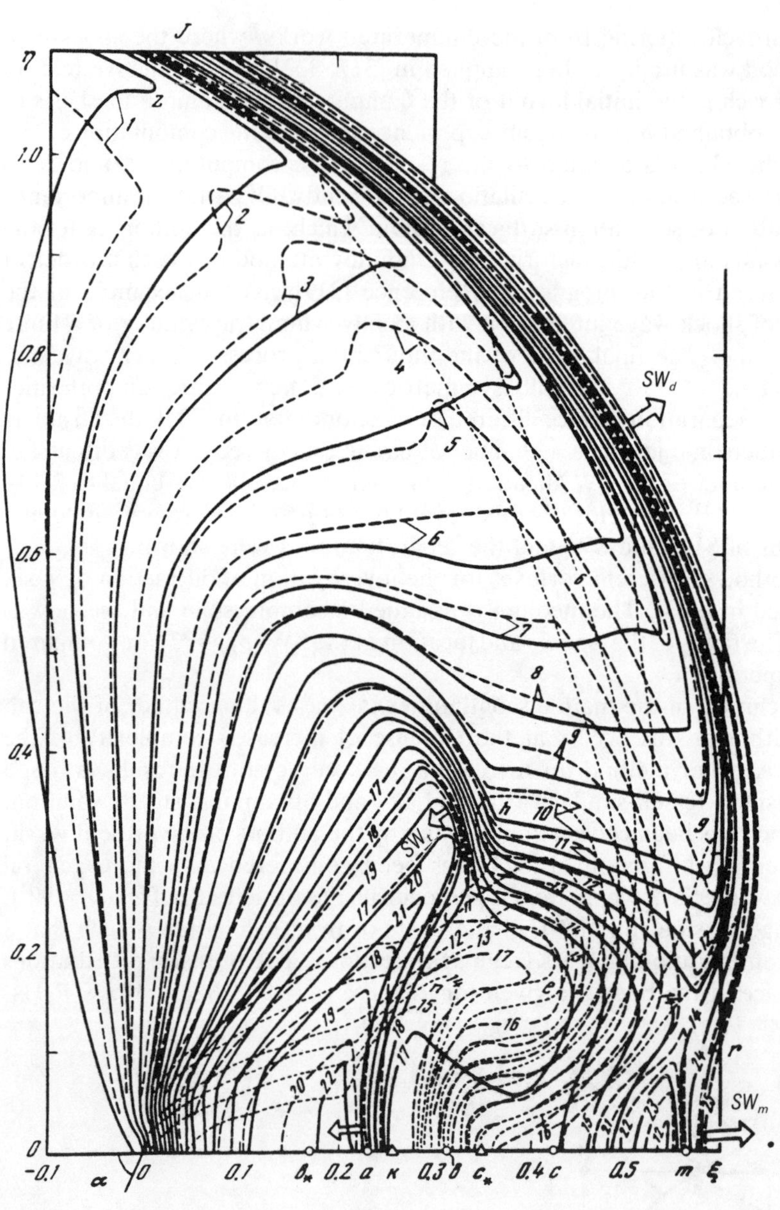

Figure 8.10. Equal-level lines for M_0 = 4, 71; γ = 1.4, and α = 65° p, ρ—solid and dashed lines, respectively, p/p^{γ}—dotted line.

and α = 10, 20, 30 and 40°; for the case α = 30°, the field of equal-level lines of p and ρ are also given. On the left boundary *de* the Prandtl-Meyer solution is given, according to which the flow turns without separation up to surface *OA*.

The problem of diffraction has been treated theoretically in the last years in [325–330].

Formulation of the problem. The numerical model under consideration is based on the use of finite difference analog (8.21) of differential Euler's equations of the dynamics of an ideal gas (inviscid, zero thermal conductivity) and an explicit finite-difference scheme of the second order of approximation (8.17) with a smoothing-out operator of the third order of the infinitesimal (8.19, 8.20). The formulation of the problem and the method of solution are analogous to the method used in section 8.2 when solving the problem on shock-wave reflection from a wedge.

Numerical [318, 319] and physical [6, 224] experiments confirm the self-similarity of the process in the presence of the effects of artificial and physical viscosity. Therefore, to increase the accuracy of the calculations it is expedient, as in section 8.2, to make the transition to the self-similar variables $\xi = x/(tu_0)$, $\eta = y/(tu_0)$, and also to introduce $\tau = \ln t$. Then the finite-difference analog of the system of the gasdynamic equations for ideal gas acquire the form (8.21).

Let us formulate the boundary conditions (Fig. 8.9). From the left the computational domain is bounded by two methods: along *ED* or *ed* (minor domain). Along *ODEJ*, parameters $\mathbf{f}_1$ behind SW_s are given, along *JBA*—parameters $\mathbf{f}_0$ ahead of SW_s are given. In the vicinity of point *J* linear transition is carried out at four or five points from $\mathbf{f}_1$ to $\mathbf{f}_0$. Calculation over the minor domain *deBA* was carried out when a non-separating flow was expected ($\alpha \lesssim 30$–$40°$). In this case, the parameters on *ed* are given taking into account the Prandtl-Meyer solution, as in [324]. When the entire domain *DEBA* is used, there is a special point in the field of calculations—the vertex of the corner *O*. The gas flow near this point must strongly change its velocity and direction. The shock-capturing method was used in the task, the artificial viscosity of this method smearing the expansion wave fan in this region. The alternative computation over the minor domain permits the exclusion of the vertex of the corner *O* and permits an exact Prandtl-Meyer solution on *de*. However, the numerical calculations give in this case, when angle α is large, a flow which is removed from reality, since when $\alpha \gtrsim 40°$ separation and even reversal flow circulation is formed in a real flow and in the shock-capturing computation in domain *DEBA* [6, 224, 318–322]. Therefore, the calculations must be conducted over the entire domain *DEBA*. Later, a comparison will be given of the results of calculations based on the entire and minor domains.

At the wall *OA* we assumed zero flow through the wall and a complete slip flow along the surface. Comparison with experimental data, however, indicated that although the velocity along the surface in the shock-capturing method becomes less than when the flow completely turns into a Prandtl-Meyer fan near the wall, it is still sufficiently high for SW_a to be formed near the surface (see Figs. 6.2*a*; 8.9), which is not observed in the experiment when $\alpha \gtrsim 50°$. In this case, it is wise not to limit ourselves with the slip condition, but assume the entire spectrum of velocities tangential to the surface: from a complete slip to a complete no slip condition. Trial calculations of various problems show that when the flow velocity along the wall is decreased, the results of the calculations tend to correlate with the physical experiment qualitatively. There is no systematic investigation so far of the quantitative characteristics in these conditions.

Initial conditions. We determine the coordinate of point J—the intersection with boundary EB of the undisturbed front of SW_s drawn at distance $\xi = 1$ from the tip of corner O (Fig. 8.9). Next, an arbitrary line is drawn from point J to the proposed vicinity of point m and on the right of this line we take parameters $\mathbf{f} = \mathbf{f}_1$ and on the left—$\mathbf{f} = \mathbf{f}_0$.

Modeling parameters, features of the solution. The numerical simulation is conducted[1] for $\gamma = 1.4$ and for the following variations of the determining parameters

$M_0 = 3$ $\quad\alpha = 5°–120°$ (Fig. 8.12, *a*; 8.13, *a*)

$M_0 = 2.2–3,$ $\quad\alpha = 30°$ (Fig. 8.12, *b*; 8.13, *b*)

$M_0 = 2.2–5,$ $\quad\alpha = 65°$ (Fig. 8.12, *c*; 8.13, *c*)

The parameters of SW_s for the majority of the calculated cases are presented in table 8.3.

Here, $\rho_{10} = \rho_1/\rho_0$, $T_{10} = T_1/T_0$, $s_{10} = s_1/s_0$ $(s = p/\rho^\gamma)$, $a_{10}^2 = p_{10}/\rho_{10}$, $M_1 = 2(M_0^2 - 1)/((\gamma + 1)\, M_0 a_{10})$. M_1 is the Mach number of the concurrent flow behind the SW.

Some features of the flow are weakly manifested in only some calculation variants, and only analysis of systematic results allows one to draw specific conclusions regarding these features.

Calculations of shock wave diffraction are mainly conducted for a condition of complete slip of the flow along the face of the corner. For purposes of illustration and comparison we shall present the results of one computation with the boundary no-slip condition on the surface at $M_0 = 4.71$ and $\alpha = 65°$. Every time this boundary condition is used, it will be mentioned separately.

The significance of using the no-slip boundary condition here consists of critically verifying the validity of its exclusion from mass computations of similar problems. For example, when the shock wave interacts with a sphere (see section 9.1), the velocity of the gas in the area of the frontal and rear points is low; therefore, the solution will be the same whether the boundary condition of slip or no-slip is used in these regions. On the lateral surface of the front part of the sphere the solution already reacts to the difference in boundary conditions; however, the discrepancy is only quantitative (up to 5%). On the lateral surface of the rear part of the sphere, on the other hand, qualitative differences are to be expected due to separation phenomena. These differences, as calculations show, do not appear at once but after some time following diffraction, when transition to a steady flow process takes place.

The numerical calculations of the current section and section 9.1 show that the use of the no-slip boundary condition leads to quantitative and qualitative

[1]The results of the calculations were obtained jointly with V.F. Yatsuk. See also: L.G. Gvozdeva, V.N. Lyakhov, "Shock-wave diffraction near an apex of a convex corner with strong flow deflection. Analysis of numerical simulation results."—In Proc. of 16-th International Symposium on Shock tubes and waves. Aachen, 1987.

Table 8.3

M_0	p_{10}	ρ_{10}	T_{10}	s_{10}	M_1
2.2	5.48	2.95	1.86	2.05	1.07
2.5	7.12	3.33	2.14	1.32	1.20
2.77	8.78	3.63	2.42	1.44	1.29
3.0	10.33	3.86	2.68	1.56	1.36
3.5	14.125	4.26	3.32	1.87	1.47
4.0	18.50	4.57	4.05	2.20	1.55
4.5	23.49	4.81	4.87	2.60	1.61
4.71	25.71	4.90	5.25	2.78	1.63
5.0	29.00	5.00	5.8	3.05	1.66

shifts toward agreement with the experiment. Also the calculations of the present section show that the change of solution occurs in the small domain, which is commensurable with the thickness of the boundary layer. Since the boundary layer begins to form behind SW_m, the parameters behind the wave are still the same in the solution of both types of boundary conditions, and discrepancy is

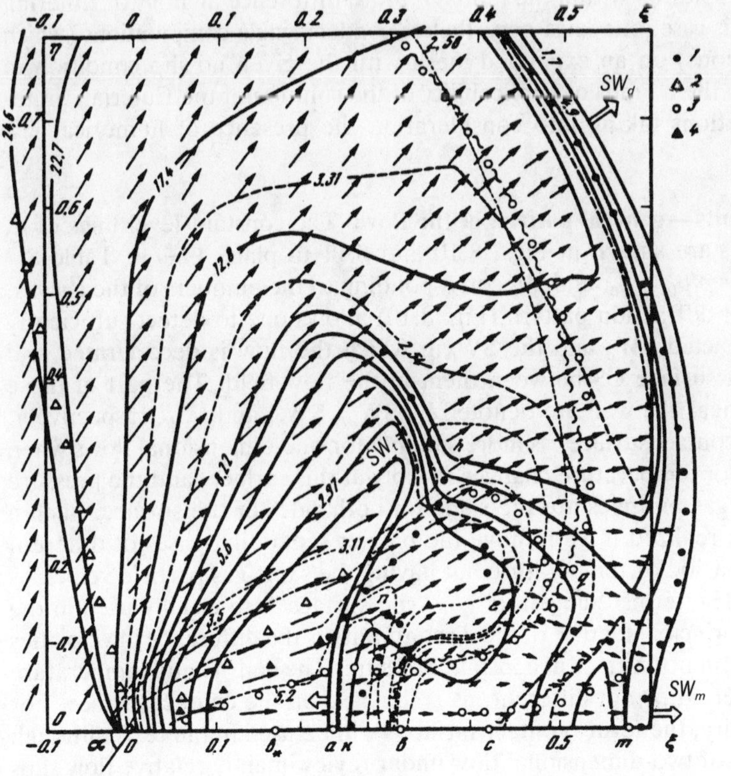

Figure 8.11. Part of the flow field from Fig. 8.10 drawn by velocity vectors and data of physical and numerical experiments at $M_0 = 4.71$, $\gamma = 1.4$, and $\alpha = 65°$.

detected, as will be shown later, in region *oec* (Fig. 8.11). Comparison with the experimental data of chapter 6 indicates good agreement of the results outside this region, which gives ground to using such a numerical model, excluding the indicated region where the effects of viscosity must be carefully considered.

Thus, examination of the solution with respect to the degree of dependence on the "viscous" boundary no-slip condition shows that qualitative and quantitative differences are observed in a limited region near the solid surface. This question will be discussed in more detail during the discussion of the results.

The difference of the nature of numerical (approximate) viscosity from physical viscosity in the present case for such verification is, in our opinion, insignificant, although this question requires detailed investigation. The artificial viscosity, as has already been mentioned, is analogous, with respect to its action, to the physical viscosity and gives qualitative and quantitative shift toward agreement with the experiment. A statement to the contrary has not been found in the literature.

At the present time, the limited capabilities of computers with respect to speed and memory do not allow one to conduct extensive computations with correct consideration of the effects of physical viscosity, but dictate the construction of the solution within the limits of finite-difference analogs of Eulerian equations. In this case, one must nevertheless conduct single computations (which are quite laborious) on an expanded mesh with the given no-slip condition in order to explain the bounds of applicability of the solution of the Eulerian finite-difference equations taking into consideration the presence of numerical viscosity.

Calculation results—general pattern of the flow. The constant-level lines of p, ρ, and $s = p/\rho^{\gamma}$ are shown in Figs. 8.10 and 8.11 in plane $\xi - \eta$. Table 8.4 shows the values p/p_0, ρ/ρ_0, and s/s_0 corresponding to the numbers of the curves.

The general diffraction pattern (Fig. 8.10) allows one to detect sufficiently clearly the diffracted SW_d and the SW_t in which the flow is deccelerated and also the contact surface CS between them in the flow field. The part of these discontinuities near the wall are denoted by SW_m, SW_a, and CS_b, respectively.

The term "contact surface" is more suitable for one-dimensional flows when the density and/or temperature changes sharply at this surface, and the pressure and velocity of gas particles are the same on both sides of the surface. Such a situation can be realized if we remove the barrier separating gases of different composition, density, or temperature but having the same gas particle velocities and pressures. The terms "tangential" or "slip surface" are usually used in the cases when the principal feature of the discontinuity is the difference in velocities along the discontinuity surface at constant pressure. The density and temperature can also be different or not different, as is the case on the contact surface. For the sake of brevity, the term "contact surface" will be used in the text, although in the situations of two-dimensional flow under review mainly relative flow slips are observed on both sides of the discontinuity.

Table 8.4

N	1	2	3	4	5	6	7	8
p/p_0	25.3	22.0	19.4		16.2	14.1	12.8	11.5
ρ/ρ_0	4.59	4.33	4.08	3.82	3.56	3.31	3.05	2.79
s/s_0				2.74		2.4	2.06	

Table 8.4 (*Continued*)

N	9	10	11	12	13	14	15	16
p/p_0	10.2	8.9	8.2	7.6	6.9	6.2	5.6	4.9
ρ/ρ_0	2.54	2.28						
s/s_0	7.57	3.11	3.46	3.81	4.52	4.87	5.22	1.53

Table 8.4 (*Concluded*)

N	17	18	19	20	21	22	23	24	25
p/p_0	4.3	3.6	3.0	2.3	1.7	1	6.9	4.9	3.6
ρ/ρ_0	0.87	0.62	0.49	0.36					
s/s_0									

In addition to the enumerated discontinuities (Fig. 8.10), we also see: a Prandtl-Meyer fan (solid and dashed lines) originating from the tip O with some smearing near it, another fan (solid lines) originating from the vicinity of point h towards point r, a fan of the dotted lines s = const originating from the area of point O and characterizing the separations zone sector which is resting with its wide part against the contact surface CS. The zones characterizing contact surfaces in the numerical solution are shown in Fig. 8.11 by the wavy broken lines. The arrows near all the enumerated flow features above in Fig. 8.9 show the direction from a higher to a lower density. Point a on the face of corner OA denotes SW_a, and point m—SW_m. Between these two shock waves there are two more points: b and c, which characterize the contact surfaces CS and rc near the wall. In the calculation with the no-slip boundary condition on the wall point b shifted to the vertex of the corner at point b_* (Fig. 8.10; $\xi \approx 0.17$), and SW_t did not reach the wall. Both these changes correspond more to the real flow pattern [6, 224] (see Fig. 6.2c, d).

Points k and c_* from the photographs in Fig. 6.2c, d denote the location of SW_k and the approximate location of the end of line rc by the triangular symbols.

The vector field of the velocities is plotted in Fig. 8.11, which is a part of Fig. 8.10. Several lines of equal levels of p, ρ, and s are reserved for comparison. The wavy broken lines, as noted already, characterize contact surfaces, the location of which was determined from the distribution of ρ and s along the various rays ξ = const or η = const. The remaining symbols *1–4* denote the location of discontinuities in shadowgraphs (see Fig. 6.2*c*, *d*).

It is evident that symbol *1*, characterizing the experimental locations of the SW, correlate well $SW_d + SW_w$, and SW_t. However, discrepancy is also observed: in the numerical calculation we have SW_a which reaches the wall, and in the experiment a λ-configuration appears instead (see Fig. 6.4*a*: SW_t + SW_f + *he*. The latter leans on the separation surface *on*. Although the constant-pressure lines in Fig. 8.11 form a certain bifurcation *hng*, SW_a disappears when the no-slip boundary condition on the wall is used; however, the flow pattern above region *oec* when $\xi \gtrsim 0.1$ remains practically unchanged (Fig. 8.11).

From the photographs of Fig. 6.2*c, d*, the location of the boundary of the Prandtl-Meyer fan is shown by the symbols *2*, and symbols *3* denote the contact surfaces, the locations of which agree with the results of the numerical calculations (wavy broken lines). Symbols *4* characterize the separation zone observed in the experiments. The dotted lines s = const enclose a sector with the top at point O, and its position agrees with the direction of the separation surface *on* in the experiment.

In chapter 6, Fig. 6.4*b* shows a comparison of the position of gasdynamic discontinuities, which are observed in the photographs of Fig. 6.2*c, d* and in the numerical calculation in Figs. 8.10 and 8.11. On the basis of the photographs of Fig. 6.2*c, d*, the following plots are made in Fig. 6.4: solid lines—position of the SW; dashed line—contact surface CS; wavy broken lines—surface of flow separation; dash and dot line—boundary characteristic of the centered rarefaction wave RW_1. The interpretation of the results of the numerical calculation is plotted in Fig. 6.4*b*, where they are denoted by the following symbols: dark dots (A)—the locations of $SW_d + SW_m$ and $SW_t + SW_a$; the light dots (B)—contact surface CS; triangular symbols (C)—boundary characteristics of the fan of rarefaction waves RW_1 and RW_2, the dotted line characterizes the position of the isobars qualitatively in Fig. 8.10. It is seen that there is good correlation of the results of physical and numerical experiments for the diffracted SW_d and its part near the wall, SW_m. The same can also be said about the upper part of the contact surface CS and boundary characteristics RW_1 and RW_2. The deviation of the results of the numerical calculation, obtained with the condition of complete slip on surface *om*, is due to the fact that SW_t reaches the wall at point a and SW_k is absent (see near point k in Figs. 6.4 and 6.2*c, d*), and the contact surface CS arrives at point b instead of vertex O (see Figs. 8.10 and 8.11). The implementation of no-slip boundary condition in the numerical solution causes the contact surface near the wall to shift from point b towards the vertex of the corner, to point b_*, and SW_a disappears.

Thus, analysis of the presented results shows that the general flow picture is reproduced sufficiently well, with the exception of region *oec* when $\xi \gtrsim 0.1$ in Fig. 8.11. The presence in the numerical solution of SW_a near the wall is an indication that the velocity of flow along the surface is very high, i.e., the discontinuity surface is not localized but strongly smeared, and there are not enough mesh points in the given calculations for its localization. It is more expedient to follow the approach [317, 322] where this separation surface is specially noted, although this complicates the algorithm. The solution obtained with the no-slip boundary condition on the wall corresponds more to reality. We note that in the numerical solution zones of reversed flow were observed at some regimes even with slip boundary condition on the surface (for example, $M_0 = 2.2$ and 2.5 at $\alpha = 65°$ and $M_0 = 3$ at $\alpha = 120°$). All this indicates that the influence of artificial viscosity in the numerical flow corresponds to the physical viscosity.

The rigorous mathematical substantiation of the use of no-slip boundary condition in the given case is not the purpose of the present work. Only factual material is presented here about the behavior of the numerical solution which is shaped with the participation of artificial viscosity. This, however, allows one to draw some conclusions about those differences which appear when the influence of physical viscosity is not taken into account. The expounded computations were mainly conducted on a mesh ~50 × 70; therefore, it is not expedient in this case to account for the physical viscosity of the laminar flow, because its effect for such a mesh is weaker than the effect of the artificial viscosity. Consideration of the transport properties of the turbulent flow also falls beyond the scope of the present work.

Results of parametric calculations. The distribution of the parameters along surface *OA* is presented in Fig. 8.12*a*. The three solid curves for $\alpha = 70$, 90,

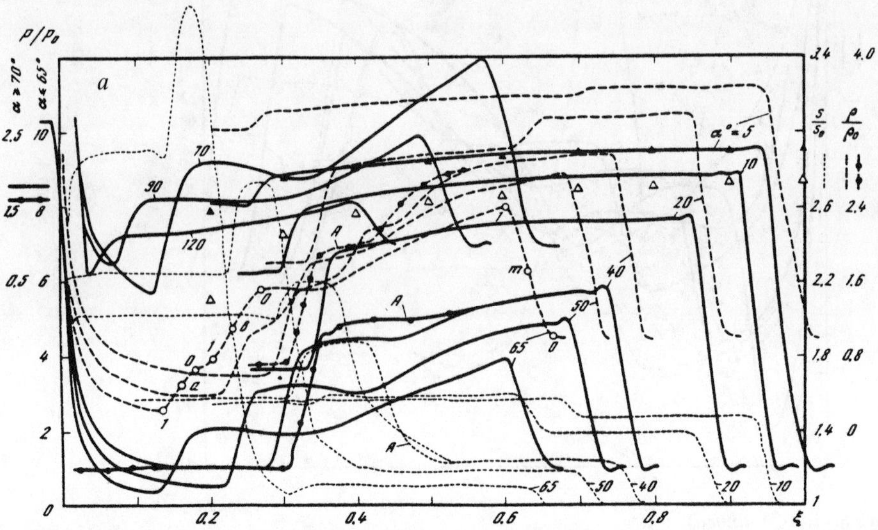

Figure 8.12. Distribution of parameters along surface *OA* (Fig. 8.9) for $\gamma = 1.4$. p, ρ, $s = p/\rho^{\gamma}$, T) solid, dashed, dotted, and dash-dot lines, respectively; *a*) $M_0 = 3$, $\alpha = 5–120°$ (calculation over the minor domain *deBA* at $\alpha \leqslant 20°$ and $\alpha = 40°$—curves *A*); *b*) $\alpha = 30°$, $M_0 = 2.2$ and 2.77 (calculation over minor domain *deBA* at $M_0 = 2.77$—curves *A*); *c*) $\alpha = 65°$, $M_0 = 2.2–4.71$ (calculation with no-slip boundary condition at $M_0 = 4.71$—curves *A*).

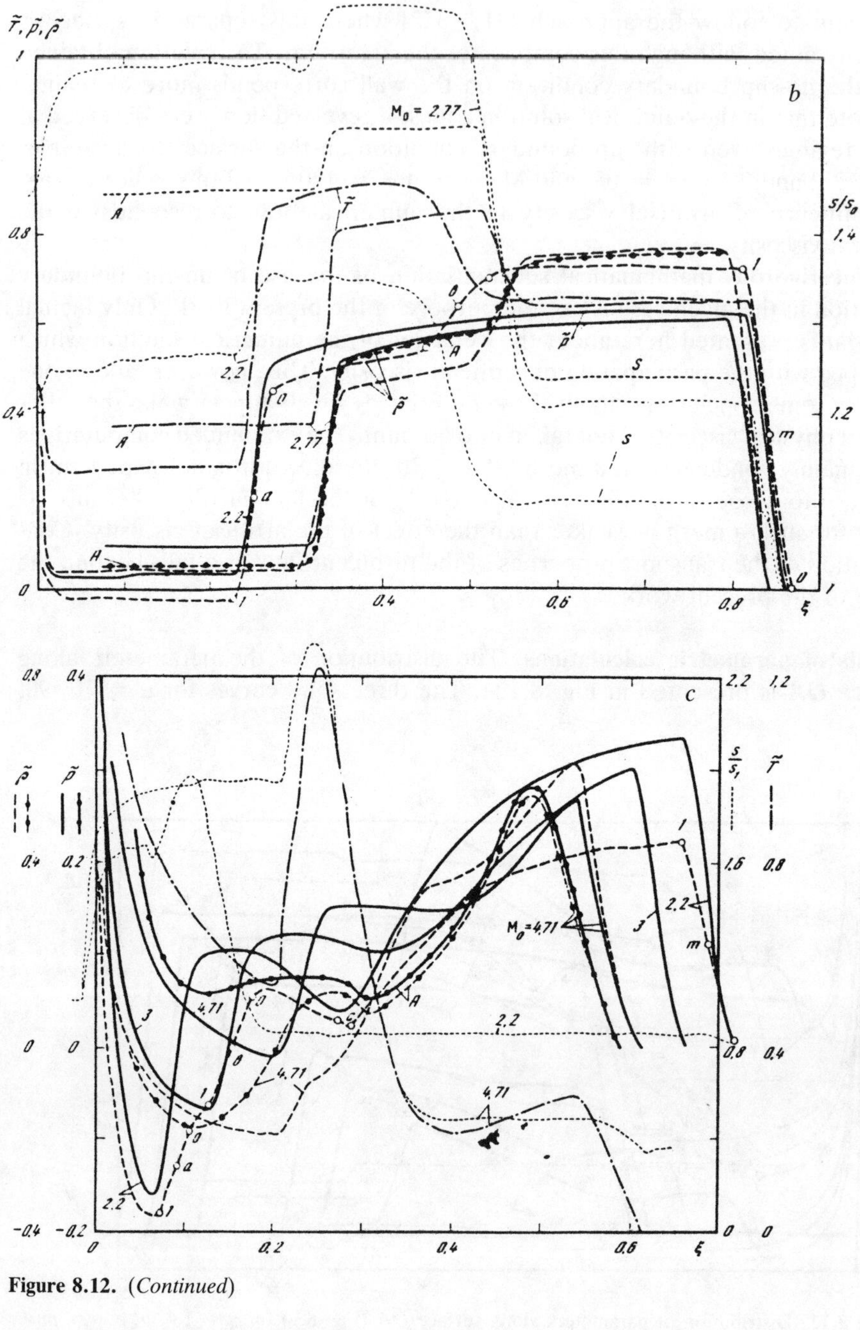

Figure 8.12. (*Continued*)

120° are lifted up and their scale is increased two times. When $\alpha \leqslant 40°$, calculation was conducted over the minor domain *deBA*. For $\alpha = 40°$ three curves *A* are plotted, which are obtained for the calculation over the minor domain (boundary *de* in Fig. 8.9). The remaining curves when $\alpha \geqslant 40°$ characterize the calculation over the complete domain *DEBA*. Some discrepancy ($\xi \leqslant 0.55$) is observed when calculated over the complete and minor domains at $\alpha = 40°$. The discrepancy decreases or increases with corresponding decrease or increase of angle α. The data on the pressure from the analytical solution of [159], which is obtained on the assumption that the flow is isentropic, are plotted using dark and light triangles for $\alpha = 5°$ and 10°, respectively. The correlation of the results is good at $\alpha = 5°$ and worse at $\alpha = 10°$, which is explained by the sufficiently large change in entropy (dotted curve at $\alpha = 10°$).

The characteristic points *m*, *a*, *b*, which are the average of the transition zones of large gradients, are plotted on the dashed curve for $\alpha = 65°$. These points characterize the position on the wall of SW_m (point *m*), SW_a (point *a*), and contact surface CS_b (point *b*). The boundary points marked by digit *1* are located closer to the vertex, i.e., to region *1* (Fig. 8.9); *O*—closer to the undisturbed region of *O*. Therefore, the points will be denoted by the letters *a*, *b*, and *m* with the corresponding subscripts, for example, a_0, a_1, m_1, and so forth. Later, the coordinates of these points will be presented in Fig. 8.13, *a*–*c* together with the values of the parameters on the boundaries of these zones of large gradients (at points *O* and *1*).

At small angles $\alpha \lesssim 20°$ the pressure diagrams have a Π-shaped form. At $\alpha \gtrsim 40°$ the pressure behind the SW_m is higher than on the remainder of the surface, and falls isentropically from point m_1 to the contact surface CS_b.

Curves analogous to the ones presented in Fig. 8.12*a* are plotted in Fig. 8.12*b*, characterizing the distribution of the parameters along the wall $p = (p - p_0)/(p_1 - p_0)$, $\sim\rho = (\rho - \rho_0)/(\rho_1 - \rho_0)$, $T = (T - T_0)/(T_1 - T_0)$, and s/s_0.

Calculations were carried out for $M_0 = 2.77$ over minor and whole domains. The values of the pressure for all these variations correlated completely. However, the densities were different—compare the dashed line with dots (*A*) and the dashed line, which corresponds to the calculations over the minor and whole domains. From the distributions of ρ (ξ), it is seen that the density differs by $\Delta\rho = 0.07$, i.e., ~8% in the range $\xi = 0.35$–0.52. There is also discrepancy with respect to the value of *s*. This comparison points to the fact that when $\alpha = 30°$, the method of calculation is sensitive to the way the boundary conditions are given on lines *ED* or *ed*.

The symbols in Fig. 8.12*c* are the same as in Fig. 8.12*b*, with the exception of two curves *A* with dots—solid and dashed, which determine the values $\tilde{p}$ and $\tilde{\rho}$ on the surface when calculating with no-slip boundary conditions of flow $M_0 = 4.71$. It can be seen that the SW_a is absent in this case and, as was explained when describing Figs. 8.10 and 8.11, point *b*, which characterizes the position of contact surface on the wall, shifts closer to the vertex of the corner to the vicinity of the point b_*—$\xi \approx 0.17$. These qualitative changes lead to better correlation of the numerical and experimental data.

It should be noted that a region with s = const is observed on the surface behind SW_m in Fig. 8.12*a–c*, up to the contact surfaces—points *c* and *b*.

In Fig. 8.13*a*, plotted are data which generalize the information from Fig. 8.12*a*. For $M_0 = 3$ and various angles α the values of the parameters at the characteristic points *O* and *1* of the surface are plotted on the boundaries of zones *a*, *b*, *c*, and *m*, which are shown in Fig. 8.12*a* and in the upper part of Fig. 8.13*a*.

The dotted lines give the coordinates ξ of the location of discontinuities near the wall: SW_m—point *m*, SW_a—point *a*, and the contact surfaces—points *b* and *c*. Point *c* marks the coordinate of a weak contact surface originating from the region of the bending point *r* (see Fig. 8.9) and seen only on some density distributions along the surface in the numerical solution. The density ratio ρ_{c0}/ρ_{c1} = 1.13–1 at α = 40–60° and 1–0.964 at $\alpha \approx$ 60–90°. This density gradient was controlled by the presence of an inverse gradient of the entropy function $s = p/\rho^{\gamma}$.

Curve m_1, which characterizes the numerical value of the pressure on the wall behind SW_m, is compared with the experimental data: *1*—approximation of the experimental data by sine curve (6.3) and exponential curve [6] for $\alpha \leqslant$ 50° and $\alpha \geqslant$ 50°, respectively; *2*—calculations by the Whitham method expounded in work [6]; *3*—approximate method [331], which is true for non-separation flow over the corner ($\alpha \leqslant$ 40°). Good agreement of these data is observed.

Curve a_1 is compared with the values of the pressure (*4*) in the flow if it turns into a Prandtl-Meyer fan at angle α up to the wall. It is obvious that this pressure (*4*) at $\alpha \geqslant$ 60° becomes less than curve a_1, which is plotted from the numerical solution.

Data (*5*), which characterize the pressure when the flow turns at angle β_s (section 6.3), are plotted above curve a_1. Angle β_s in the experiment [6] determines the deflection of the flow up to the separation surface on (see Fig. 6.4*a*). The comparison implies that the numerical flow is overexpanded relative to the flow in a physical experiment, even though it gives higher pressure than that obtained from a complete turning before the wall (according to the Prandtl-Meyer solution). Thus, the effect of the numerical viscosity with the boundary condition of complete slip of the flow along the wall is not adequate for the used mesh (60 × 70) to obtain a flow pattern which is adequate to the real flow. The boundary condition of an incomplete slip or an exclusion of a separation surface can be used as a means for decreasing velocity, as was done in [317, 322].

Curve *A* in Fig. 8.13*a* characterizes the pressure ratio p_{a0}/p_{a1} on the surface. It is evident that when angle α is increased, this ratio first increases, then drops, reaching a maximum at $\alpha \approx$ 50°. Despite the fact that in the experiment instead of this SW_a a λ-configuration forms with flow separation, the noted tendency can be transferred to the upper part of λ-configuration for SW_t. This is in agreement with the fact that when $\alpha \approx$ 50°, the maximum pull towards the vertex of the corner and deformation of the part of the SW_m near the wall is observed in the numerical and physical experiments, since the flow loses more momentum in a strong SW_t at $\alpha \approx 50°$.

The solid curve a_0 (with a discontinuity at $\alpha = 40°$) represents the pressure behind the SW_a. The discontinuity is obtained as a result of conducting the computations over the minor domain *deBA* (see Fig. 8.9) when $\alpha \leqslant 40°$ and over the whole domain *DEBA* when $\alpha \geqslant 40°$ (see solid curves with dots (*A*) and without dots in Fig. 8.12*a* for $\alpha = 40°$).

The values of densities ρ/ρ_0 are presented by the dashed curves at the stated points b_0, b_1, c_0—near the contact surfaces.

The dash-dot curve determines the values of the temperature T/T_0 at point a_0. It is seen that the temperature is maximum at $\alpha \approx 60°$. The temperature rise in the range $\alpha = 30$–$60°$ agrees with the data of [331], which are marked by symbol *6*.

Fig. 8.13*b*, *c* gives a comparison of well-known experimental and calculated data with the results of systematic computations given in Fig. 8.12*b*, *c* ($\tilde{p}$, $\tilde{\rho}$, and $\tilde{T}$ are the same as in Fig. 8.12*b*). The meaning of the curves and the symbols

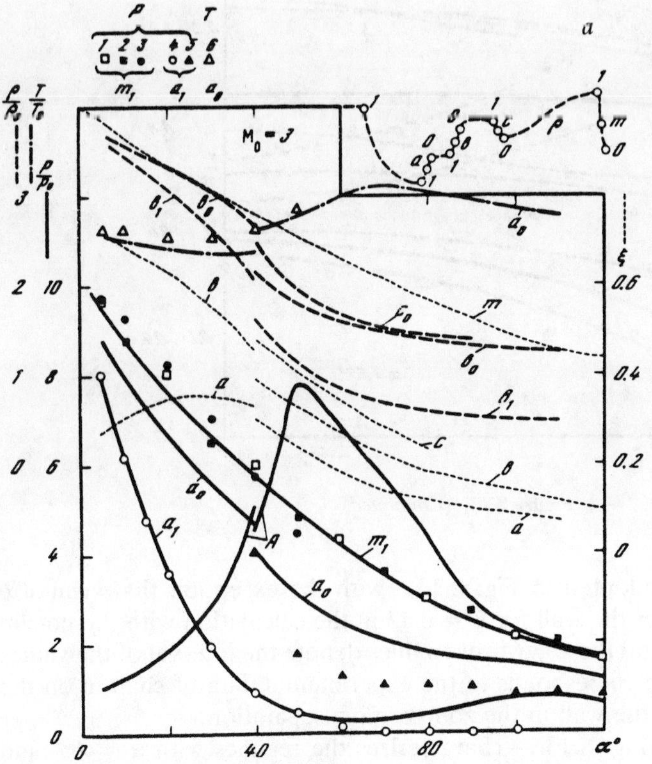

Figure 8.13. Flow parameters on the wall at characteristic points for $\gamma = 1.4$. p, ρ, T, ξ) solid, dashed, dash-dot, and dotted lines, respectively; *a*) $M_0 = 3$, $\alpha = 5$–$120°$; *b*) $\alpha = 30°$, $M_0 = 2$–3 (upper); *c*) $\alpha = 65°$, $M_0 = 2$–5 (lower).

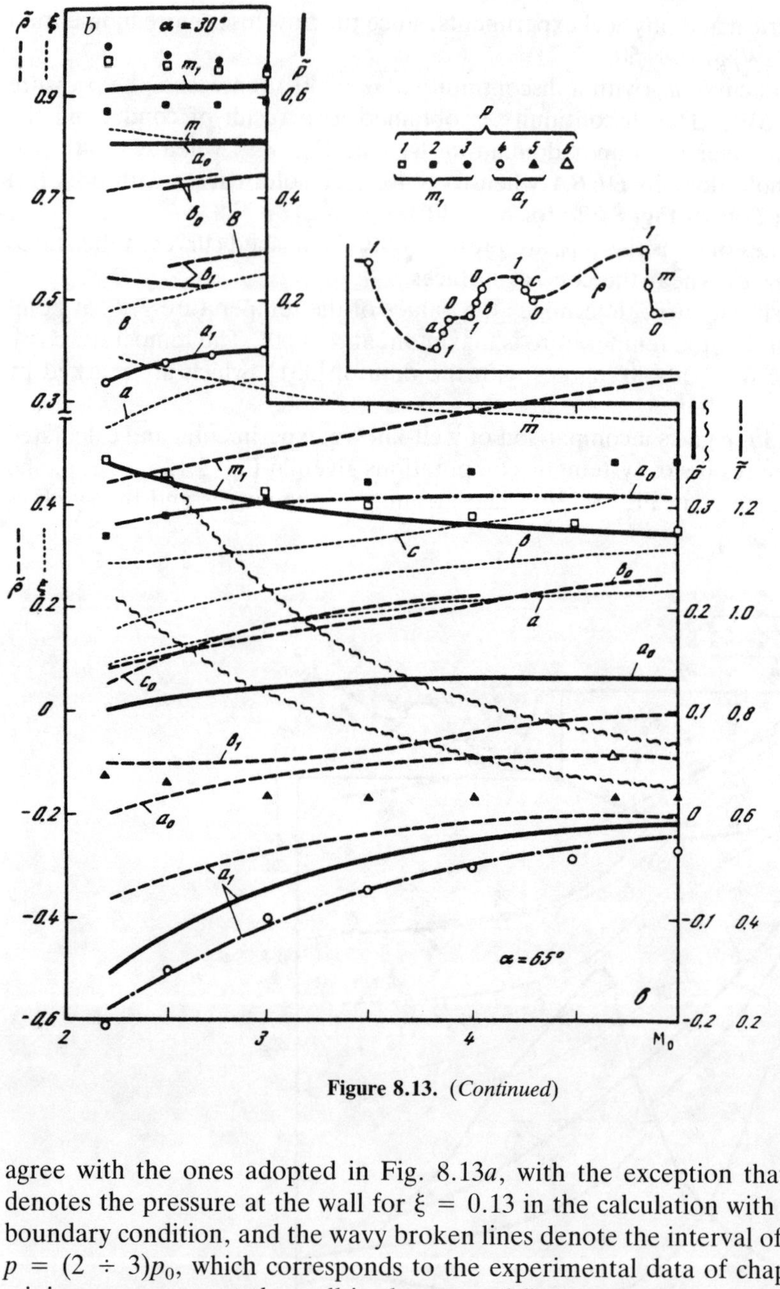

Figure 8.13. (*Continued*)

agree with the ones adopted in Fig. 8.13*a*, with the exception that symbol *6* denotes the pressure at the wall for $\xi = 0.13$ in the calculation with the no-slip boundary condition, and the wavy broken lines denote the interval of the values $p = (2 \div 3)p_0$, which corresponds to the experimental data of chap. 6 on the minimum pressure at the wall in the zone of flow separation.

The lower part—Fig. 8.13*c*—characterizes the regimes with $\alpha = 65°$ and $M_0 = 2$–5; the upper part—Fig. 8.13*b*—the regimes with $\alpha = 30°$ and $M_0 =$ 2.2–3, for which computations were conducted in the minor (line *B*) and whole domains as a result of which a small overlapping of the dashed lines occurs, characterizing the values of densities at points b_0 and b_1. The pressure at $\alpha = 30°$ in these two cases is the same as was stated already when discussing Fig. 8.12*b*.

Comparison of curve m_1 (pressure behind the front of SW_m) for $\alpha = 65°$ (Fig. 8.13*c*) with the data of the experiment (*1*)—approximation by sine curve (6.3)—indicates good correlation, whereas data (*2*) by the Whitham method [6] intersect curve m_1 when $M_0 = 3$.

As in Fig. 8.13*a*, the values of the solid curve a_1 in Fig. 8.13*c*, characterizing the pressure at point a_1 in the numerical simulation were higher than the values of pressure (*4*) when the flow in the Prandtl-Meyer fan turned by an angle α, and lower than pressure (*5*) when the flow similarly turned by an angle β_s, which is observed in the experiment [6, 224]. Thus, here, too, is an indication of overexpansion and excessively high flow velocity in the numerical simulation as compared to the experimental data when the no-slip boundary condition is used.

Boundary conditions for numerical integration. Let us discuss the question about boundary conditions and the effect of numerical viscosity with respect to a system of four finite difference equations (8.21). These are first order equations; therefore, in order to obtain continuous smooth solution for each of them only one boundary condition can be given. For example, to give function f on one of the two boundaries (see Fig. 8.9): η = const (*DA* or *EB*) or ξ = const (*DE* or *BA*). On the remaining boundaries the functions will be computed. The number of boundary conditions must be equal to the number of characteristics arriving at the boundary under consideration from outside.

When integrating the gasdynamic equations (8.21) at a solid boundary only one condition is given, condition of impermeability, i.e., equality of the normal component of flow velocity and the body. In this case, on the opposite boundary three functions can be given and one must be computed. Three functions must be computed on the solid boundary. Algorithmically it is more convenient to conduct the computations at the boundaries using the same difference scheme as at the interior points. For this purpose, additional rays are drawn along the boundary of the computation region, thus completing one more row of cells along the entire perimeter. Assuming the equivalency of the solid wall and the plane of symmetry in the sense that in both cases the condition of impermeability is fulfilled, the boundary conditions on the wall can be the same as near the plane of symmetry. Therefore, the parameters are transferred from the interior points of the computation field to the nodes of the additional rays, this being accompanied by a sign change of the normal component of the velocity. Such a realization of boundary conditions is equivalent to the computation at the boundary using one-sided differences with a first order of approximation. Experience with computations shows that these algorithms allow one to conduct steady calculation and obtain results with sufficient accuracy. This is confirmed by comparisons with analytical solutions and good convergence of the solution when computed with various mesh intervals.

If the number of the boundary conditions is larger than the number of equations (four in the present case), then a discontinuous solution can be obtained in the general case. In particular cases the solution will either correspond to reality (shock wave, contact surface) or be non-physical.

For the system of gasdynamic equations (8.21) to be numerically integrated, it must be written in a finite-difference form (8.21). This finite difference analog contains the highest derivative with small parameters of the order $\Delta\xi$, $\Delta\xi^2$, . . . , $\Delta\eta$, $\Delta\eta^2$, . . . , or Δt, Δt^2 . . . , or their combinations. In this case, if more than four boundary conditions are given a smooth solution is always obtained, because the discontinuities arising in the ideal gas in this case are represented as narrow transient zones with large gradients. The form of these additional terms is established by analyzing differential approximations [275, 332], and their influence is similar to the effect of physical viscosity, namely, instead of the ideal shock wave a narrow zone is formed at which energy is dissipated and the Rankine-Hugoniot condition [266] is implemented. Instead of the ideal contact discontinuities or slip surfaces, mixing zones with characteristic velocity profiles are formed. In contrast to and by analogy with the properties of the physical viscosity, the combination of artificial viscosity (or the smoothing operator) and remaining terms of the differential approximations are called the numerical viscosity. The character and specific properties of the numerical flow in the indicated transient zones are determined by the numerical viscosity which, in its turn, depends on the dimensions of the calculation cells and the employed finite difference scheme. Thus, one has to factually deal with a simulated numerical flow in which the forces of numerical viscosity act. These forces, as already mentioned, are qualitatively analogous to a great extent to the effect of the forces of real physical viscosity. The quantitative characteristics of the numerical viscosity on real difference meshes are almost unresearched presently. As calculations show, separation and reversal flows are observed in the numerical solutions even in spite of the stipulation of the traditional condition—complete flow slip along the wall.

It can be positively stated that in many works in which the finite difference analogs of the system of gasdynamic equations are numerically integrated and the flow behind the body examined, a certain contradiction of the "order of boundary layer thickness" exists. This happens when a condition of flow slip is given on the surface of the body in addition to the condition of impermeability, thereby ignoring the presence of a boundary layer, i.e., virtual decrease of velocity near the solid boundary in the real flow. Therefore, the solutions obtained with the condition of complete slip can differ noticeably sometimes from the physical flow picture not only quantitatively but also qualitatively as, for example, the presence of SW_a in the numerical simulation (Figs. 8.10–8.12), whereas they do not exist in the real flow.

The results of the numerical calculations (for example, a jet or flow over a sphere or cylinder) show that the action of the numerical viscosity forms a flow which is qualitatively similar to turbulent flow with $Re \approx 10^5 - 10^6$; however, no systematic investigations whatsoever have been carried out in this direction. Comparison of results of numerical and physical experiments show that the presence of the boundary layer and action of the numerical viscosity at the interior points of the computation field can be ignored in many cases, for example, when the SW is reflected from a stationary wedge [305] or from the frontal part of blunt bodies [254–256]. However, when a SW is diffracted at a convex corner or at the rear section of a sphere or cylinder both quantitative and qualitative differences are observed, which makes it imperative to approach the question of using the boundary slip condition more carefully.

As was already pointed out, when the condition of a complete slip flow along the wall is assumed the flow velocity turns out to be so high that a SW, which is not observed in the experiment, is formed. Calculations conducted with the additional condition of no-slip at the surface of the corner (mesh 70×60) indicated that this "superfluous" boundary condition within the bounds of the ideal gas theory exerted influence at a distance of 6% along the normal to the wall with respect to the dimension of the entire disturbed region, and furthermore, the qualitative correlation with the experiments improved. It can be assumed that for a reasonable mesh, when nearly ten cells are provided for a sufficient solution of the developing features in the flow, such an additional condition deccelerating the flow near the solid surface (incomplete slip in the general case) makes the simulation more physical in a qualitative sense. When the space steps are decreased, the numerical boundary layer also decreases and we obtain, in the limit, a non-physical discontinuous flow near the wall. This, however, does not, as calculations show, cause the deterioration of the solution at the remaining interior points.

The use of the condition of incomplete slip is required to verify the numerical solution with respect to its degree of dependence on boundary conditions for a specific formulation of the problem. If the no-slip boundary condition gives a solution that differs strongly from the case with the slip condition near the wall, then the formulation of the problem must be changed or the results of the numerical calculations used only in the region where the discrepancy is insignificant. This verification implies that the importance is in the presence of the boundary layer and not in the exact computation of its thin structure which, certainly, is not described within the limits of the Eulerian equations.

At the present time, one can already use the results of the shock-capturing method, in which SW or contact surfaces are regarded as transient zones at several intervals of the mesh. A specific number of cells is reserved at these zones—about ten in all. Similarly, sufficient number of cells are provided for mixing zones and boundary layers appearing in the flow when the no-sip condition is used.

What will happen with solution convergence in this case? Since the remaining terms determining the numerical viscosity depend on the space intervals of the difference mesh, then when they tend to zero the influence of the numerical viscosity will decrease. The thickness of the numerical boundary layer, which forms near the body when the boundary no-slip condition is given, decreases (in the limit to zero). An example of such a model solution with discontinuity at the boundary is given in [267, p. 249]. Thus, the finite difference solution will converge to the solution of differential equations by ignoring (with the formation of small oscillations near the boundary) the no-slip condition. By selecting the intervals of the mesh, the thickness of the numerical boundary layer can be regulated in accordance with the solutions of the Navier-Stokes equations or experimental data.

Based on the conducted analysis of the results of systematic calculations, it can be concluded that the employed technique of the shock-capturing method allows one to obtain results which correlate well with the experiment in the entire diffraction region at small angles $\alpha \lesssim 40°$ when the flow over the corner is without separations. When $\alpha \gtrsim 40°$ correlation is violated in the region bounded by region *oec* for $\xi \gtrsim 0.1$ (Fig. 8.11) as we get closer to the wall. The boundary condition of incomplete slip gives a shift of the flow picture in this region toward agreement with the experiment; however, the quantitative effect of such a boundary condition has not yet been resolved completely. Of extreme necessity in regimes with separation is the exclusion of a separation surface or a contact surface which starts at the vertex of the corner *o*, or a significant concentration of points in this region during the shock-capturing calculation, and also taking into consideration the effects of viscosity in the region *oec* (Fig. 8.11).

8.4 ENGINEERING ESTIMATE OF PRESSURE LEVELS UPON EGRESSION OF SHOCK WAVES INTO A DIVERGING SECTION OF A FLAT CHANNEL

The data presented in section 8.3 on the diffraction of SW near the vertex of a convex corner are obtained as a result of laborious computer calculations using numerical integration of the finite-difference analog of the system of differential equations of unsteady gasdynamics. In addition to conducting such works, more simplified techniques are required which allow one to quickly conduct an engineering estimate of pressure levels at individual points of the surface. Discussed below is such a technique [331], which was constructed on the assumption of non-separation of flow during shock wave diffraction and is valid in the stage when the process is still self-similar. Comparison with the experimental data of chapter 6 and reference [6] show satisfactory agreement of the results for $0 \lesssim \alpha \lesssim 40°$. Discussed are the regimes when the medium in front of the SW is at rest or moves at a supersonic velocity, forming a centered expansion fan near the vertex of the corner.

Formulation of the problem and method of calculation. Let us discuss the plane-parallel flow of an ideal gas in the case when the SW (Fig. 8.14*a*) is egressing from channel *ao* into its diverging section *oc*. The shock wave gas motion resulting from this depends on four dimensionless parameters: $p_{21} = p_2/p_1$ is the ratio of the pressure behind and ahead of the shock wave front SW_s (from now on the double-digit subscripts will denote ratios of the corresponding values); $M_1 = u_1/a_1$ are Mach numbers of the flow ahead of SW_s in the undisturbed region *bo* (u_1, a_0 are the velocity of this flow and speed of sound in it); α are

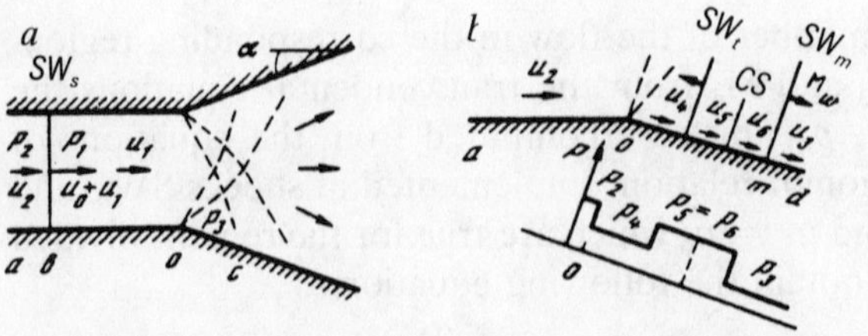

Figure 8.14. Schematic diagram of SW_s egression into the diverging section of the channel. *a*) moment of time before egression; *b*) moment of time after the SW passes the vertex of the corner—formation of two SW and contact surface CS between them.

the angles of inclination of channel walls; $\gamma = c_p/c_v$ is the Poisson adiabatic index. Subscripts *1* and *2* pertain to the parameters ahead of and behind the SW_s, respectively.

Case $M_1 = 0$ has been discussed in chap. 6 and section 8.3. For this very regime, computations were conducted in [288] using a simplified one-dimensional scheme of discontinuity decay at $\alpha = 90°$. At a sufficiently large distance from the corner tip *o* the decrease in wave intensity can be approximately described by a dependence on the cross sectional area of the channel. The techniques for obtaining similar dependences and review of corresponding works are presented in [6].

A method for the evaluation of pressure distribution on the wall will be discussed later in the homogeneous region *oc* (Fig. 8.14*a*) at the moment of passage of the diffracted SW_m for various values of the determining parameters.

The proposed calculation method includes limitations and assumptions. *a*) $M_1 = 0$ or $M_1 > 1$ and $M_2 > 1$ ($M_2 = u_2/a_2$ is the Mach number of the flow behind the SW_2; *b*) $\alpha \leqslant \alpha_* \sim 30–40°$, where α_* is the angle at which the real flow begins to separate from the wall of the channel *oc*; *c*) the gas is perfect, the effect of viscosity and thermal conductivity is not taken into account; *d*) the shock wave motion of the gas in the homogeneous region *oc* near the wall is plane and self-similar. The latter assumption is true when the size of the channel is much larger than the region *oc* under consideration, which is bounded by the boundary characteristics of rarefaction waves (see dashed lines in Fig. 8.14*a*).

Analyses of the oscillograms in [6] and the pressure diagrams along the wall (see Fig. 8.12*a*, *b*) allow one to introduce the following flow scheme. After the wave bends around the corner a complex system of gasdynamic discontinuities is formed (see Fig. 8.14*b*): SW_m moves downstream of the flow which has a velocity u_3; SW_t is oriented opposite to the flow, which has a velocity u_4, and deccelerates it; contact surface CS, located between SW_m and SW_t, moves with a velocity $u_5 = u_6$. The numbering of the regions coincides with the subscripts of *u* and *p* (see Fig. 8.14*a*, *b*).

For the determination of the parameters in regions *3* and *4* the Prandtl-Meyer solution is used

$$\alpha = \omega(M_3) - \omega(M_1), \quad \alpha = \omega(M_4) - \omega(M_2)$$

$$\omega(M_i) = \operatorname{arctg}\sqrt{\nu(M_i^2 - 1)}/\sqrt{\nu} - \operatorname{arctg}\sqrt{M_i^2 - 1} \tag{8.23}$$

$$i = 1, 2, 3, 4, \quad \nu = (\gamma - 1)/(\gamma + 1)$$

where $M_i = u_i/a_i$ is the Mach number of the flow in the corresponding region. After the determination of M_3 and M_4 from the transcendental equations the remaining parameters (p_3, ρ_{31}, p_{42}, ρ_{42}) are computed from the equations of isentropic flow. Using the Hugoniot relation, implemented at shocks SW_m and SW_t, and conditions $p_5 = p_6$ and $u_5 = u_6$ which are true for the regions divided by the contact surface CS, we obtain the following equation

$$M_3 + \frac{p_{53} - 1}{\gamma\sqrt{(p_{53} + \nu)/\mu}} = a_{43}\left(M_4 - \frac{p_{54} - 1}{\gamma\sqrt{(p_{54} + \nu)/\mu}}\right), \quad \mu = \frac{2\gamma}{\gamma + 1} \qquad (8.24)$$

Pressure p_{53} is numerically determined from this equation.

Calculation results. Curves *1–3* in Fig. 8.15*a*, *b* characterize the change of Mach number $M_2 = \sqrt{(p_{53} + \nu)/\mu}$ of the wall part of the diffracted wave SW_m (p_{53} determined from (8.24)). Curve *4* represents the data obtained from the empirical equation (6.2). The discrepancy between the results is considerable when $\alpha \geqslant 45°$, which is explained by the separation of the flow over the vertex of the corner *o* in the physical experiment; separation of the flow begins at $\alpha \geqslant 20$–$30°$. As was mentioned earlier, a non-separation flow is assumed in the suggested scheme. Curves *2* and *3* in Fig. 8.15*b* are plotted for the case $M_1 = 1$ and $M_1 = 2$, respectively. Unlike the case $M_1 = 0$, the value of M_w here increases with increasing α. At constant angles α, the value of M_w rises with increasing

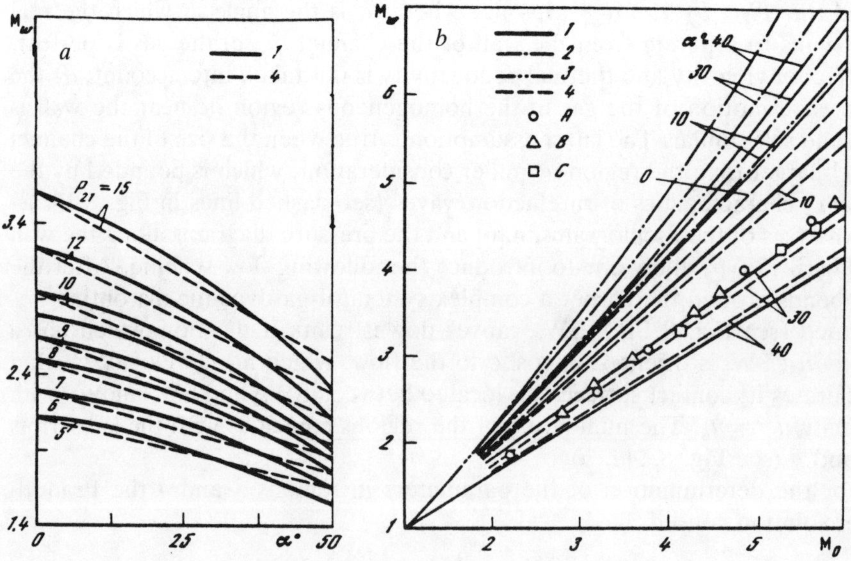

Figure 8.15. Comparison of calculated and experimental data (for $\gamma = 1.4$) from dependence M_w (α) at $p_{21} = 5$–15 (*a*) and M_w (M_0) at $\alpha = 0$–$40°$ (*b*). *1–3*, calculation for $M_1 = 0, 1, 2$, respectively; *4*) approximation of experimental data for $M_1 = 0$ (6.2); *A, B, C*) experimental data [6] for the diffraction of SW in N_2, air and CO_2, respectively.

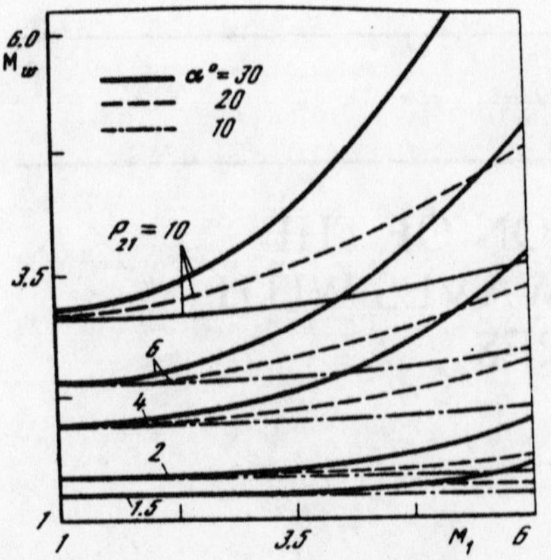

Figure 8.16. Calculated dependence M_w (M_1) for various α, p_{21} at $\gamma = 1.4$.

M_1. For $\alpha = 30°$ and $M_1 = 0$, symbols A, B, and C denote experimental results [6]. As in Fig. 8.15*a*, good correlation of the results is observed. We note that in the case $\alpha = 90°$ and $M_1 = 0$ the data of [288] exceed the experimental values by 1.1–1.35 times in thc range of $M_0 - 2$–7.

Dependence M_w (M_1), obtained from (8.24) for $M_1 \geqslant 1$ at various p_{21} and α, are depicted in Fig. 8.16 in which one also sees the increase of M_w when M_1 and α are increased.

The values of M_w are presented below as a function of the adiabatic index γ at $p_{21} = 10$, $\alpha = 20°$ and different M_1.

M_1	$\gamma = 1.1$	1.2	1.4	5/3	2.0	2.5
0	2.73	2,72	2,69	2.64	–	–
1	2.78	2.89	3,07	3.31	3.65	4.29
2	2.77	2.90	3,14	3.51	4.16	5.97

Here, it is evident that M_w (γ) decreases at $M_1 = 0$ and increases at $M_0 \neq 0$.

The described computation scheme and presented results can be used at $\alpha \leqslant 35$–$40°$ for the evaluation of the pressure on the wall behind the front of the wall part of the diffracted wave SW_m when the medium ahead of the wave is at rest or moving. The utilization of this technique at $M_1 \neq 0$ requires that the limits of applicability be specified by comparing with the results of physical or numerical experiments, which are not as yet available in the published literature.

CHAPTER

NINE

NUMERICAL INVESTIGATION OF THE INTERACTION OF SHOCK WAVES WITH BODIES OF VARIOUS SHAPES

9.1 INTERACTION OF A SHOCK WAVE WITH A STATIONARY CYLINDER OR SPHERE

Flow scheme in the calculation region. Figure 9.1 illustrates the specific features of the process of interaction. A shock wave, which is depicted prior to interaction at $t = t_0$, moves toward a cylinder or a sphere of radius R from the left. The shock wave front is parallel to the axis of the cylinder. The velocity of the wave is u_0, velocity of the gas behind it is υ_1, the pressure ahead of and behind the front is p_0 and p_1, respectively, $p_{10} = p_1/p_0$.

The upper section of the figure pertains to the unsteady phase of the interaction: t_2 is the regular, t_3 is the Mach (irregular), t_4 is the diffraction at $\omega > 90°$, t_5 is the moment the wave reaches point $\omega = 180°$ when the shock waves, emanating from different sides, collapse, t_6 is the wave motion after the collapse.

The lower section of the figure illustrates a steady flow over the body following the passage of the shock wave (SW), with wave d remaining ahead of the body when $t \to \infty$ if $M_1 = \upsilon_1/a_1 > 1$, and shock wave c closing the supersonic zone. The angular coordinate of the point on the surface at which this wave c arrives is denoted by ω_c.

By limiting ourselves with the bounds of the theory of an ideal gas (inviscid, zero thermal conductivity), it can be assumed that the flow under consideration has symmetry. In this case only half the flow field is supposed to be computed.

The computation field is bounded by the upper half of the surface of the body *gf*, lines *ag* and *fe* (axis or plane of symmetry), and the external boundary *aje* which is located sufficiently far from the body so that at small flow numbers M_1 the disturbance, reflected from it, could not have reached the body during the interaction duration.

Review. The analytical study of the process of shock wave interaction with a body involves considerable difficulties, since the process of interest to us is described by a nonlinear system of differential equations of unsteady gasdynamics (8.1). Several special cases have been discussed analytically in [333, 334] for weak waves ($p_{10} \to 1$). The time dependences of the pressure at the critical point and the unsteady departure of the SW along the stagnation streamline *ag* have been obtained analytically for a SW impinging on a cylinder, a sphere [335], and an ellipsoid [336] when the adiabatic index $\gamma = c_p/c_v$ is close to unity. Time dependences of the departure of SW of moderate intensities [337] have been found with the help of approximations of the derivatives with respect to velocity and curvature of the reflected shock. The parameters on the surface of the body were determined in the regimes of regular reflection by expanding into series relative to the frontal point in [338], and in the entire flow field in [250], the expansion being conducted relative to the reflection point. In [339], the experimental results are compared with the data of the Whitham theory [340, 341].

Experimental studies (since 1950) of the process of shock wave interaction with a cylinder or a sphere have allowed one to obtain photographs of the shock wave pattern [339, 342–346] and determine the change as a function of time of the shock wave departure from the critical point [342–347]. Measurements of density and pressure were also conducted at various points on the surface [344, 347, 227, 348, 237].

Numerical simulation of shock wave interaction with a sphere or a cylinder began with [349] in which calculations were conducted using the Godunov method of the first order of accuracy [350] and constant-pressure and constant-volume

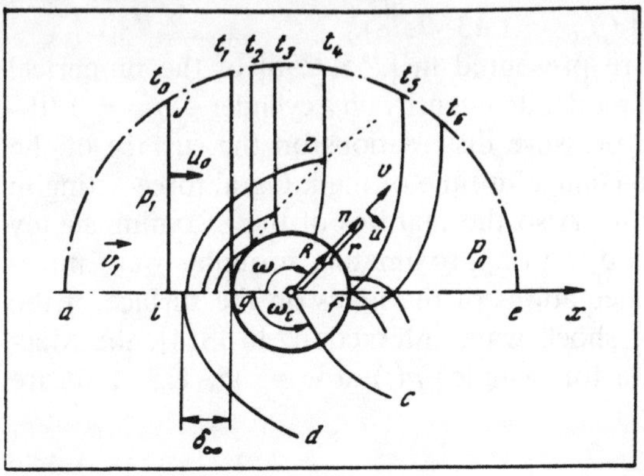

Figure 9.1. Schematic of the computation field; stages of shock wave reflection from a sphere or cylinder.

fields presented for the interaction of a SW with a cylinder or a sphere—p_{10} = 1.35–3.75; γ = 1.4. The values of the unsteady force F_x acting in the direction of shock wave travel were given and the question of the separation zone, forming behind the body for a steady flow *1*, was discussed.

Some results of the calculations of the action of SW (M_0 = 5, γ = 1.4) on a wedge and solid or elastic cylindrical shells were discussed briefly in [304]. Considered were the features of the pressure diagrams in the range $\omega_m \lesssim \omega \lesssim \omega_e$. The interaction of SW (M_0 = 5, γ = 1.4) with an elastic cylindrical shell is discussed in more detail in [351–353]. The integration of the system of unsteady gasdynamic equations was carried out by the shock-capturing method without the exclusion of discontinuities in the flow field with the help of a two-step difference scheme of the second order of approximation (8.17). Instead of artificial viscosity a smoothing operator (8.20) was used with the inclusion of diagonal differences, which were particularly necessary for oblique shock wave reflection from the surface of a body. It is seen from expression (8.20) that the smoothing operator has a third infinitesimal order on smooth solutions; therefore, the order of the difference approximation of the equations remains a second. For every type calculation, the constant coefficient $\Omega(p_{10})$ in the operator is so chosen that the condition $p_m = p_n$ is fulfilled at the critical point $\omega = 0$ (see 8.22).

Some results of the calculation of the action of a SW on a solid sphere and cylinder are given in work [254]—p_{10} = 1.5–100, ω = 0–150°, and the problem of transition of a regular reflection to a Mach regime during the travel of a SW along a convex surface is discussed. This same problem was discussed in work [253] where the method of integral ratios with the exclusion of gasdynamic discontinuities was employed. This method is more accurate than the shock-capturing method; however, it is more laborious and requires an *a priori* assumption of the shock wave configuration. The shock-capturing method, on the other hand, selects steady and feasible configurations for the given boundary conditions. The authors of [304, 254, 253] conclude that Mach reflection at a convex surface begins before the limiting angle—when $\omega \lesssim \omega_e$. It is possible that this happens at $\omega \approx \omega_m$ [253, 254]. Reference [258] presents the results of numerical calculations of the interaction of SW at a cylindrical body which were obtained by the Godunov method (p_{10} = 1.13–7.78).

Sufficiently detailed results are presented in [255, 256] of the numerical calculations of the action of a SW of moderate intensity on a cylinder—p_{10} = 1.01–5, α = 1.1–3, ω = 0–180°. The pressure distributions on the surface of the body are presented along with the change in time of the integral force acting in the direction of shock wave motion. Also the features of the incoming steady flow over a cylinder were discussed, and approximating formulas were introduced, suitable for engineering calculations of the loads on the surface of the cylinder in the unsteady phase of shock wave interaction. In [354], the MacCormack method is used and data for cylinder $p(t)$ at ω = 45, 135, 180° are given for regime M_0 = 1.17.

Formulation of the problem. It is convenient to study the scope of the phenomena under consideration in spherical or polar systems of coordinates. Since we are concerned with the detailed discussion of the process of shock wave reflection near the surface of the body, it is expedient to condense the mesh in this region introducing the transformation: $n = 1 - R/r$, where $R = R(\omega, t)$, which is a function characterizing the surface of the body and depending, in the general case, on the angular coordinate ω and time t. This transformation allows one to examine various surface forms by condensing the mesh near the body.

The system of gasdynamic equations (8.5) has the following form with respect to the variables ω, n, t

$$\frac{\partial \mathbf{f}r}{\partial t} + \frac{\partial \mathbf{A}}{\partial \omega} + (1-n)\left[\frac{\partial \mathbf{B}}{\partial n} - \frac{1}{R}\left(\frac{\partial R}{\partial \omega} \cdot \frac{\partial \mathbf{A}}{\partial n} + \frac{\partial R}{\partial t} \cdot \frac{\partial \mathbf{f}r}{\partial n}\right)\right] + \mathbf{C}_1 = 0 \tag{9.1}$$

For a nondeformed shell $\delta R/\delta t = 0$, and for a true sphere or cylinder $\delta R/\delta\omega = 0$.

Boundary conditions. If one limits himself with the cases of plane or axial symmetry, then the calculation field will occupy the upper part of Fig. 9.1 ($\omega = 0$–$180°$). The external boundary *aje* must be positioned sufficiently far from the body so that the errors in the assumed boundary conditions would not reach the body surface during the interaction. Choosing $r_{max} = 50R$, we obtain $n_{max} = = 0.98$. For this value of $n = n_{max}$ parameters $\mathbf{f}_1$ are given behind the shock wave on the external boundary layer at the left of the SW front, and the parameters $\mathbf{f}_0$ of the gas coming to rest at the right. When the undisturbed section of the shock wave lies within the range $x = (0$–$0.1)R$, the flow adjustment condition $\delta\mathbf{f}/\delta n = 0$ must be assumed at the external boundary for $\omega \gtrsim 90$–$180°$ in order to prevent the reflection of the strong disturbances from this section of the boundary (x is measured from the center of the sphere).

At the boundaries *ag* and *fe* the symmetry condition must be fulfilled

$$u = \frac{\partial \rho}{\partial \omega} = \frac{\partial v}{\partial \omega} = \frac{\partial e}{\partial \omega} = 0 \tag{9.2}$$

This condition can be achieved by introducing additional rays of the mesh outside the calculation region, to which the values of the functions form the symmetrical, relative to the axis, interior points of the calculation region are transformed. In this case, the component of the velocity vector u, which is normal to the wall, changes its sign.

When computing rays *ag* and *fe* the components of vector $\mathbf{C}_1$ has an indeterminate of type 0/0, since the values entering b (8.5) are equal to zero ($u = \operatorname{tg}\omega = 0$). In this case, L'Hospital's rule is used to evaluate the indeterminate form, and term $u/\operatorname{tg}\omega$ is replaced by $\delta u/\delta\omega$ in $\mathbf{C}_1$.

A similar procedure of "mirror image" transfer is used when implementing

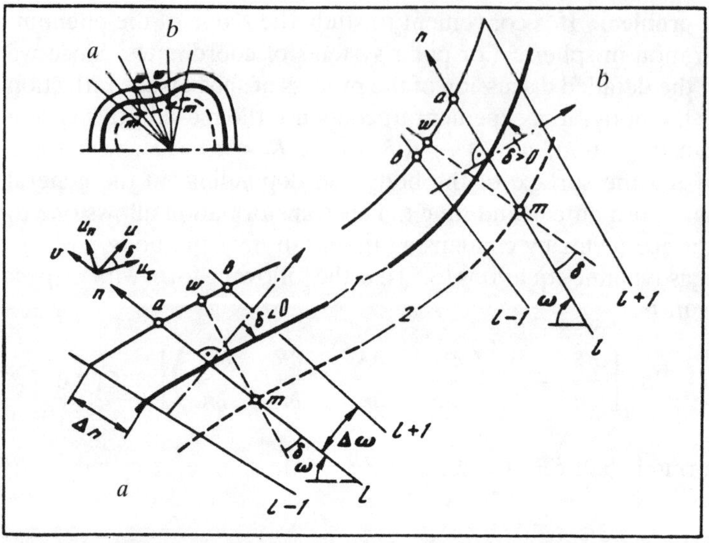

Figure 9.2. Schematic diagram of the implementation of boundary conditions on the surface of a deformable shell for $\delta < 0$ (*a*) and $\delta > 0$ (*b*). The values of parameters at point *m* are transferred from point *w* by mirror imaging.

the boundary conditions near the body, and if the surface deforms and shifts, the transfer of the parameters is carried out along the normal to the surface element taking into account the surface velocity $\delta R/\delta t$ and the derivative $\delta R/\delta\omega$. Two probable types of deformation of the shell elements in the computation are illustrated in Fig. 9.2. First, the parameters at the interior point *w* of the computation region are calculated by linear interpolation from the values of the functions at two nearest nodes of the mesh (ignoring the terms of the order Δn^2 $\Delta n \Delta r$)

$$\mathbf{f}_w = \mathbf{f}_a + (\mathbf{f}_b - \mathbf{f}_a)c_*, \quad c_* = \frac{aw}{ab} = \frac{2\Delta n \mid \sin\delta \mid \cos\delta}{(1+\Delta n)\cdot\Delta\omega}$$

Here, δ is the angle between the positive direction of the change of coordinate ω and the surface of the body ($\delta < 0$, if the body surface deflects from the positive direction of coordinate ω clockwise during deformation). The value $\mathrm{tg}\delta$ is then determined

$$\mathrm{tg}\,\delta = \frac{1}{R}\frac{\partial R}{\partial\omega} = \begin{cases} (R_{l+1} - R_l)/(R_l\Delta\omega) & \text{if } \mathrm{tg}\,\delta < 0 \text{ (Fig. 9.2, } a) \\ (R_l - R_{l-1})/(R_l\Delta\omega) & \text{if } \mathrm{tg}\,\delta \geqslant 0 \text{ (Fig. 9.2, } b) \end{cases}$$

Finally, the values of the parameters at point *m* are computed as follows

$$\rho_m = \rho_w, \quad e_m = e_w$$

$$u_m = u_\tau \cos\delta + (-u_n + 2D)\sin\delta, \quad v_m = -u_\tau \sin\delta + (-u_n + 2D)\cos\delta$$

where

$$u_n = v_w \cos\delta + u_w \sin\delta, \quad u_\tau = -v_w \sin\delta + u_w \cos\delta,$$

$$D = \frac{\partial R}{\partial t}\left[1 + \left(\frac{1}{R}\frac{\partial R}{\partial \omega}\right)^2\right]^{-1/2} = \frac{\partial R}{\partial t}\,|\cos\delta\,|$$

is the velocity of motion of the surface element in the direction of the external normal.

The implementation of the boundary conditions on the body is described in the present section, since the accuracy of the obtained solution depends on this to a great extent. The procedure under consideration was tested in the steady case for the calculation of a body whose shape $R(\omega)$ was a cone (wedge) made blunt by a sphere (circle). Comparison with the tables of [19] showed agreement of magnitudes of the pressure within 2–3%.

Initial conditions. The given conditions behind the SW at time t_0 inside contour *aji* are: $\rho = \rho_1$, $e = e_1$, $u = v_1 \sin\omega$, $v = -v_1 \cos\omega$ and in the remaining region the parameters of the undisturbed flow $\mathbf{f} = \mathbf{f}_0$.

The system of equations (9.1) together with the boundary conditions are integrated using a finite difference scheme of the second order of approximation (8.17) with a smoothing operator with diagonal differences (8.19, 8.20). The SW and slip surfaces appear in this case as zones of large gradients which are smeared through several computation cells. The time step Δt is chosen with respect to the condition of stability.

Results of numerical calculations. The unsteady process of shock-wave interaction with a nondeformable body within the limits of the ideal gas theory (inviscid gas, zero thermal conductivity, no real-gas effects) is determined by the body geometry $R(\omega)$ and two dimensionless parameters M_0 or p_{10} and $\gamma = c_p/c_v = \text{const}$. When the deformation of the shell is accounted for, the number of dimensionless parameters increases [304, 351–353].

Later, the results of systematic calculations of shock-wave interaction in an ideal gas with a stationary solid sphere and a cylinder will be discussed in a wide range of determining parameters: $\gamma = 1.4$ and $p_{10} = 1.01–100$; $\gamma = 1.1–3$ and $p_{10} = 1.5$.

Figure 9.3*a* shows the instantaneous pressure diagrams (*1*) $p = p(\omega)$ on the surface of a sphere at various moments of time t_s, the values of which are shown near the curves in units R/u_0. Figure 9.3*a* also shows envelope (*2*) of the local maxima of the pressure p_m in the vicinity of the reflection. The SW touches the body at $t_s = 0.5$. From the results of [305, 284], in which the shock-wave reflection from a wedge was studied, it follows that during Mach reflection when weak waves are reflected, the local pressure maximum p_m on the body for Mach lies near the Mach wave, and when strong waves are reflected, p_m shifts to the region of contact surface (see section 8.2). It can be assumed that a similar tendency exists in the case of shock wave reflection from a curvilinear surface.

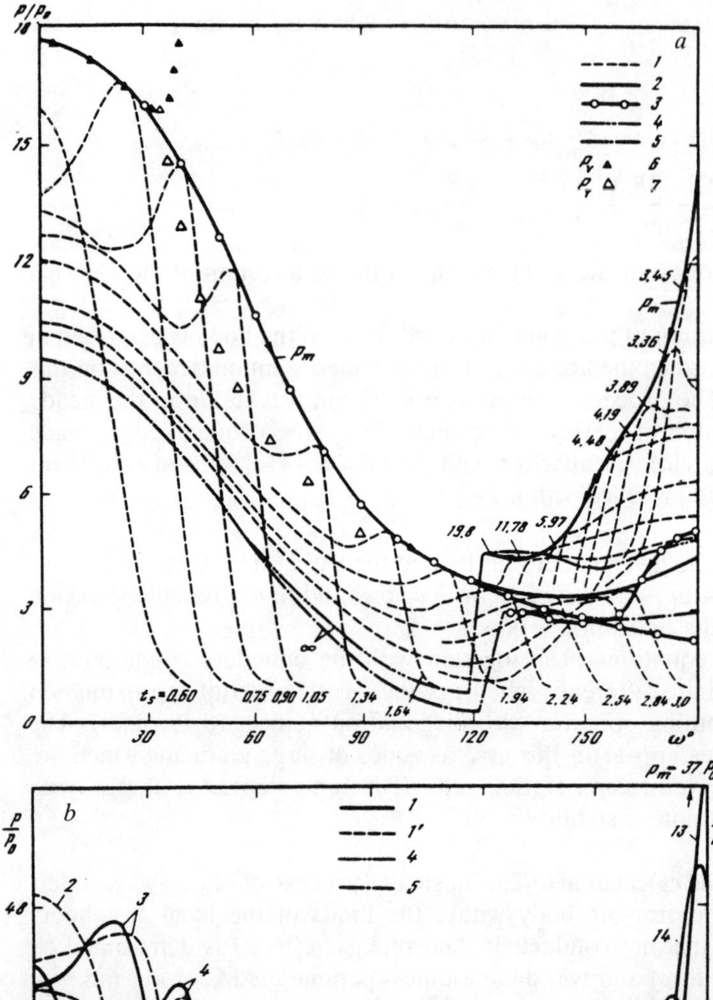

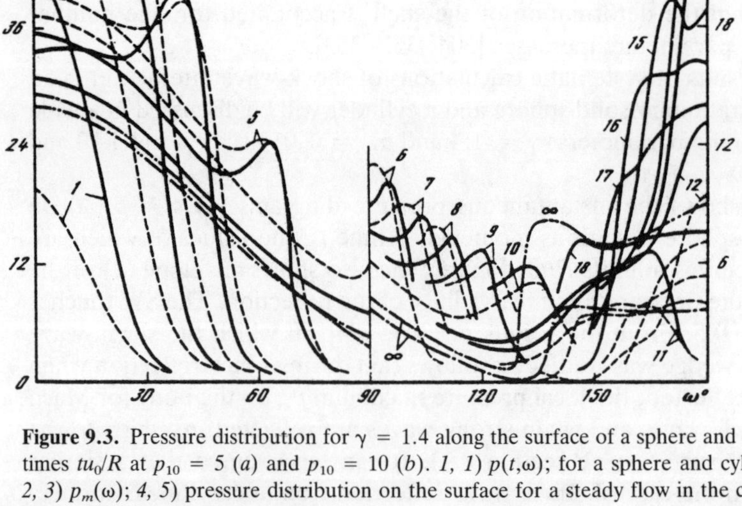

Figure 9.3. Pressure distribution for $\gamma = 1.4$ along the surface of a sphere and cylinder at different times tu_0/R at $p_{10} = 5$ (*a*) and $p_{10} = 10$ (*b*). *1, 1)* $p(t,\omega)$; for a sphere and cylinder, respectively; *2, 3)* $p_m(\omega)$; *4, 5)* pressure distribution on the surface for a steady flow in the case of a sphere and a cylinder, respectively ($t_s \sim 20$).

This is evidenced by the presence of local maxima of curves *1* near the large gradient which characterizes the transition zone of the smeared shock wave front. The direction of motion of the SW is indicated on curve p_m by an arrow. The envelope for the cylinder is shown by curve *3*. It is evident that the values of p_m, in the range of angles $\omega = 0–120°$ when the SW moves to the right, for a sphere and a cylinder coincide with an accuracy of up to 2%. Thereafter, curve p_m for the cylinder dips lower, and for the sphere it rises higher due to adverse cumulative effect for the wave converging at point $\omega = 180°$ (see Fig. 9.1, following $t = t_4$). The values of p_m in the region of $\omega = 170–180°$ were determined by extrapolation, since the frontal part of the smeared shock-wave front reaches point $\omega = 180°$ prematurely. Following their collapse, the waves move against the flow (on the envelope p_m the arrows to the left) and at a particular value of $M_1 = \upsilon_1/a_1$ and $\omega \simeq \omega_c$ they come to a halt, enclosing the formed supersonic zone for the sphere at $M_1 \gtrsim 0.6$, and for the cylinder of $M_1 > 0.4$ [255] (see Fig. 9.1, wave *c*). Curves *4* and *5* in Fig. 9.3, *a*, which characterize a steady flow, coincide at $\omega = 0–60°$ with an accuracy of up to 2%. The location of the enclosing shocks observed in the steady regime ($\omega_c = 120–125°$) characterizes the turbulent flow (at $M_1 \approx 0$) over the body with high Reynolds numbers $Re = 10^5–10^6$ [355]. The symbols *6* indicate the solution of the theory of regular reflection (theory V)—p_υ. Symbols *7* denote the values behind the Mach wave from the elementary theory of Mach reflection (theory *Y*)—p_y, if the wave is oriented perpendicular to the reflecting surface. It is evident that the values of p_m are first close to the solution of theory V ($\omega = 0–30°$), then they go down, whereas symbols *6* "go" up. The envelope of the numerical values p_m goes down, but remains above the values p_y of the solution of theory *Y*.

The local maximum p_m on the body for Mach reflection of strong shock waves shifts to the vicinity of the contact surface originating at the triple point *z*, whose trajectory is shown in Fig. 9.1 by the dotted curve. The density distribution ρ (ω) in the numerical shock-capturing solution with smeared fronts is similar to the distribution of pressure (Fig. 9.3, *a*, *b*). Their nonmonotonous character agrees with the experimental data of [250].

Thus, curves *2* and *3* in Fig. 9.3*a* in the region $\omega = 30–40°$ are not shifted upwards beyond solution *6* of the theory V, but downwards. In discussing this fact, the following can be noted:

- The numerical method cannot "capture" the narrow peak zone of large gradients, which was obtained on a wedge experimentally [165] and numerically [305, 284] in the vicinity of $\omega = \omega_e$, despite the fact that special computations were carried out with increasing number of points over the surface of the sphere. These computations, however, did not yield any results for the rising branch of p_m (ω) near $\omega = \omega_e$;
- Mach reflection on a curvilinear surface begins earlier than $\omega \approx \omega_e$ (for example, $\omega_e \approx 41°$ for $p_{10} = 5$, $\gamma = 1.4$), and the behavior of p_m (ω) is characterized by a curve similar to the one obtained for shock-wave reflection from a wedge when a regime of Mach reflection has already been observed

[305, 307] (this assumption, however, contradicts the experimental data of chap. 7);

- The results of theory V cannot be used for a curvilinear surface, since the reflection point moves along the trajectory with deceleration, which contradicts the assumptions of theory V.

Fig. 9.3*b* compares the instantaneous pressure diagrams obtained in numerical simulation for shock-wave reflection from a sphere (*1*) and a cylinder (*1'*). It is seen that the local pressure maxima p_m (ω), characterized by the envelopes, are close to each other at $\omega < 120°$; however, the instantaneous diagrams p (ω) over the cylinder lie predominantly higher than over the sphere. On the contrary, when $\omega \geqslant 120°$, the instantaneous diagrams p (ω), and also p_m (ω) on the cylinder lie lower than on the sphere due to the cumulative effect when $\omega \to 180°$. The magnitudes of the moments of time $t_s = tu_0/R$, corresponding to the curve numbers, are presented below.

№	2	4	6	8	10	12	14	16	18	∞
Sphere		0.89	1.63	1.93	2.38	3.27	3.57	4.17	5.36	20
Cylinder	0.67	0.93	1.60	1.87	2.42	3.75	4.56			18

The pressure change as a function of time at individual points of the body—the numerical analog of the oscillograms when the signal is picked up by the pressure transducers—is given in Fig. 9.4. It is evident that the pressure rises rapidly up to $p = p_m$, then it drops. The curves for $\omega = 140°$ and 160° denote the second rise caused by the incidence of the wave, which travels in the opposite direction to the flow following its collapse at point $\omega = 180°$. From these curves, the moment of SW incidence at a given point of the surface can be determined. For the points other than $\omega = 0$ and 180°, when the SW experiences a normal reflection, the time corresponding to the middle of the curve rise is selected as the time of wave incidence at the given point. For points $\omega = 0$ and 180°, the point at which the first quarter of the rise ended was used. This algorithm ensures the uniform measurement of moments of time with a scatter of ± (1–2)%. A narrow pressure peak p (t) is observed at point $\omega = 180°$ for a sphere, and the maximum pressure holds for some time in the case of a cylinder. The small nonmonotonicities at points $\omega = 0$, $t_s \approx 1$ (sphere, cylinder) and $\omega = 180°$, $t_s \approx 3.7$ (sphere) have no physical meaning and are the result of the numerical effects of the shock-capturing method.

The values of τ_s, characterizing the time interval between the moments of signal appearance at points $\omega = 0$ and 180°, are plotted for a sphere and a cylinder in Fig. 9.5. Point *3* at $M_0 = 1$ represents the value $\tau_s = 2.57$ for a sonic wave which traverses the first half of the body (length R) during the period R/u_0, and then its wall part, turning around the body, traverses a path of a quarter circle $0.5\ \pi\ R = 1.57\ R$. Symbol *4* denotes the results of analyzing (± 2%) shadowgraphs for a cylinder ($p_{10} = 1.5$; air [345]). Points *5* show experimental data given by the authors of [342], where shadowgraphs are presented for a cylinder (air). For comparison with these data, two series of computations

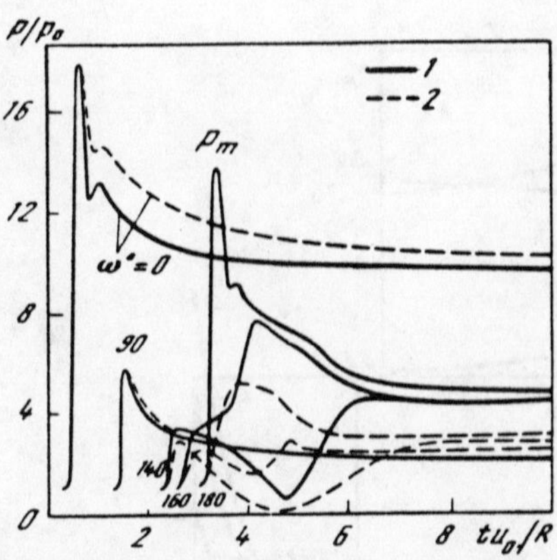

Figure 9.4. Pressure change in time at individual points of the surface. *1*) sphere; *2*) cylinder.

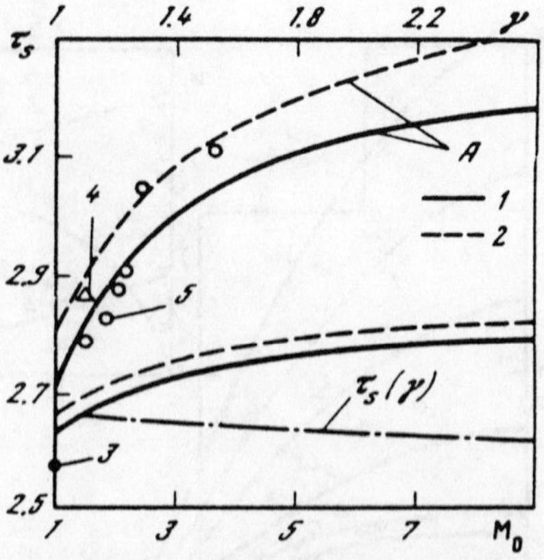

Figure 9.5. Dependence on γ and M_0 of the value $\tau_s = \Delta t_s u_0/R$—time interval between the appearance of signals at points $\omega = 0$ and 180°. *1*) sphere; *2*) cylinder; *A*) calculation with no-slip boundary condition; *3*) $\tau_s = 2.57$ (for a sonic wave); *4, 5*) experiment [345, 342].

were conducted with different boundary conditions for the tangential component of velocity u_w near the surface of the body: with slip (the lower pair of curves *1* and *2*) and with no-slip—$u_w = 0$ (upper pair—*A*). The value $u_w = 0$ in the given case simulated the boundary conditions used when calculating the flow of viscous gas near the surface of the body. It is assumed that for such a small time interval ($\tau_s \approx 3.3$) the no-slip boundary condition is valid for a qualitative investigation, since the formed boundary layer is still thin and does not exert significant influence on the shock wave process. It is interesting to note that such a non-traditional statement of boundary conditions for Eulerian difference equations had little bearing on the distribution of pressure on the surface of the sphere (see Fig. 9.7, solid curve with slip condition for $p_{10} = 3$ and symbols *1* with no-slip). This had even less bearing on the integral characteristics—axial and body forces (up to 2–3%). However, intervals τ_s increased by 5–15% with regard to the calculated data with slip condition ($M_0 = 1$–7) and, hence, correlation with the physical experiment improved (see Fig. 9.5, upper curve *1* and symbols *4, 5*). Based on this, one can come to the conclusion that taking into account the boundary layer, influence increases intervals τ_s, and the boundary-layer flow near the body must be taken into consideration to correctly determine these intervals quantitatively.

During calculations with no-slip condition the coefficient Ω in the smoothing operator had to be increased by 40–50% (compared with Ω in the calculation with slip condition) in order to obtain $p_m = p_n$ at the frontal point (see 8.22). Increasing Ω has almost no influence on the values of intervals τ_s ($\lesssim 2\%$). Methodical calculations showed that when Ω changes by 100%, the values of the pressure or density changes by only 10–20%. In the calculation with slip con-

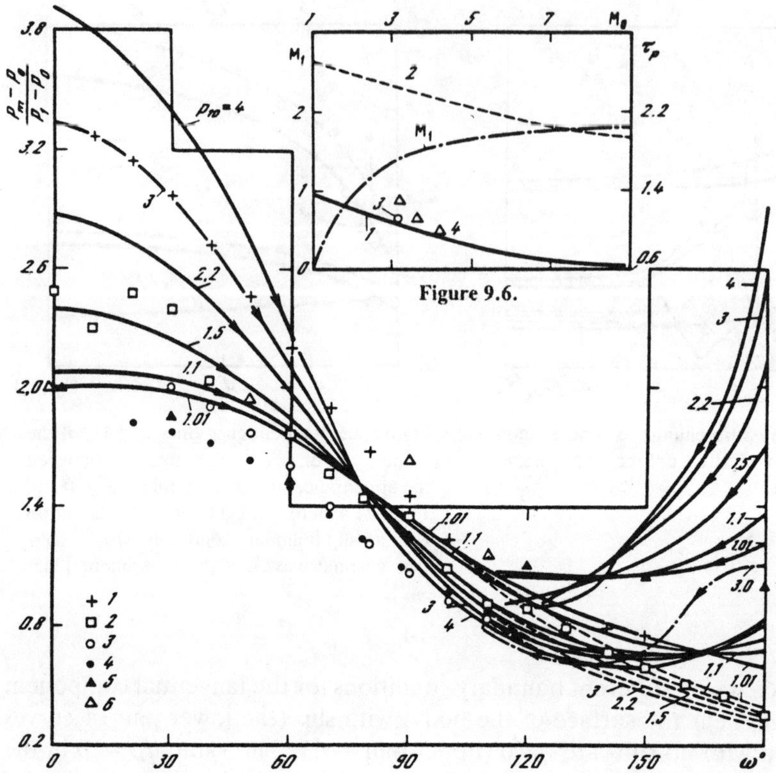

Figure 9.6. Dependence of time interval $\tau_p = \Delta t_p u_0/R$ on Mach number of shock wave for $\gamma = 1.4$. *1, 2*) numerical calculation for a sphere and cylinder, respectively; *3*) experiment [344]; *4*) results of the method of [337].

Figure 9.7. Dependence of the maximum pressure p_m on angle ω for $\gamma = 1.4$.

dition when $M_0 = 1.5–9$ $\Omega \approx 0.8–0.9$ for a cylinder, and $\Omega \approx 0.5–0.6$ for a sphere. When $M_0 \to 1$ $\Omega \to 6$ for a cylinder, and $\Omega \to 0$ for a sphere. Such values of Ω ensure that $p_m = p_n$ when $\omega = 0$.

The dash-dot curve in Fig. 9.5 shows how τ_s changes as a function of the adiabatic index $\gamma = 1.1–2.6$ for a cylinder when $p_{10} = 1.5$ (with slip condition). A weak dependence of τ_s on γ is observed, which allows one to use this circumstance for the verification and interpretation of the results of gashydraulic analogy [356], and also for the adjustment of physical or numerical experiments.

Figure 9.6 shows the dependence τ_p (M_0), which characterizes the time interval between the moment of signal appearance at point $\omega = 0$ and the moment of time when condition $(p(t)-p_{t1})/(p_n-p_{t1}) = 0.2$ is satisfied at this point, where p_{t1} is the pressure at a frontal point of the body $\omega = 0$ when a stream with Mach number M_1 flows over the body. Point (*3*) is a plot of the value from the experiment [344] for a cylindrical body with spherical head, the triangles (*4*) represent the data on τ_p from [337] for a sphere, recalculated with help of the coefficient $5.6 = \tau_\delta/\tau_p$. The value of τ_δ characterizes the time interval from the start of interaction t_1 to the moment when the magnitude of the separation $\delta(t) = 0.8\ \delta_\infty$, where δ_∞ is the magnitude of steady separation. This relation was found experimentally in [344] for $M_0 = 3.2$, $\gamma = 1.4$. In the present calculations this coefficient for sphere amounts to about 5. The SW reflected from the frontal section of the body travels in the opposite direction to the flow. Its intensity gradually decreases and it comes to a stop at a distance δ_∞ in front of the body for a specific magnitude of the SW (Fig. 9.1, wave *d*). To carry out the process until $\delta = \delta_\infty$, the computation time required is 4–5 times longer than for τ_p. The curves in Fig. 9.6 have an uncertainty of $\pm 15\%$, which is caused by the probable variations of the intervals of the calculation mesh, the Courant coefficient which determines the time step ($C_r = 0.9$–1.1), and the smoothing coefficient Ω (see section 8.1). The other data exhibit less scatter—around $\pm$ 5%, since they are determined at shorter calculation times. For ease of comparison, dependence M_1 (M_0) at $\gamma = 1.4$ is presented.

Plotted in Fig. 9.7 are the values of the local pressure maxima p_m enveloping the instantaneous diagrams p (ω, t) for $p_{10} = 1.01$–4; $\gamma = 1.4$; $\omega = 0$–$180°$ (see Fig. 9.3*a*). The solid curves, constructed for a sphere, coincide with the dashed curves (for a cylinder) in the range from $\omega = 0$ to $\omega \approx 100$–$110°$, then the dashed curves drop lower. After the reflection of the SW at point $\omega = 180°$, the envelopes are plotted with an arrow to the left (for a cylinder—one dash-dot curve—$p_{10} = 3$). The results for $\omega \to 180°$ show that the effect of shock-wave collapse increases the pressure at point $\omega = 180°$, the increase for a sphere being stronger than for a cylinder. For example, for $p_{10} = 3$ the ratio p_m at point $\omega = 180°$ amounts to $2.58/1.1 = 2.34$ (sphere/cylinder). Symbol *1* for a sphere denotes the values of p_m for $p_{10} = 3$ obtained in the calculation with no-slip condition (the remaining—with slip condition). As has already been mentioned, the coefficient Ω had to be increased by approximately 30% so that $p_m = p_n$ at point $\omega = 0$. Here, nearly a complete agreement was obtained with respect to the values of p_m at the frontal section ($\omega = 0$–$60°$), and at $\omega = 80$–$140°$ with values of p_m in the calculation with no-slip condition were respectively higher by 16–25% relative to the results without slip conditions). The symbols *2* denote the experimental values of p_m for a cylinder at $p_{10} = 1.5$ [237]. Good correlation of the results is noticed despite the fact that there was a blast wave of variable intensity in the experiment. Points *3* and *4* give the results of the numerical calculations[2] for a sphere (analogous to [349, 350]), obtained by the Godunov method for $p_{10} = 1.5$ and 1.1, respectively. It can be seen that these values of local maxima p_m (ω) lie below the data of the stated calculations in the compared range $p_{10} = 1.1$–10. In Table 9.1, for the range of angles $\Delta\omega$, the difference ε is shown in percent of the values of the solid curves. From the table, it follows that the difference increases with increasing p_{10}. The values of the integral forces

[2]Calculations were conducted by V.V. Podlubny

differ by only 2–4% during the unsteady phase ($tu_0/R \lesssim 2$–3). The results of the linear theory for the cylinder are designated by the symbols *5* and *6* [333, 334]. Symbols *5* correspond to the solution of [333], in which eight terms are left in the expansion, and symbols *6* correspond to the results of [334]. In the range $0 \lesssim \omega \lesssim 90°$, the data of [333] lie somewhat below the numerical curves ($p_{10} = 1.01$–1.1), and the results of [334] lie above. The linear solutions on the shadow side ($\omega > 90°$) give values higher than the numerical calculation.

From the discussion of the curves in Fig. 9.7 and the data which are analogous to the data presented in [256, see Fig. 2], the following nodal points can be singled out together with the regions where the lines are densely clustered:

for spheres and cylinders:

$\omega = 0°$, $\quad p_n^* = 1\ (\pm 1\%)$. $\quad p_{10} = 1$–100

$\omega = 77°$, $\quad p_1^* = 1.51\ (\pm 2\%)$, $\quad p_{10} = 1.01$–5

$\omega = 98°$, $\quad p_m = p_1\ (\pm 1\%)$, $\quad p_{10} = 4$–100

$\omega = 104°$, $\quad p_n^* = -0.03\ (\pm 1\%)$, $\quad p_{10} = 4$–100

for spheres:

$\omega = 145°$, $\quad p_n^* = -0.23\ (\pm 1\%)$, $\quad p_{10} = 1.01$–1.5

$\omega = 164°$, $\quad p_1^* = 0.68\ (\pm 2\%)$, $\quad p_{10} = 1.01$–5

for cylinders:

$\omega = 180°$, $\quad p_1^* = 0.35\ (\pm 15\%)$, $\quad p_{10} = 1.01$–5

where $p_1^* = (p_m - p_0)/(p_1 - p_0)$, $p_n^* = (p_m - p_1)/(p_n - p_1)$

The values in the brackets are the percentages of uncertainty due to the change of spatial and time computation intervals and Ω. These data can be utilized when constructing approximation $p_m(\omega)$ for a sphere using the method described in [256] for a cylinder. The indicated values for a cylinder differ somewhat from the results given in [256] because of the variety of techniques used for the interpretation of the results of numerical experiments. However, this does not introduce significant changes in the values of p_m. The nodal point for a cylinder in air is obtained at $\omega = 80°$ and $p_1^* = 1.4$ in the experiment of [237] ($p_{10} = 1.35$–1.9).

The dependence of the Mach number of the wall part of the diffracted wave M_w on the number M_0 when $\omega > 90°$ is almost linear: $M_w = \lambda M_0$ (with the exception of $\omega = 180°$ for a cylinder). The values of λ ($\pm 1\%$) as a function of ω are presented in table 9.2, and the values of M_0 and M_w for a cylinder at $\omega = 180°$ are given in table 9.3. We note that a similar linear dependence is observed when a SW is diffracted at convex corners (see [254] and section 8.4).

It must be emphasized, for the sake of definiteness, that in the present section 9.1, the results of calculations with no-slip condition are presented by the two upper curves *A* in Fig. 9.5 and by the symbols *1* in Fig. 9.7; the remaining data are computed with slip condition. The results for various $\gamma = 1.1$–2.6 and $p_{10} = 1.5$ are indicated by the dash-dot curve in Fig. 9.5 (γ = const for each calculation). The rest are obtained for $\gamma = 1.4$.

The stated results of the numerical simulation of the action of a SW on a sphere or a cylinder allow one to assess, in the range $p_{10} = 1.01$–100, the

Table 9.1

p_{21}	$\Delta\omega^\circ$	ϵ	p_{21}	$\Delta\omega^\circ$	ϵ
1.1	20–100	2	5	30–90	15–12
1.5	30–90	7–6	10	60–90	18–35

Table 9.2

	Sphere				Cylinder		
ω°	90	115 180	160	140	140	160	180
λ	1.125	0,935	0.691	0.679	0.595	0.522	–

Table 9.3

Cylinder, $\omega = 180^\circ$					
M_0	1	3	5	7	9
M_w	1	1.7	2.2	2,7	3.1

instantaneous and maximum values of the pressure on the surface of the body and some time characteristics—τ_s, τ_p. From the results of the calculations it can be concluded that taking into account the effect of the boundary layer near the body exerts influence on the magnitude of time interval τ_s, which is, in its own way, an integral characteristic of the process. The value of τ_s can be easily measured in physical and numerical experiments and characterizes the unsteady phase in its entirety. Therefore, this quantity can be used for verifying, adjusting, and comparing various experiments with each other, including data of hydraulic analogy.

9.2 INTERACTION OF A SHOCK WAVE WITH A BLUNT BODY IN SUPERSONIC FLOW (SHOCK-ON-SHOCK PROBLEM)

Comparative analysis has been conducted of the results obtained using one-dimensional models (OM) of the flow and the finite-difference shock-capturing method of two-dimensional unsteady flow. For the OM, the boundaries of applicability of their results, in which they agree with the experimental data and also with the results of the numerical calculation, are indicated. However, a qualitative difference is observed in the behavior of the curves characterizing the maximum reflection pressure. Analysis of the obtained material points to the fact that OM do not give sufficiently correct description of flow and there is no basis to recommend their use in a wide range of determining parameters. The main analysis of the results will be based on information about pressure on the surface of the body.

In the last fifteen years, many authors have discussed the process of unsteady shock-wave interaction with a blunt body in supersonic flow. Investigations have been conducted in three directions: obtaining experimental data (for example, [357–359]); constructing approximate OM [357, 360–363]; calculation of two-dimensional unsteady flow using the method of finite differences [269, 364–366].

One-dimensional models. Fig. 9.8 shows the main features of the process under discussion and introduces the notations for the parameters and regions. At the initial moment $t = t_0$ (Fig. 9.8*a, b*), we have in front of the blunt body *cc'* a bow shock *b* which is caused by the incidence of a supersonic flow *1* on the body with velocity u_1. A SW_a travels in the gas of region *1* toward the body with a velocity u_0 relative to the gas *1* (the parameters of the gas behind the SW_a are denoted by subscript *2*). Following the intersection of waves *a* and *b* at point *d* two waves—*ed* and *cd* (see diagram *x-t* in Fig. 9.8*a*) and a contact surface

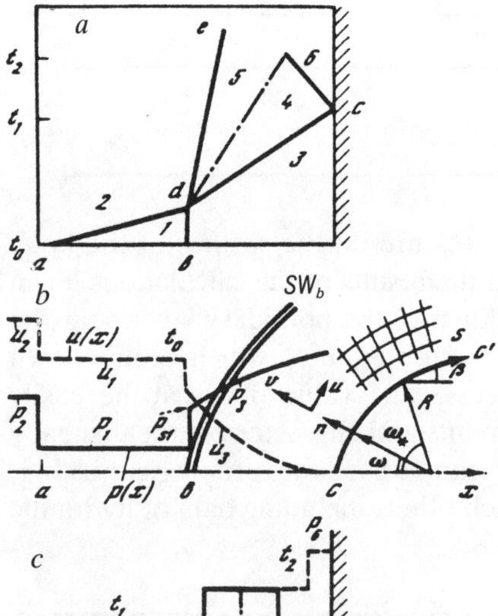

Figure 9.8. Schematic of one-dimensional flow along the stagnation streamline. *a*) *x* is the *t*-diagram: *bd* is the position of bow shock ahead of body prior to interaction; *ad* is the trajectory of incoming SW; solid and dash-dot lines are the trajectories of SW and contact surface, respectively: *b*) position of shock wave prior to interaction (t_0; *cc'* is the contour of body surface *n*, *s* are the natural coordinates; *c* is the pressure distribution after intersection of incoming and bow shock waves; solid and dashed lines prior to (t_1) and following (t_2) interaction of SW with body, respectively.

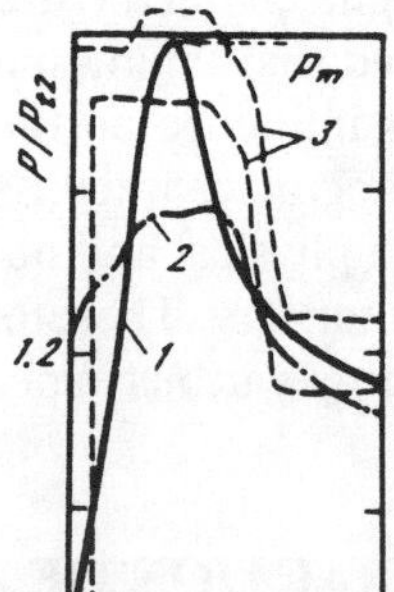

Figure 9.9. Pressure change in time at the stagnation point of the body for $M_1 = 5.16$; $M_0 = 4.46$, $\gamma = 1.4$. *1*) the result of the described numerical calculation [366]; *2*) two-dimensional calculation [364]; *3*) experimental data taking scatter into account [361].

between them are formed. The case at time $t = t_1$ is depicted by the solid line in Fig. 9.8*c*, and time $t = t_2$ (following the reflection of wave *cd* from the body) by the dashed line.

When obtaining the OM only the flow in the vicinity of the stagnation streamline *abc* is taken into account (Fig. 9.8*b*) and the pressure is evaluated at the stagnation point *c* of the body. The value of this pressure depends on three dimensionless parameters: $M_1 = u_1/a_1$, $M_0 = u_0/a_1$, and $\gamma = c_p/c_v$. Here, a_1 is the speed of sound in region *1*; γ is the Poisson adiabatic index, which has the value 1.4 in all the calculations in the present section.

Subscript *s1* denotes the parameters of the gas at a point behind the bow shock *b* on the stagnation streamline during a flow over a blunt body at Mach number M_1. Subscript *s2* denotes the gas parameters behind the shock which is formed by the incidence of the gas at velocity u_1 along the normal to a plane impermeable wall. Let us write the relation linking the values of pressure ahead of and behind the shock [254, 307]:

$$p_{s1} = p_1(\mu M_1^2 - \nu), \quad u_{s1} = a_1\left(M_1 - \frac{p_{s1}/p_1 - 1}{\gamma M_1}\right)$$

$$p_{s2} = p_1\left[1 + \frac{z}{2} + \sqrt{z\left(\mu + \frac{z}{4}\right)}\right], \quad u_{s2} = 0$$

$$z = \gamma^2 M_1^2/\mu, \quad \mu = 2\gamma/(\gamma+1), \quad \nu = (\gamma-1)/(\gamma+1)$$

Let us denote by subscripts *t*1 and *t*2 the values of the pressure at the stagnation point for a flow over a body at Mach numbers M_1 and $M_2 = u_2/a_2$— prior to and following shock-wave interaction, respectively.

$$p_{ti} = p_i\left(\frac{\gamma M_i^2}{\mu}\right)^{\gamma/(\gamma-1)} (\mu M_i^2 - \nu)^{1/(1-\gamma)}, \quad i = 1, 2$$

In the OM under consideration the real distribution of the parameters **f** (x) in the bow shock layer *bc* (Fig. 9.8*b*) is replaced by the constant values **f** = const. We shall consider three methods of selecting these values when $t = t_0$ (table 9.4). It is assumed in [357, 362, 363] that the pressure in the bow shock layer $p_3 = p_{s1}$. Physically, this corresponds to the model of permeable plane wall (OM-2,3). It is assumed in [362] that $p_3 = p_{s2}$—model with impermeable wall (OM–1). In [366] one more natural case with $p_3 = p_{t1}$ is discussed (OM–4). Practically, this is also a model of permeable wall; however, permeability differs from the first case. The subsequent discussion of the process for all models is the same, with the exception of OM–3 [362, 363], where it is assumed that the flow velocity u_3 tends to zero directly prior to the moment (t_*) of reflection from the body. This circumstance, as will be seen from the presented results, lowers the maximum pressure p_m considerably.

In Table 9.4, the value V_b of the initial velocity of bow shock *b* relative to the body prior to interaction is qualitatively indicated. This velocity is different from the velocity of the body due to the assumptions made.

Table 9.4

Model	Initial physical assumptions, $t = t_0$		V_b $t = t_0$	u_3 $t = t_*$	References
	p_3	u_3			
OM-1	p_{s2}	0	<0		[362]
OM-2	p_{s1}	u_{s1}	0	Same	[357, 362, 363]
OM-3	p_{s1}	u_{s1}	0	0	[362, 363]
OM-4	p_{t1}	$\neq u_{s1}$	<0	Same	[366]

When considering the decay of the gasdynamic discontinuity between the adjacent regions *2* and *3* (Fig. 9.8*c*), condition $p_4 = p_5$ and $u_4 = u_5$ is used at the contact discontinuity, and from the resulting transcendental equation p_5, $M_4 = u_4/a_4$, and $M_5 = u_5/a_5$ are determined

$$u_5 = u_3 + a_3 \frac{p_{5\,3} - 1}{\gamma\sqrt{(p_{5\,3} + \nu)/\mu}} = u_2 - a_2 \frac{p_{5\,2} - 1}{\gamma\sqrt{(p_{5\,2} + \nu)/\mu}} = u_4$$

$$p_{5\,3} = \frac{p_5}{p_3}, \quad p_{5\,2} = p_{5\,3}\frac{p_3}{p_2}$$

Furthermore, collision of the flow with Mach number M_4 with the wall takes place and the flow is decelerated through the unsteady SW which is directed against the flow. The pressure on the wall rises. As a result of repeated shock-wave reflections from the wall and interaction with contact surface, the pressure on the wall will rise from p_3 to p_m and after dropping it will reach a fixed level which is determined by a flow over a body at Mach number M_2. Presented below are the results of the calculation with single (principal) flow decceleration with Mach number M_4 through the unsteady SW (see dashed line in Fig. 9.8*c*)

$$\frac{p_6}{p_4} = 1 + \frac{z}{2} + \sqrt{z\left(\mu + \frac{z}{4}\right)}, \quad z = \frac{\gamma^2 M_4^2}{\mu}, \quad M_4 = \frac{u_4}{a_4} \tag{9.3}$$

To be more precise, in reference [366] M_5 in equation (1.4) one should read $M_4 = U_4/a_4$, and in (2.1) in the equation for $h\ \partial\ (h_s h_n)$ should be replaced by $\partial\ (h_s h_\phi)$.

Numerical calculation of axisymmetric dimensional unsteady flow. Figure 9.8*b* shows section of the calculation mesh near a sphere (radius *R*) with conical afterbody of angle β. The system of unsteady gasdynamic equations, describing the flow of an ideal gas in natural coordinate system *n-s* (8.2), was integrated with the help of an explicit finite difference scheme (8.17) of the second order of approximation

$$\frac{\partial \mathbf{f}}{\partial t} + \frac{1}{h_s}\frac{\partial \mathbf{A}}{\partial s} + \frac{1}{h_n}\frac{\partial \mathbf{B}}{\partial n} + \frac{\mathbf{C}}{h_s} = 0$$

For a plan-parallel flow $h_s = h_n = h_\phi = 1$, for axially symmetric flow (in table 8.1, h_1, h_2, h_3 are replaced by h_s, h_n, h_ϕ)

	h_s	h_n	h_φ
$s < s_*$	$1 + n$	1	$h_s \sin s$
$s > s_*$	1	1	$h_s \sin s_* + (s - s_*) \cos s_*$

Here, pressure p and density ρ are ascribed to p_1 and ρ_1. The components of the velocity vectors u and υ, ascribed to $\sqrt{p_1/\rho_1}$, as directed along the independent variables s and n. Coordinate s is measured from the stagnation point along the body, n—from the surface perpendicular to it. At $s = s_*$ the spherical head and conical afterbody are joined. The linear dimensions are ascribed to R, time t to $R\sqrt{\rho_1/p_1}$. Henceforth, the results of calculations will be presented as a function of time $t_s = t(u_0 + u_1)/R$. As result of the selected standardization $s \equiv \omega$ on the spherical part of the body, and $1 + n = r/R$.

The axial symmetry of the flow allows one to examine the range $0 \leqslant s \leqslant s_k$, where s_k is the coordinate of the end of the conical surface of the body.

"Mirror image" transfer of the parameters to an additional ray of the mesh is carried out on the axis of symmetry and the solid surface. On the external boundary for $n = n_{max}$ the parameters of the incoming flows $\mathbf{f}_1$ and $\mathbf{f}_2$ are given depending on the displacement of the SW, and for $s = s_k$—condition $\delta\mathbf{f}/\delta s = 0$. In the first stage of the calculations a stationary flow field is established with the Mach number of the incoming flow M_1. In this case, the stagnation parameters and also the pressure distribution over the surface of the body differed from the data of [19] with respect to p and ρ by 2–4% at the most. In the second stage—when the unsteady phase of shock-on-shock interaction is considered—the flow with parameter $\mathbf{f}_2$ was given ahead of the body. The intersection point of the shock front with the external boundary for $n = n_{max}$ moved along the boundary with velocity $u_1 + u_0$ toward the axis of symmetry. The process of integration continued until a new steady flow over the body was established with Mach number M_2.

The stability of the calculation and damping of oscillations were ensured by the smoothing operator (8.19, 8.20) with diagonal differences having an accuracy of the third order which did not hinder the second order of approximation of the difference scheme. The constant coefficient Ω in the operator, on the one hand, is selected to be small so that the solution would not depend on its value and, on the other hand, it is selected large enough to ensure the stability of the calculation and monotonicity of the results. The selection of the magnitude of Ω will be made when analyzing Fig. 9.10*a*.

The time step during integration was limited by the simplifying condition which ensures the stability of the calculation (see section 8.1).

When investigating the flows under consideration, the need arises to study the special effects which are considerable when analyzing the interaction process. These, for example, include the question of wave intersection, reflection of refracted wave from the surface of the body, transition from regular to Mach type reflection, effect of the boundary layer and so forth. The formation of Mach type reflection and conditions of transition from the regular type had been investigated in detail when a wave is reflected from rectilinear and curvilinear surfaces at $M_1 = 0$ in sections 8.2 and 9.1. This can be regarded as substantiation of accuracy and applicability of the used shock-capturing method to the unsteady process. The conducted comparison with exact solutions shows a discrepancy of no more than 3–5% with respect to p and ρ. If the boundary layer is not taken into account in the steady flow state, inaccuracy increases further by 10% [255]. Generally, taking into account the conducted comparisons, the accuracy of the results of the numerical simulation under consideration can be evaluated to be approximately equal to 15–20%.

Analysis of the results of OM and numerical calculation. Figure 9.9 depicts dependence $p\ (t_s)$ obtained numerically at the stagnation point. Curve *1* agrees well with the experiment (*3*) with respect to the maximum pressure p_m, although the front of the solid line is stretched. In that sense, curve *2*, computed by the Godunov finite difference method of the first order of accuracy, correlates better, since contact surface between regions *5* and *4* has been clearly indicated (discontinuity-fitting method) in the calculation (see Fig. 9.8*c*). However, the level of the maximum pressure p_m is lower in this case by ~30% which apparently can be explained by the large numerical viscosity of the method employed in [364]. The parameters of the steady flow at the critical point (for $t_s \approx 10$) obtained in the present calculations $p/p_{t2} = 0.98$, and in [364] are $p/p_{t2} = 0.86$ (i.e., the difference from 1.0 is larger). Having some experience with calculations the stretching of the front of the solid line *1* poses no obstacle to the correct interpretation of the results of the computations.

Figure 9.10*a* shows in dashed lines the instantaneous pressure diagrams on the surface of the spherical part of the body for various moments of time. The values of quantity $s \equiv \omega$ in degrees are plotted along the abscissa. The solid line with the smoothing coefficient $\Omega = 1.2$ represents the envelope of the maximum values p_m of the dashed diagrams. For the purpose of comparison the envelopes for other values of $\Omega = 0.8$–1.5 are also presented. It is evident that for a significant change in Ω (up to 100%) the magnitude of p varies within the bounds ± 10% (when $\omega = 0$) which is an indication of the weak dependence of the solution on Ω. When $\omega \gtrsim 50°$ this dependence is almost absent. Within the feasible variation of Ω its value was selected so that the maximum pressure p_m at the stagnation point did not differ from the exact solution p_n (8.22) by more than 2% when the wave was reflected from the stationary body ($M_1 = 0$). This condition is satisfied by $\Omega = 1$–1.2.

Analysis of the results allows one to conclude that the observed non-monotonicity of the dashed curves is caused not by the specific features of the numerical calculation but as a consequence of secondary reflections. The latter are indicated by the additional rise in curves *2* and *3* in Fig. 9.9. In the described computations measures were not taken to indicate shock *ed* and the contact surface (see Fig. 9.8*a*) as was done in [364]; however, later along the surface it appears that stronger separation of the flow in homogeneities takes place and the stated non-monotonicities appear. In the described computations this non-monotonicity has also been observed for some regimes ($M_1 >> M_0$) at point $\omega = 0$ as by curve *2* in Fig. 9.9.

Figure 9.10*b* presents time dependences of the axial force (solid lines) acting on the blunt cone with $\beta = 15°$ and $M_1 = 4$ for various M_0. The integral force $F_x(t)$ is calculated as follows

$$F_x(t) = F_1(t) + F_2(t)$$

$$F_1(t) = \frac{2\pi}{p_{t2}} \int_0^{s_*} p(t, s) \cos s \cdot \sin s \, ds, \quad 0 \leqslant s \leqslant s_*$$

$$F_2(t) = \frac{2\pi}{p_{t2}} \int_{s_*}^{s_k} p(t, s) \cos s_* \left[\sin s_* + (s - s_*) \cos s_*\right] ds, \quad s_* < s \leqslant s_k$$

Since only the frontal part of the blunt cone is being considered, the trend of the curves does not drop as time t_s progresses. For the sake of convenience of comparison the curves in the figure are slightly shifted in order to superpose the initial gradient section. It is obvious that the unsteady phase and the features of reflection are realized in the interval $t_s \approx 0.3–1.3$; furthermore, an increase

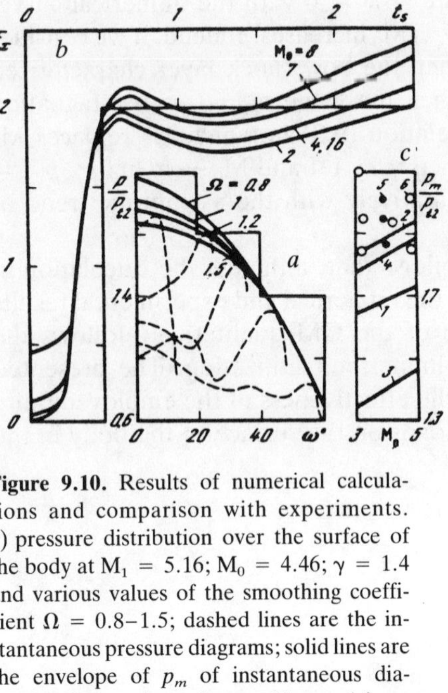

Figure 9.10. Results of numerical calculations and comparison with experiments. *a*) pressure distribution over the surface of the body at $M_1 = 5.16$; $M_0 = 4.46$; $\gamma = 1.4$ and various values of the smoothing coefficient $\Omega = 0.8–1.5$; dashed lines are the instantaneous pressure diagrams; solid lines are the envelope of p_m of instantaneous diagrams; *b*) time dependence of integral force $F_x(t)$ acting in the direction of axis x when $M_1 = 4$, $\gamma = 1.4$; *c*) comparison of maximum pressures at the stagnation point: *1-4* are the one-dimensional models; *5* is the numerical calculation [366]; *6* is the experiment [357].

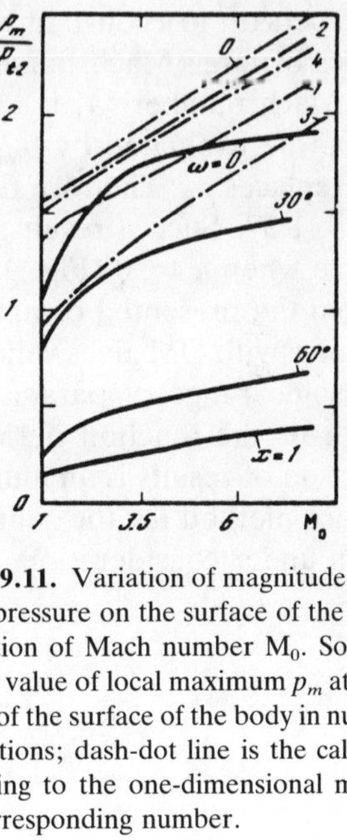

Figure 9.11. Variation of magnitude of maximum pressure on the surface of the body as a function of Mach number M_0. Solid lines are the value of local maximum p_m at various points of the surface of the body in numerical calculations; dash-dot line is the calculation according to the one-dimensional model of the corresponding number.

in $F_x(t)$ takes place with the increase in the surface area behind the front of the acting wave.

Lines 1–4 in Fig. 9.10*c* represent the results of calculations according to the OM at $M_1 = 5.16$. The numbers of the curves correspond to the numbers of the OM. Points *5* represent the values of the maximum pressure p_m at the stagnation point from Fig. 9.10*a* obtained by the finite difference method at $\Omega = 1$, 1.2, and 1.5 (from top to bottom, respectively); points *6* show experimental data from [357]. Agreement is observed between the results of numerical calculation at $\Omega = 1$–1.5, the experimental data, and the data obtained form OM–2 and OM–4. The curve obtained from OM–3 lies below all values.

The results of calculations for $M_1 = 4$ are plotted in Fig. 9.11. Curve *0* represents the magnitudes of the maximum pressure p_m obtained in [361], where a quasi one-dimensional computation is conducted taking into account the change of γ in the bow-shock layer. Curves *1–4* show the data corresponding to OM. The solid curves are the values of maximum pressure p_m at various points on the surface of the blunt cone with $\beta = 15°$ which were obtained by using the finite difference method. Comparison of curves *0–4* with the solid lines at $\omega = 0°$ shows that in the range of $M_0 = 2$–5, the results of OM do not differ greatly (~20%) from the data of numerical calculation. However, none of the OM under consideration qualitatively correlate at $\omega = 0$ with the numerical curve, which tapers off to a constant value ~1.89 as M_0 increases. Indeed, if we consider the cases of large Mach numbers M_0 when the bow shock layer characterized by the Mach number M_1 is neglected, then the value p_m/p_{t2} at the stagnation point can be evaluated by p_6/p_{t2} from relation (9.3) in which M_4 replaces M_2 and p_4 replaces p_2. Then we find that when $\gamma = 1.4$ and $M_0 \to \infty$ $\lim p_m/p_{t2} = 8/5.09 = 1.57$. Such a result agrees qualitatively with the asymptotic trend of the curve when $\omega = 0$ (Fig. 9.11).

From the presented comparison it follows that, although the calculation in accordance with OM gives values close to the numerical and experimental results within some range of parameters, none of the OM qualitatively follows the behavior or the function $p_m(M_0)$ which limits their utilization. The presented comparison of results is an indication of the effectiveness of the employed finite difference method for the simulation of loads on the surface of the body in the problem under consideration.

CHAPTER

TEN

UNSTEADY INTERACTION OF DETONATION WAVES

10.1 FEATURES OF THE STRUCTURE OF DETONATION WAVES

Shock waves, behind which physiochemical reactions take place in a gas with a negative thermal effect, have been discussed in the previous chapters. When the thermal effect is positive, a special type of waves is formed—detonation waves, which represent a combination of a shock wave and combustion zone. The fundamentals of the theory of detonation were laid down by Zel'dovich [367–369]. The problems of detonation are discussed in monographs [49, 148, 370–383]. The principal feature of detonation waves is the existence of a self-supporting regime—the so-called Chapman-Jouguet detonation.

The structure of the detonation wave in a one-dimensional approximation, taking into account the finite rate of chemical reactions in the combustion zone, is discussed by Zel'dovich [367]. In the coordinate system, moving with the wave, the detonation wave in the one-dimensional theory has the following form. Behind the shock wave, in which translational and rotational degrees of freedom of the gas molecules are excited, a relaxation zone exists in which excitation of molecular vibrations and chemical reactions take place. The velocity of the gas in the zone is subsonic relative to the front. It gradually increases with release of heat of reaction. The subsequent state of the gas depends on flow conditions behind the detonation wave. If there is a piston maintaining gas flow, the velocity of the gas remains subsonic as in the case of a shock wave. These waves are called overdriven waves. In the absence of a driving gas the detonation wave is converted into self-supporting detonation. The velocity of gas increases up to the speed of sound in the plane in which chemical reactions come to an end—the Chapman-Jouguet plane. Then an unsteady expansion wave follows, which does not weaken the detonation wave because its front propagates with the speed of sound.

In the laboratory coordinate system at the Chapman-Jouguet plane the velocity of detonation is equal to the sum of the speed of sound and gas flow velocity behind the detonation wave (Chapman-Jouguet condition). Since sound can propagate at various speeds in a relaxing medium such as a hot gas, the question becomes: relative to which sound speed is the Chapman-Jouguet condition fulfilled? Since the front of the expansion wave propagates with a frozen sound speed and the main part of the energy is transferred with equilibrium sound speed, a conclusion is made in several works that the Chapman-Jouguet condition must be fulfilled relative to the equilibrium sound speed [384, 385].

Another essential feature of detonation waves is related to their instability to transverse perturbations in the gas behind the wave front and their complex three-dimensional structure [373, 374]. The reason for the instability is the strong temperature dependence of the reaction rate in the wave [386]. The instability of the one-dimensional system: shock-combustion plane can cause the disturbance, arising in the Chapman-Jouguet plane, to increase in intensity, transfer to the shock plane, distort its surface, and create transverse waves. The entire front surface is divided into cells, the dimensions of which are in the order of the length of the reaction zone in the one-dimensional model. The dimensions of the cells depend on the reaction rate and are characteristic of the given explosive mixture. When the number of cells in the detonation front is large, each triple-shock configuration travels in a spiral path, collides with other configurations, collapses, and then returns to its original form again.

A model of a detonation cell is constructed in [387], and analysis of a multi-headed structure is made in [388–390]. The formation of a multi-headed detonation from a one-dimensional complex as a result of random initial disturbances was successfully followed numerically in [391]. An approximate schematic of multi-headed detonation is illustrated in Fig. 10.1. The combustion zones in some regions border with a shock front, in other regions they lag. Combustion revival takes place upon collision of transverse waves (points O_1, O_2), and also as result of adiabatic flashes at the end of the induction zone [388]. It was found that the triple configurations, moving along the front, can be of a simple Mach reflection type (Fig. 10.1) or a double Mach reflection with second triple points and contact surfaces which are curled into a vortex [392].

A structure other than one-dimensional detonation front is observed practically in all investigated mixtures for Chapman-Jouguet waves and for several overdriven detonation waves. Since highly overdriven waves are analogous to shock waves, so far as the heat released during the combustion process does not play a noticeable role in the energy balance, their structure is close to the one-dimensional structure [393, 394]. By analogy with the instability of detonation waves, a model is proposed explaining the instability of very strong shock waves

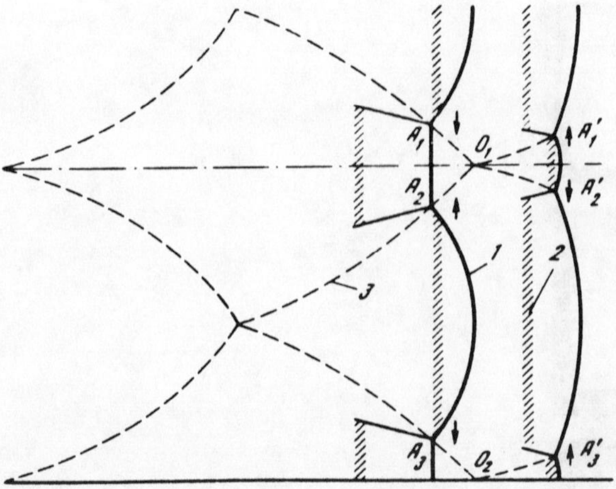

Figure 10.1. Schematic of the structure of a multi-headed detonation. *1*) shock waves; *2*) combustion waves; *3*) trajectories of motion of the triple points A_1, A_2, A_3.

in a relaxing gas when a delay in part of the reaction with heat release is possible despite the general negative thermal effect of the reaction [395]. The multi-dimensional structure of detonation leads to oscillation of local values of gas parameters and location of the Chapman-Jouguet plane. The model used for calculations is one-dimensional model which gives sufficiently accurate results in the majority of cases for the wave parameters averaged across the cross section of the channel, although it does not correspond to the real structure [396–398].

The internal instability inherent in the detonation wave is manifested not only in the multi-headed detonation structure but also in all its interaction with obstacles. In the motion of a detonation wave in tubes, when the tube diameter is comparable with the dimensions of a cell, spin detonation arises instead of multi-headed detonation. When the diameter is further decreased, the detonation wave propagates in some mixtures in the so-called "galloping regime" [399]. In this situation, periodic separation of the shock wave front from the combustion zone takes place with subsequent step-wise adhesion. The transition of a detonation wave into an expanding channel and transition of the detonation wave from one combustible mixture to another takes place with the disintegration of the complex shock wave-combustion zone and its subsequent re-establishment [400, 401]. At the stage of detonation initiation in transient processes behind the reflected shock waves in a combustible mixture, the formation of detonation takes place with unstable appearance of ignition spots behind the wave [402, 403].

Investigation of unsteady interaction of detonation waves has been conducted in three tubes of square cross section: 3 × 3 cm^2, length 5 m; 5 × 5 cm^2, length 6 m; 7.2 × 7.2 cm^2, length 8.3 m (4 ea and vol). Stoichiometric mixtures by volume of commercial methane and hydrogen with medical oxygen was prepared in gasometers above water and stirred for several days. The mixtures were ignited by a 180 mF capacitor charged to a voltage of 2 kV. The Schlieren photographs of the detonation waves in a mixture of $CH_4 + 2O_2$ at an initial atmospheric pressure are shown in Fig. 10.2. The wave propagates

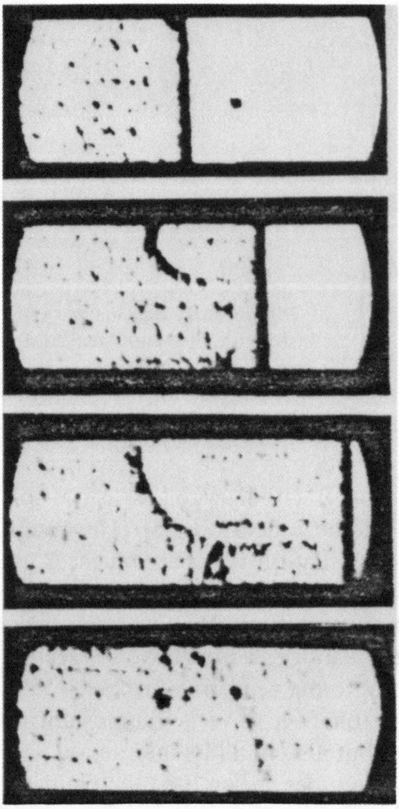

Figure 10.2. Detonation wave in mixture $CH_4 + 2O_2$, $p_0 = 1$ atm.

from left to right. The width of the dark strip in the pictures corresponds to the thickness of the reaction zone with transverse shock waves. Large scale disturbances behind the front are seen at the upper and lower walls of the channel. These disturbances are caused by the frames into which thin films were mounted when investigating refraction. The entire structure of the flow appears discontinuous and a multitude of small scale disturbances are observed. At a first glance, these appear to be turbulent disturbances which occur as a result of contact surfaces in triple configuration curling into vortexes at the wave front. Figures in chap. 5 showing triple configurations also show how vortexes are formed that appear only a considerable distance away from the triple points (of the order of one centimeter). Furthermore, in the experiments on the reflection of shock waves from an end with opening, which are described in the chapter on starting supersonic nozzles (see [6]), it is noted that vortexes disappear in the flow in a nozzle during transition through the speed of sound. Therefore, it is difficult to expect that vortexes could be created and preserved at the front of the detonation wave in the flow behind the Chapman-Jouguet plane. The observed small-scale perturbations are sound waves.

Figure 10.3 shows the streak record of a detonation-wave refraction during its transition from a combustible mixture into air. An expansion wave is reflected back in the mixture with its front travelling in the combusted gas with the speed of sound relative to the flow a_1-u_1. Comparison of the streak record with the picture frames shows that small-scale perturbations behind the front inscribe lines parallel to the expansion front in the streak record. Therefore, the perturbances in the photographs are sound disturbances due to flashes in the cells, the mechanism of which was observed in [388]. By measuring the velocity of such detonation sound waves or the velocity of the expansion waves fronts, the values of the quantity (a_1-u_1) can be found, where a_1 is the sound speed, and u_1 is the gas velocity behind the detonation wave in the laboratory coordinate system [396, 401].

The detonation wave in tubes appeared at a distance of 40–60 cm from the ignition point at an initial pressure of 1 atm. The velocity of detonation wave in the narrowest tube at a distance of 1.5 m from the ignition point is greater than the velocity of self-supporting detonation which is an indication of the wave being overdriven. The wave velocity remained constant within the bounds of experimental error starting from a distance of 2.5 m. Investigation of the structure of the disturbances in the flow behind the detonation wave showed that the wave is overdriven. Let us go back to Fig. 3.6*a* which depicts the wave structure appearing when the shock wave travels along the surface with a discontinuity at a point O. An analogous picture is obtained when the detonation wave travels with speed $D_1 = u_0$. Had the Chapman-Jouguet regime been implemented, the velocity of the detonation wave would have obeyed the condition $D_1 = u_1 + a_1$, then the sonic circle *1* would have come into contact with the wave at a point *m* and angle χ_1 would have been equal to zero. If, on the other hand, the triple point moves at angle $\chi_1 > 0$, this indicates that the disturbance behind the wave catches up with the wave, i.e., the wave is overdriven.

The propagation of the disturbances from the roughnesses in the upper and lower sides of the tube is seen in Fig. 10.2. They have a shape similar to the one presented in Fig. 3.6, which indicates that the wave is overdriven. The secondary waves propagating in the gas flow behind the detonation wave at a velocity lower than the detonation wave is one further evidence of the wave being overdriven (Fig. 10.3). Waves like these appeared at increased ignition energy and at lower initial pressures. Photographs of detonation and secondary waves are presented in Fig. 10.4.

The magnitude a_1-u_1 turns out to be extremely sensitive to the state of the gas behind the detonation wave. Simultaneous measurement of other values, for example the angle of motion of the triple point for elementary perturbation χ (see Fig. 3.6), which is dependent on a_1 and u_1, allows one to determine a_1, u_1, and then γ_{eff}. The latter is the isentropic index, a value which is required for all calculations of the interaction of detonation waves [404 and 405].

The results of measurements of a_1-u_1 near the detonation front in tubes of various dimensions at various distances from the ignition point are given in Table 10.1.

The measurements in the narrowest tube were made at a distance 3.5 m from the ignition place. The values of a_1-u_1 were obtained as the average from

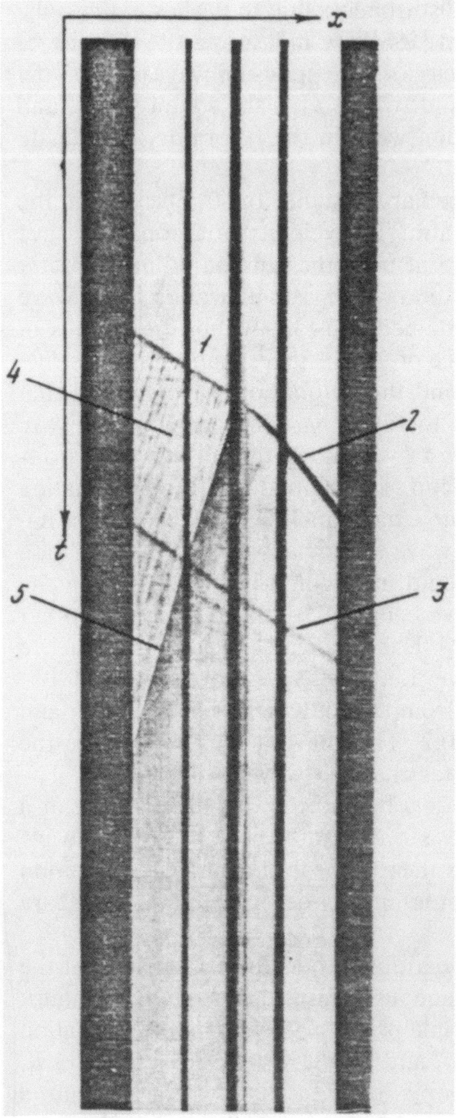

Figure 10.3. Streak-recording of refraction of a detonation wave during reflection from a mixture of CH_4 + $4O_2$ to air. *1–3* are incident, refracted, and secondary waves, respectively; *4* are sound perturbations; *5* is the front of an expansion wave.

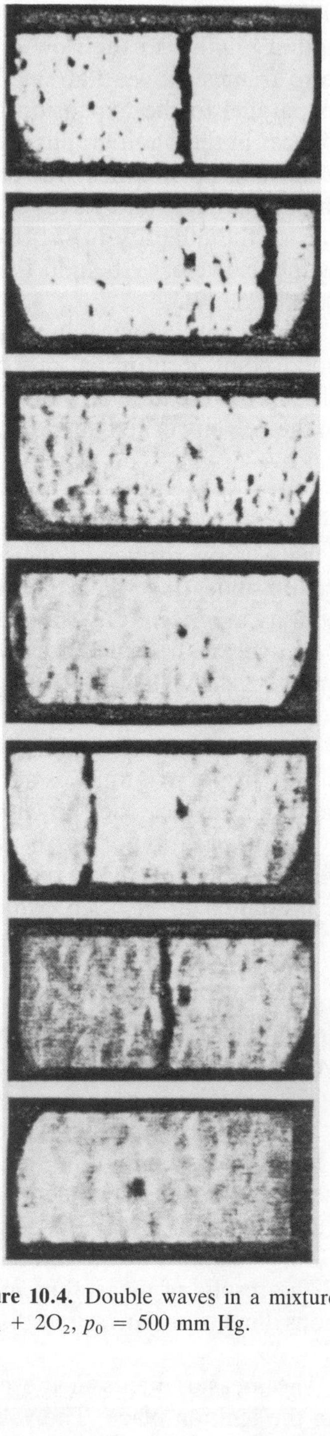

Figure 10.4. Double waves in a mixture of $CH_4 + 2O_2$, $p_0 = 500$ mm Hg.

Table 10.1

Mixture	Parameters of experiment	D, m/s	$(a_1 - u_1)$, m/s
$CH_4 + 2O_2$	3 × 3 см; l = 4 м	2290	440 ± 70
	7.2 × 7.2 см; l = 8 м	2350	370 ± 40
	l/d = 0	2400	320
$2H_2 + O_2$	5 × 5 см; l =5 м	2750	520 ± 30
	7.2 × 7.2 см, l = 8 м	2820	505 ± 20
	l/d = 0	2850	450

10–20 tests. The presented magnitudes of the errors correspond to the scatter from one test to another and to the inaccuracies of determining $(a_1\text{-}u_1)$ in each test. The error in determining the detonation velocity is equal to ± 50 m/sec. The values of D and $(a_1\text{-}u_1)$ were linearly extrapolated to an infinite tube diameter and compared with calculations. The detonation adiabat was computed from the laws of conservation for one-dimensional flow using the method of successive approximation without accounting for the losses from the equations of Chap. 1. The enthalpies of individual components of the mixture were taken from the tables of [5]. The obtained results are given in Table 10.2.

Here, D is the wave velocity; p_1, T_1, ρ_1, μ_1 are pressure, temperature, density, and molecular weight of the combustion products; u_1 is the gas velocity behind the wave in the laboratory coordinate system; a_e is the equilibrium sound speed in the gas behind the wave. The first row shows the parameters behind the Chapman-Jouguet wave in the case when the gas velocity is equal to the equilibrium sound speed. The following rows give the parameters of the overdriven detonation waves.

In addition to the equilibrium sound speed a_e, the frozen sound speed a_f (during the passage of the sound wave the chemical equilibrium is frozen and the vibrations are excited) and a_{ff} (during passage of the sound wave there is no shift in chemical equilibrium and no excitation of vibrations) are computed at the Chapman-Jouguet state. As is generally known, the velocity of the per-

Table 10.2

D, km/s	p_1, atm	T_1, K	ρ_0/ρ_1	u_1, km/s	a_e, km/s	$(a_e - u_1)$, km/s	μ_1
2.392	29.7	3725.7	0.541	1.097	1.295	0.198	21.09
2.395	30.0	3729.8	0.536	1.108	1.296	0.188	21.08
2.473	40.0	3829.8	0.417	1.441	1.320	−0.121	20.87
2.615	50.0	3917.8	0.345	1.712	1.349	−0.372	20.64
2.771	60.0	4005.8	0.298	1.946	1.363	−0.583	20.39
2.926	70.0	4082.3	0.263	2.156	1.387	−0.769	20.154
3.225	90.0	4234.3	0.218	2.523	1.432	−1.091	19.66

turbation front in the immediate proximity of the source is determined by the frozen sound velocity, then transition to equilibrium sound speed occurs. The distance at which this transition will take place depends on the ratio between the period of sound vibrations and relaxation times. The expansion wave front in a relaxing gas must propagate with a speed frozen relative to the excitation of vibrations during a period of time of the order of the vibrational relaxation time. In the course of experimental investigation of the velocity of weak perturbations in a relaxing gas it was established in [49] that the weak perturbations in carbon dioxide behind the shock wave travel a distance of no less than 2 cm from the source of the disturbance when $\gamma = 1.4$ and the shock wave Mach number is 6–12.

The equilibrium sound velocity was determined for the given composition of the burning gas from the equation

$$a_e = \sqrt{\frac{\gamma_1 R T_1\left[1 + T_1 \dfrac{1}{N}\left(\dfrac{\partial N}{\partial T}\right)_v\right]}{\mu_1\left[1 + T_1 \dfrac{1}{N}\left(\dfrac{\partial N}{\partial T}\right)_p\right]}}$$

where

$$\frac{1}{N}\left(\frac{\partial N}{\partial T}\right)_p = \sum_i \frac{1}{N}\left(\frac{\partial N_i}{\partial T}\right)_p, \quad \frac{1}{N}\left(\frac{\partial N}{\partial T}\right)_v = \sum_i \frac{1}{N}\left(\frac{\partial N}{\partial T}\right)_v$$

γ is the ratio of specific heats c_p/c_v for the reacting gas

$$c_p = \frac{1}{\mu_1}\left[\sum_i c_{p_i} x_i + \Sigma H_i \frac{1}{N}\left(\frac{\partial N_i}{\partial T}\right)_p\right]$$

$$c_v = \frac{1}{\mu_1}\left[\sum_i c_{p_i} x_i - R + \Sigma H_i \frac{1}{N}\left(\frac{\partial N_i}{\partial T}\right)_v - RT\sum_i \frac{1}{N}\left(\frac{\partial N_i}{\partial T}\right)_v\right]$$

where H_i is the molar enthalpy of the components; N_i is the number of atoms of a given type; x_i is the $P_i/p = N_i/N$. Summation was conducted with respect to all types of atoms. The magnitudes of a_f and a_{ff} were determined from equations

$$a_f = \sqrt{(\gamma_i R T_1)/\mu_1}, \qquad a_{ff} = \sqrt{(\gamma_{ff} R T_1)/\mu_1}$$

where γ_f and γ_{ff} are ratios of specific heats in the combustion products for a known frozen composition. The values of enthalpy of components in the calculations were taken from the tables in [5].

Table 10.3 presents the experimental and calculated values of D and (a_1-u_1). The parameters of the Chapman-Jouguet state for mixture $2H_2 + O_2$ are taken from [406], where the Chapman-Jouguet condition was assumed satisfied for the equilibrium sound speed; the experimental values are constant at a distance of up to 2 cm from the wave front.

Table 10.3

Calculated and experimental values	Mixture		Calculated and experimental values	Mixture	
	$CH_4 + 2O_2$	$2H_2 + O_2$		$CH_4 + 2O_2$	$2H_2 + O_2$
D	2.392	2.839	$(a_1 - u_1)_{exp}$ $(5 \times 5\ cm^2)$	–	0.520±30
$(a_e - u_1)$	0.198	0.247	$(a_1 - u_1)_{exp}$ $(3 \times 3\ cm^2)$	0.440±70	–
$(a_f - u_1)$	0.239	0.307	D_∞	2400±50	
$(a_{ff} - u_1)$	0.331	0.411	D_{exp} $(7.2 \times 7.2\ cm^2)$	2350±50	
$(a_1 - u_1)\ (1/d = 0)$	0.320	0.450	D_{exp} $(3 \times 3\ cm^2)$	2290±50	
$(a_1 - u_1)_{exp}$ $(7.2 \times 7.2\ cm^2)$	0.370±40	0.505±20			

Comparison of calculated and experimental value indicate that:

- First, the Chapman-Jouguet state is not implemented in the tubes used which are not sufficiently wide. Detonation speed D is less than calculated, and the experimentally determined value $(a_1\text{-}u_1)$ is much higher than all the calculated values. It must be noted that this effect cannot be the result of the fact that the wave in tube $3 \times 3\ cm^2$, for example, was overdriven. As evident from the calculations, the magnitude of $(a_1\text{-}u_1)$ on the overdriven part of the adiabat decreases even more. This means that the point which represents the state behind the wave does not lie on the equilibrium adiabat or, in other words, for the determination of the state one cannot use the conservation equations without losses. In either case, this can be shown as follows. If we take the values of D and $(a_1\text{-}u_1)$, obtained experimentally for tube $3 \times 3\ cm^2$, and make use of the laws of conservation for one-dimensional flow, then the effective value of the adiabatic index entering into the determination of the speed of sound can be found. The obtained value is $\gamma = 1.57 \pm 0.1$, which does not correspond to logical assumptions
- Second, if linear extrapolation is applicable, the comparison of the experimental data, reduced to an infinite diameter, with the calculated results shows that velocity $(a_1\text{-}u_1)$ for a tube having an infinite diameter is close to the velocity computed with the assumption that the speed of sound is frozen relative to dissociation and vibrational excitation. This corresponds to the magnitude of the effective adiabatic index $\gamma = 1.39$ for mixture $CH_4 + 2O_2$.

Tests on the interaction of detonation waves were conducted mainly in tube $3 \times 3\ cm^2$. Since it follows from the experiments that the effect of heat losses in the wall is felt in this tube, then to determine the state of the gas behind the detonation wave a larger number of parameters must be determined. Angle of motion χ of the triple point upon disturbance of the detonation front has been determined in [400] and the following magnitudes of the parameters for the detonation wave in $CH_4 + 2O_2$ obtained: $D = 2300 \pm \pm 100$ m/sec, $\mu_1 = 1100 \pm 90$ m/sec. $a_1 = 1540 \pm 160$ m/sec. From these data, it follows that a_1-$u_1 > D$, i.e., the wave is overdriven. The larger error in the given case are related to the fact that the presented values are the results of analysis of several experiments which were conducted at distances 2500 and 3500 mm from the end.

The problem of propagation of shock waves in tubes, taking losses into account, has been discussed by Zel'dovich [367], where accounting for losses has been carried out in one-dimensional approximation. Taking losses into account in this case changes all conservation equations except for the continuity equation, which when applied up to the present case results in $\rho_0/\rho_1 = 0.521$. For the determination of the pressure behind the wave the equation for the determination of speed of sound can be used if it is known which sound speed exactly was determined experimentally. The value a_1 obtained in [409] is the velocity of propagation of the front of the expansion wave. If we assume that the expansion wave travels with the frozen speed of sound, relative to the vibrational excitation and chemical equilibrium, then in the expression for the sound speed we must take $\gamma_{ff} = 1.39$ ($p_1 = 35.6$ atm, $T_1 = 3980$ K). The equilibrium value of the adiabatic index $\gamma_e = 1.245$. If the velocity of propagation of the expansion wave is determined by the frozen sound velocity, relative only to chemical equilibrium, then the following regime is obtained: $p_1 = 39.4$ atm, $T_1 = 4140$ K, $\gamma_e = 1.265$. Both obtained values were used in subsequent computations.

10.2 DIFFRACTION OF DETONATION WAVES

Numerous experimental investigations [407–422] and works conducted by using numerical methods [330, 423] have been devoted to the transition of a detonation wave from a channel of a certain section to a channel of larger section or infinite volume. It is established that there exists a critical channel diameter d_{cr} so that when $d > d_{cr}$ detonation turns into spherical detonation in an infinite volume, and when $d < d_{cr}$ it decays [407]. It was revealed that the determination of the critical diameter has great theoretical and applied significance, since by determining the critical diameter one can compute the critical initiation energy for the development of detonation in an infinite volume [420–422].

Systematic determination of the critical diameter was made for various mixtures upon transition of detonation from a narrow tube into a large volume [375, 407, 411, 416–418] and upon its transition into a wide tube through a conical nozzle [408, 419]. It was found that the critical diameter is related to the magnitude of the cell in a multi-headed detonation, and the dependences linking the diameter with the mean number of transverse waves of the same direction were determined [410, 411, 414, 415, 418].

At the initial stage of transition of the detonation wave from a tube of one size into another diffraction of the detonation wave at the convex angle takes place. The wave pattern of the diffraction of the detonation wave was investigated in [402, 409, 410, 414, 415], and the main feature of this process was found: wave decay, occurring at the initial stage of the process in all cases and for all diameters d. Then at $d > d_{cr}$ the detonation wave is restored, and at $d < d_{cr}$ it decays. Using numerical techniques, it was confirmed in [330] that the effect of strong wave expansion upon diffraction is felt in the decay of the detonation wave into a shock front and reaction zone.

Investigation of diffraction of a detonation wave at corners 65, 90, and 115° in $CH_4 + 2O_2$ was conducted in a tube 3×3 cm^3 ($d > d_{cr}$) [400, 409] (Fig. 10.5).

Upon diffraction, the detonation wave weakens as a result of appearance of expansion waves which travel upstream of the flow (Fig. 10.6). The shape of the front of an expansion wave is the same as that formed upon interaction with a single disturbance at the wall (Fig. 3.6). Consequently, the angles of motion of points A and A', which are the points of intersection of the unaltered part of the detonation wave AA' and its weakened parts AM and $A'M'$, must be determined by the equation

$$\mathrm{tg}\,\chi = \sqrt{a_1^2 - (D - u_1)^2 / D}$$

For the Chapman-Jouguet wave this angle must be equal to zero. The deviation of angle χ from zero is an indication that the detonation wave is overdriven. The measurements of angle χ assist in finding the gas parameters behind the detonation wave (section 10.1). As a result of weakening by the expansion wave the detonation wave decays into a shock wave AM and $A'M'$ and combustion zone on the contact surface AL and $A'L'$ (see Fig. 10.6). The decay is well observed in Schlieren pictures of the process (see Fig. 10.5), in which a white region appears behind the shock wave near the wall and increases from one frame to another as the distance between the contact surface and shock wave increases and becomes accessible to optical resolution. As a confirmation of the fact that front AM is a shock wave, we can cite, first, the self-similarity of the diffraction front and contact surface which was proven in [409]; secondly, the results of comparison of experimentally obtained profiles with the profiles of diffracting shock waves with the same flow Mach numbers.

Experiments show that weakening of waves upon diffraction is the greater the larger the angle of diffraction α_0 (Fig. 10.7).

We present the ratio of the velocities of parts of the front moving along the wall D_w and the detonation velocity D as a function of the diffraction angle α_c, and similar data for shock waves with corresponding adiabatic indexes:

α_0	65°	90°	115°
D_w/D	0.51 ± 0.05	0.47 ± 0.02	0.23±0.06
u_w/u_0	0.54	0.46	0.31

The profiles of detonation waves agree satisfactorily with the profiles of shock waves for the same Mach numbers.

Front decay leads to a sharp drop of temperature along the front. The graph

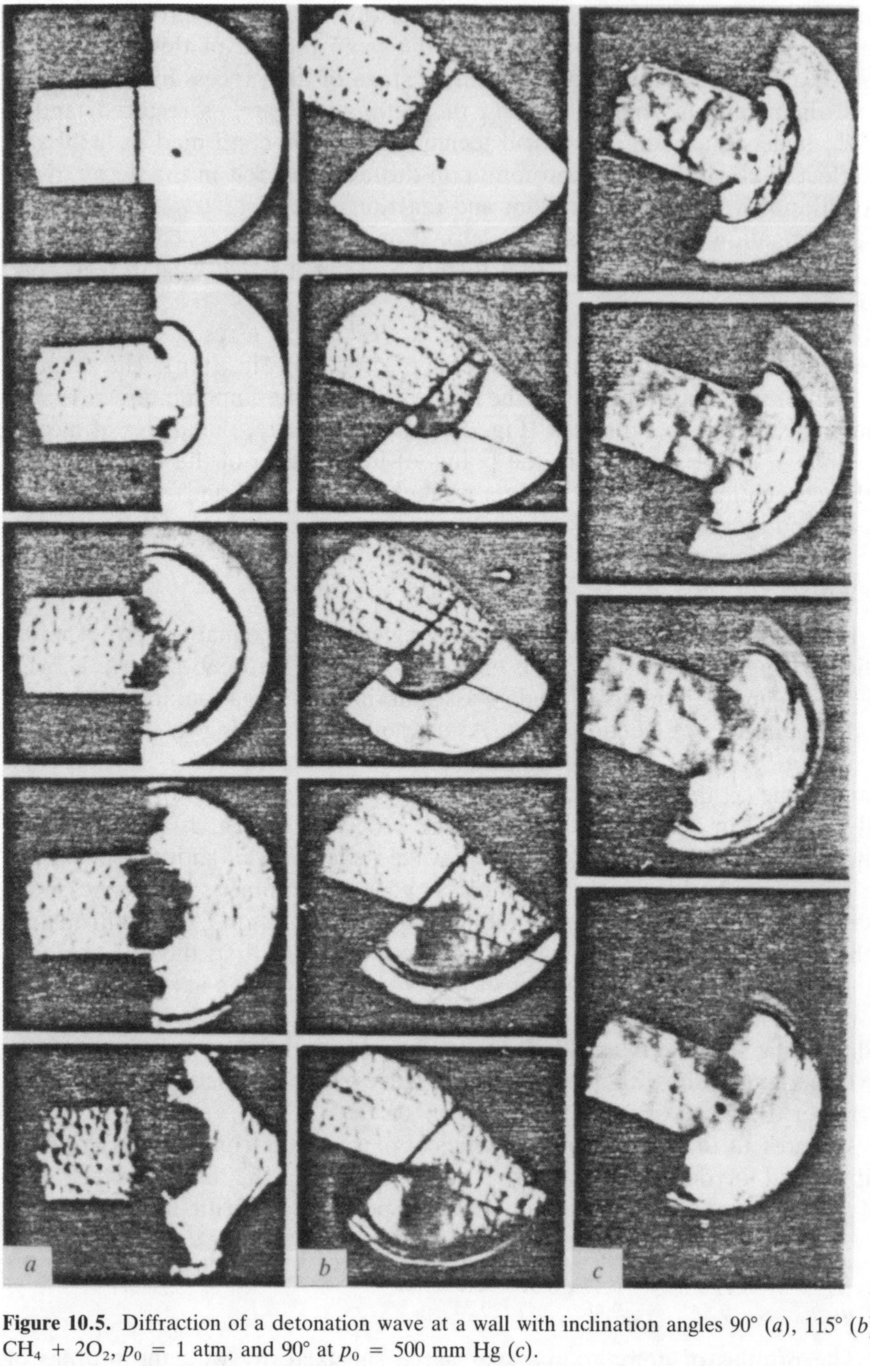

Figure 10.5. Diffraction of a detonation wave at a wall with inclination angles 90° (*a*), 115° (*b*) in $CH_4 + 2O_2$, $p_0 = 1$ atm, and 90° at $p_0 = 500$ mm Hg (*c*).

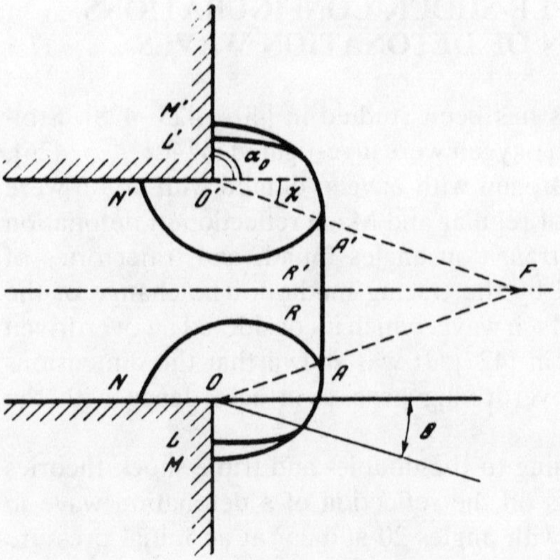

Figure 10.6. Detonation-wave diffraction. $M'A'$, AM is the front of diffracted wave; ARN and $A'R'N'$ are the fronts of reflected expansion waves; χ is the angle of motion of points A and A'; θ is the current angle.

of temperature behind the wave front, computed from the experimentally measured velocity D_w for diffraction at corner 90°, is given in Fig. 10.8. When diffraction occurs at corner 115°, the temperature drops even lower, the temperature on the wall reaches 590 K. The temperature between the shock wave and the contact surface is not constant. However, with time the gradients are smoothed out and the distance increases. The time of residence of the particles in the hot region becomes sufficient for ignition. Restoration of the front begins to take place at the points with the highest temperature, i.e., near the parts of front AA' which remained unaltered. This process has been investigated with the help of the tracing method [415].

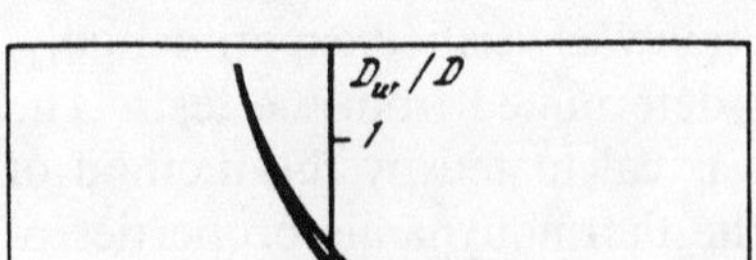

Figure 10.7. Ratio of the front velocity moving along the wall D_w to the velocity of the detonation wave D as a function of angle θ upon diffraction at corners *1*) α_0 = 65°; *2*) 90°; *3*) 115°.

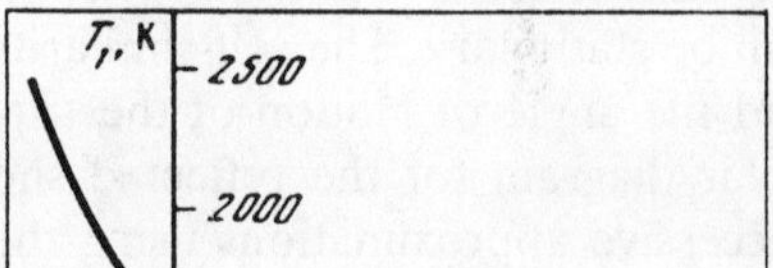

Figure 10.8. Temperature distribution behind the front of the detonation wave during diffraction at a right-angle corner (initial temperature T_0 = 290 K).

10.3 CALCULATION OF TRIPLE-SHOCK CONFIGURATIONS ARISING FROM REFLECTION OF DETONATION WAVES

The reflection of detonation waves has been studied in [401, 425–428]. Stoichiometric mixtures of methane with oxygen were investigated in [401, 425, 426], and acytelene with oxygen and hydrogen with oxygen diluted with argon were investigated in [427]. The patterns of regular and Mach reflection of detonation waves were obtained, the critical transition angles found, and trajectories of motion of triple points determined by the tracing method. The change of the dimensions of the cells behind the Mach wave, which is considered an overdriven detonation wave, was investigated in [427]. It was shown that the dimensions of the cells decrease during the overdriving process in accordance with the acoustic theory.

In [425] the calculations according to the double- and triple-shock theories are compared with the experiments on the reflection of a detonation wave in mixture $CH_4 + 2O_2$ from wedges with angles 20 and 35° at an initial pressure of 1 atm. The calculation of the triple-shock configuration upon reflection of detonation is conducted by using the method of shock polars, analogous to the calculations of shock wave configurations [49]. Detonation and shock polars were constructed in plane (p, θ), where p is the pressure behind the wave, θ is the angle of flow deflection during passage through the wave. The Mach wave is a recompressed detonation wave in a mixture $CH_4 + 2O_2$; the values of the parameters for such waves are presented in Table 10.2. The polar diagrams represent all the possible states which can exist behind various overdriven waves if we are given the gas parameters up to the wave and velocity and direction of the incoming flow. In order to determine the direction of the incoming flow the angle of motion of the triple point must be known. Thus, when the motion is self-similar the flow pattern in the coordinate system moving with the triple point will be stationary. The self-similarity of the motion was verified experimentally, and the angle of motion of the triple point χ_1 determined from the tests. The polar diagram for the reflected shock wave was calculated by the method of successive approximations using the data on the thermodynamic properties of the combustion products of mixture $CH_4 + 2O_2$. The points of intersection of two polar diagrams represent the possible solutions of triple-wave Mach configuration. The Mach configuration was computed on the assumption that a Chapman-Jouguet (Ch.-J.) state exists behind the incident wave. It was also supposed that two states exist which are determined by taking into account the losses with the assumption that the propagation of sound disturbances takes place with: completely frozen velocity (state A), and frozen velocity only relative to the shift of the chemical equilibrium (state B). The results of the calculation are presented in Table 10.4.

Here α_0 is the angle at the tip of the wedge; D, p_1, ρ_0/ρ_1 are the velocity of detonation wave, pressure behind it, and density ratios; ω_1, ω_3, ω_4 are the angles of incidence, reflection, and angle of Mach wave θ_2 is the angle of flow deflection at the reflection wave; $\overline{M}_2$ is the flow Mach number in the region behind the

Table 10.4

Parameters	A	B	C.J.	A	B	C.J.
α_0	35°	35°	35°	20°	20°	20°
D, m/s	2300	2300	2392	2300	2300	2392
p_1, atm	35.6	39.4	29.7	35.6	39.4	29.7
ρ_0/ρ_1	0.521	0.521	0.541	0.521	0.521	0.541
ω_1	52° 30′	52° 30′	52° 30′	63°	63°	63°
ω_2	48° 42′	49° 12′	regular	52° 36′	55° 18′	65° 48′
ω_4	87° 00′	87° 00′	reflection	80° 36′	78° 36′	84° 12′
θ_2	6° 00′	8° 48′	–	14° 42′	16° 24′	11° 24′
p_2/p_1	1.91	1.70	–	1.19	1.10	1.81
$\bar{M}_2$	0.925	0.938	–	1.003	1.018	1.018
ω_{1exp}	52° 30′ ± 1°		–	63 ± 1°	–	–
ω_{2exp}	31 ± 2°		–	52 ± 3°	–	–
ω_{4exp}	85 ± 2°		–	81 ± 2°	–	–

reflected wave in the coordinate system moving with the triple point; $p_0 = 760$ mm Hg, $T_0 = \; = 293$ K.

The computed results were compared with the experimental data. Experiments were conducted in a tube having a section 3×3 cm^2. The Schlieren photographs of reflection from a wedge with a 35° angle are presented in Fig. 10.9.

Comparison of angle of reflection ω_2 with the experimental data indicates that for both wedge angles the calculation, based on the assumption that the Chapman-Jouguet condition is fulfilled, does not correlate with the experiment, particularly, in the case of the angle $\alpha_0 = 35°$ where the calculation gives regular reflection. The results of the calculations conducted with assumptions A and B are close to each other. For angle $\alpha_0 = 35°$, the experimentally determined angle is much less than both computed values. For angle $\alpha_0 = 20°$, the value of ω_{2exp} agrees with the value computed in version A. The disagreement of the results of the calculation with the experiment in the case of the angle $\alpha_0 = 35°$ and their agreement for the angle $\alpha_0 = 20°$ is explained by the fact that, similar to the case for shock waves, the triple-shock theory is not implemented for weak shock waves with subsonic flow behind the wave. For $\alpha_0 = 20°$ the flow is supersonic and the triple-shock theory is valid. A regime of complex Mach reflection arises in this case with a kink at the reflected wave as can be seen in Fig. 10.9. In such a regime in carbon dioxide a double Mach configuration appeared with negative angle ω_2, since the gas was compressed more due to the endothermic processes behind the shock wave [401].

10.4 REFRACTION OF DETONATION WAVES

When a detonation wave is reflected, an overdriven detonation wave is created as a result of pressure increase behind the wave. If a process is established

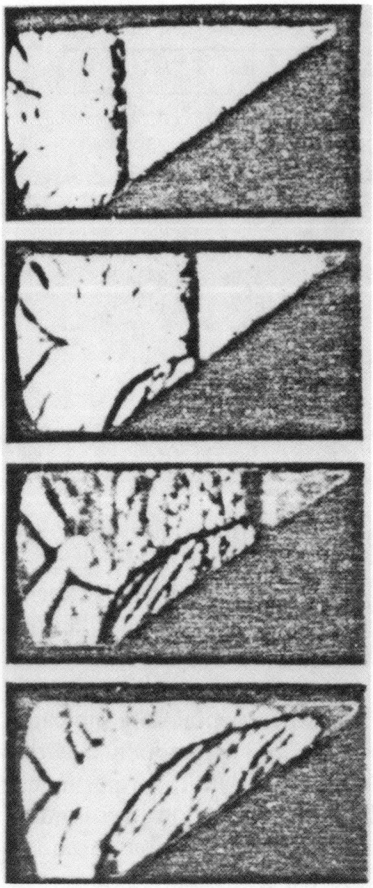

Figure 10.9. Mach reflection of detonation waves in mixture $CH_4 + 2O_2$, $p_0 = 1$ atm, $\alpha_0 = 35°$.

whereby the pressure drops sharply, then a decay of the detonation wave can take place, as in the case of diffraction. Such a process is created upon diffraction of a detonation wave and its transfer from one explosive medium to another. In this case a triple configuration arises: detonation wave, shock wave, and expansion wave.

The layout of the tests on the diffraction of detonation waves is as follows [49, 429]. A detonation tube of a square cross section 3 × 3 cm^2 was divided into two parts in the test section by a thin film stretched at an arbitrary angle to the axis of the tube. The detonation wave propagated along the tube and refracted at the interface of the two mixtures. A similar experimental layout was used in [430–433]. The film was prepared from a solution of nitrocellulose (specific weight 1.68 g/cm^3). The composition of the solution: 100 ml 2.6% solution of nitrocellulose in amyl acetate with acetone (in 2:1 ratio), 70 ml toluene, and 10 ml diamyl ftalat.

The film thickness did not exceed 800 Å. Evaluation by the formulas of [431] indicated that such films should not influence the refraction process. In order to verify this, tests were conducted with the tube filled by the same mixture on both sections. When a film was used that is three times thicker than the one normally employed, a weak compression signal, propagating at sonic speed was reflected upstream of the flow. There were no other changes in the flow and in the velocity of the transmitted detonation wave. The films were sufficiently thin so that their mass would not have an effect on refraction. On the other hand, they were sufficiently elastic so that they would not burst from moderate bending.

The tube was filled with explosive mixtures by simultaneously purging both sections of the tube; the film in this case serving as a sensitive manometer. Visual observation using the Schlieren method established that the gases from both sides of the film do not mix, at least during the first 5 minutes.

Detonation-wave refraction tests were conducted for angles of incidence 0, 30, 58° upon transition of waves from $CH_4 + 2O_2$ into air, oxygen, and methane-oxygen mixtures of various compositions. Fig. 10.10 depicts the wave systems appearing upon refraction into air and combustible mixtures for an angle of incidence $\omega = 58°$.

In Fig. 10.10*a* the wave underwent a transition from a mixture of $CH_4 + 2O_2$ into air. The inclined line dividing the picture is the trace of the frames which were holding the film; *1–6* represent the incident wave, the front of the reflected expansion wave, the refracted wave, the Mach wave appearing upon reflection of a refracted wave from the upper wall, the rear front of an expansion wave, and the contact surface, respectively.

In Fig. 10.10*b* the wave underwent a transition into a mixture of $CH_4 + 4O_2$. It was revealed that the refracted waves, upon transition of detonation waves into a reacting mixture, were shock waves no matter whether the transition was into a mixture with a higher or a lower detonation velocity. Indeed, the magnitudes of the detonation velocities used in the mixtures were as follows

$CH_4 + O_2$	2637 m/s	$CH_4 + 3O_2$	2046 m/s
$CH_4 + 1.5\,O_2$	2531 m/s	$CH_4 + 4O_2$	2067 m/s
$CH_4 + 2O_2$	2392 m/s		

Had a Chapman-Jouguet wave been initiated immediately upon refraction of the detonation wave in the second medium, its inclination could have been easily determined from the sine law. The refraction angles ω_4 should have been equal to the following values (at an angle of incidence 58°)

Mixture.	$CH_4 + 4O_2$	$CH_4 + O_2$
ω_4	47° 12′	61° 02′

Had an overdriven wave been refracted, the angles ω_4 would have been even greater. In all the tests the angles of refraction were significantly less (Fig. 10.11). Thus, as in diffraction, the detonation wave, after getting into a mixture of another composition (even a weaker one), first degenerates into a shock wave and a combustion front. The transition into detonation in the second medium takes place by igniting the mixture behind the shock wave. Unlike diffraction, the parameters behind the detached shock wave, in this case, are constant up

a

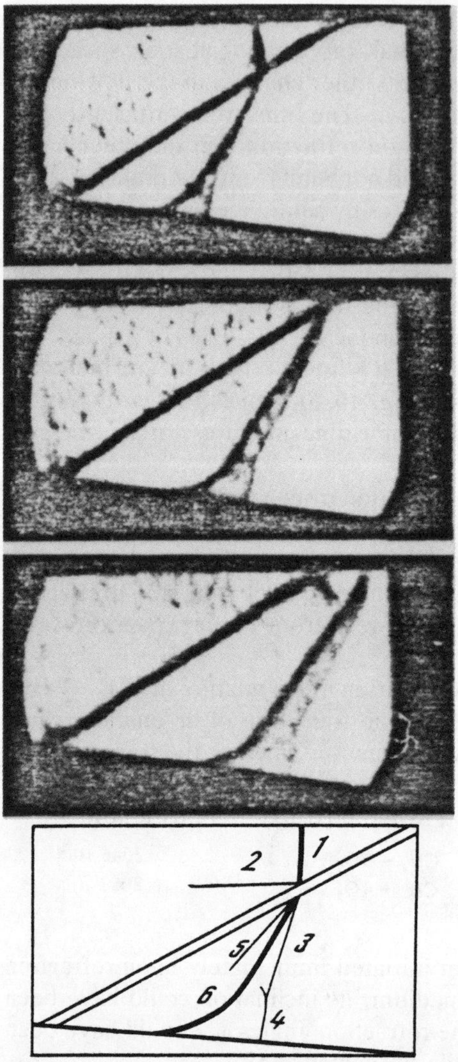

Figure 10.10. Refraction of detonation waves from mixture Ch_4 + $2O_2$ into air (*a*), into mixture CH_4 + $4O_2$ (*b*) and into mixture CH_4 + $3O_2$ (*c*).

to the contact surface. Ignition occurs when the layer of the heated gas behind the shock wave becomes comparable with the magnitude of the ignition delay at the given temperature and pressure [49]. The ignition process leads to an increase of angle ω_4. If the refracted wave is reflected from the wall irregularly, then the process of angle increase becomes quicker.

Fig. 10.10*c* shows a picture of wave refraction in mixture CH_4 + $3O_2$ when the upper plane is so inclined that the refracted wave travels normal to it. In this case, the reflection of the refracted wave does not take place. The perturbations, seen in the picture, characterize the process of transition to detonation in the refracting medium. The process of detonation development and the re-

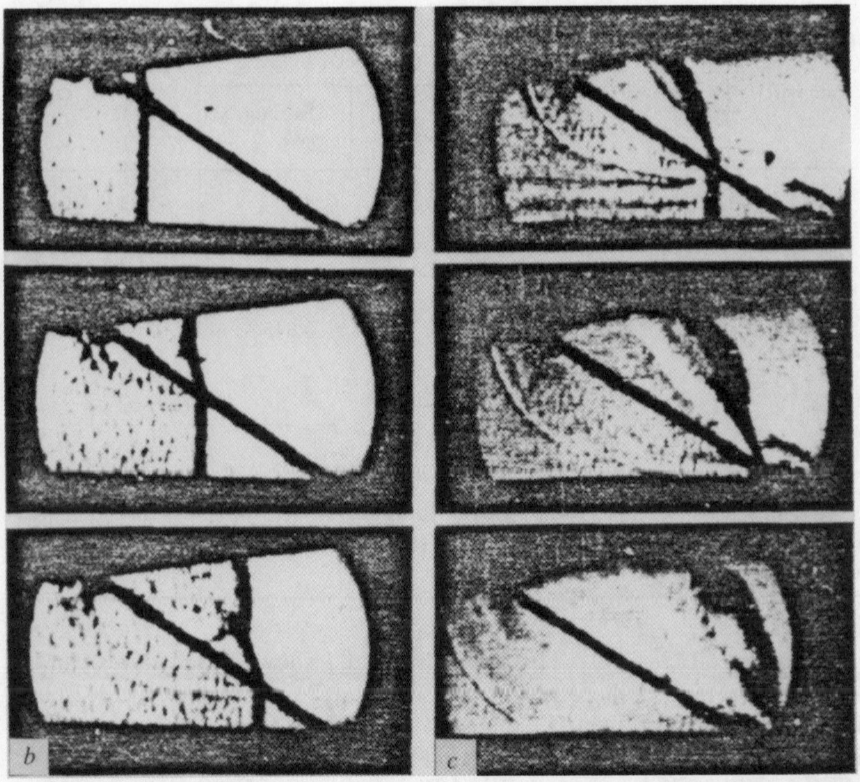

sulting change of inclination of the refracted wave can be seen. The velocity of the refracted wave in the second medium tends to a steady value of the detonation velocity.

The calculation of the triple-shock configuration, occurring upon refraction of a detonation wave from mixture $CH_4 + 2O_2$ into mixtures of methane with oxygen of various composition, was conducted by the method of shock polars in [401]. The polar diagram for the reflected wave was determined as in the calculation of Mach reflection. When calculating the expansion wave in the burnt gas it was assumed that γ in the Prandtl-Meyer wave is constant and equal to 1.39. The refracted shock waves travelling in explosive mixtures were computed by the method of successive approximations assuming that the vibrations of gas molecules without dissociation took place behind the waves. Most assumptions are made when calculating overdriven detonation waves in mixtures $CH_4 + 3O_2$ and $CH_4 + 4O_2$. The polar diagram for a wave in mixture $CH_4 + 2O_2$ was calculated taking into account the thermodynamic properties of the combustion products at different temperatures and pressures.

A schematic diagram of the triple configuration is given in Fig. 10.12. The polar diagrams of the case of refraction of a detonation wave from mixture $CH_4 + 2O_2$ into mixtures $CH_4 + 3O_2$ and $CH_4 + 4O_2$, i.e., into "slower" mixtures are illustrated in Fig. 10.13. Point A depicts the state behind the detonation

Table 10.5

Parameters	Air	$CH_4 + 4O_2$		$CH_4 + 3.5\,O_2$
		Shock wave	Overdriven wave	
ω_2	37°15′	37°15′	37°15′	37°15′
ω_2'	−11°34′	−10°24′	−5°	−10°36′
ω_4	35°36′	33°06′	52°36′	33°12′
θ_3	26°36′	26°48′	21°48′	27°
$x = \frac{P_2}{P_1}$	0.61	0.645	0.845	0.64
ω_{1exp}	58 ± 3°	—	—	—
ω_{2exp}	37 ± 3°	37°30′ ± 20°	—	—
ω_{2exp}'	−15°30′ ± 1°	−16 ± 3°	—	—
ω_{4exp}	32°30′ ± 1°	33°30′ ± 1°30′	—	31°30′ ± 2°
θ_{exp}	24 ± 1°	26°30′ ± 30′	—	27 ± 1°

Note to table 10.5: Angle of incidence $\omega_2 = 58°$; $p_0 = 760$ mm Hg; ω_2 and ω_2'—angles formed with the interface by the frontal and rear sides of the expansion front, θ_3 is the angle of surface inclination.

wave in mixture $CH_4 + 2O_2$. The polar diagram *1* shows which states can be realized in the reflected shock wave, part of polar *2* corresponds to the states which can arise in the reflected expansion wave. It is evident that for the refraction in a mixture with lower detonation velocity the method of shock polars gives two possible solutions: point *B* and *B′* for mixture $CH_4 + 4O_2$ and points *C* and *C′* for mixture $CH_4 + 3O_2$. In both cases the expansion wave is reflected; however, one cannot say which solution will be established. In any case, upon transition from a "quicker" medium with higher detonation velocity into a "slower" one, it is natural to expect that the reaction does not come to an end at the boundary and that overdriven detonation wave occurs in the second medium. However, comparison of experimental data with the calculated results indicates that a case corresponding to the lower point of the intersection with the refracted shock wave is carried out.

In Table 10.5 experimental data and calculation results are presented for the refraction of a detonation wave from mixture $CH_4 + 2O_2$ into mixtures of various composition when a shock wave propagates in the second medium. The

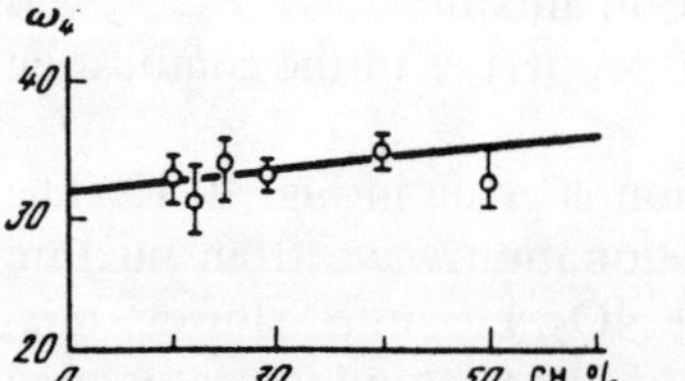

Figure 10.11. Dependence of refraction angle ω_4 on the percentage of methane content in the second mixture upon refraction of detonation wave from mixture $CH_4 + 2O_2$ into a mixture of methane and oxygen.

$CH_4 + 3O_2$		$CH_4 + 2.5O_2$	$CH_4 + 1.5O_2$	$CH_4 + O_2$
Shock wave	Overdriven wave			
37°15′	37°15′	37°15′	37°15′	37°15′
−11°	−3°12′	−11°	−11°36′	−12°
33°12′	59°06′	33°24′	34°06′	34°48′
26°06′	19°48′	27°24′	28°	28°30′
0.635	0.915	0.63	0.61	0.58
–	–	–	58°30′	
38°20′ ± 1°	–	34 ± 3°	33°30′ ± 3°	37 ± 2°
−15 ± 2°	–	−14°	−14°30′ ± 1°	–
34 ± 2°	–	33 ± 1°	34 ± 1°	32°30′ ± 2°
31 ± 1°	–	27 ± 1°	28 ± 1°	27°30′ ± 2°

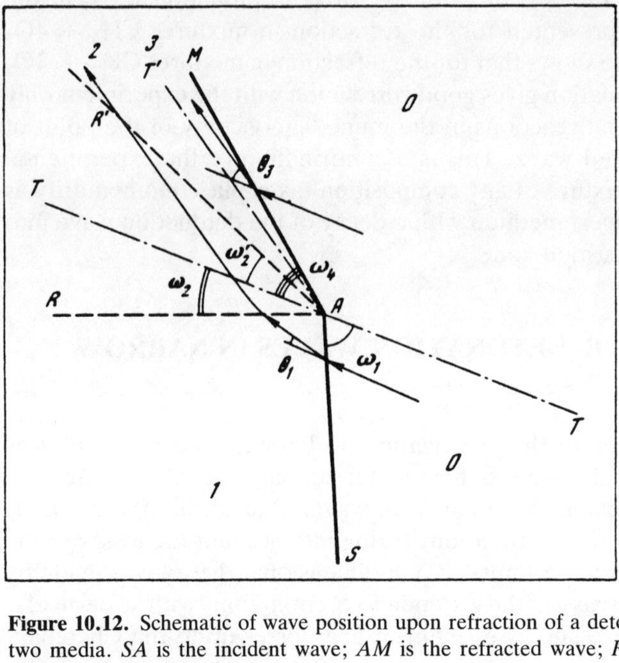

Figure 10.12. Schematic of wave position upon refraction of a detonation wave at the interface of two media. *SA* is the incident wave; *AM* is the refracted wave; *RAR′* is the reflected expansion wave; *AT* is the contact surface.

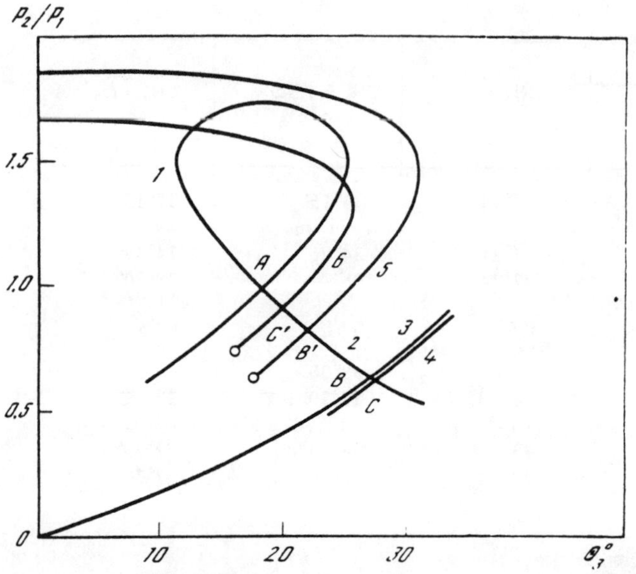

Figure 10.13. Polar diagrams computed for refraction of a detonation wave from mixture $CH_4 + 2O_2$ into mixtures $CH_4 + 4O_2$ and $CH_4 + 3O_2$ (angle of incidence 58°, $p_0 = 1$ atm) *1* is the reflected shock wave; *2* is the reflected expansion wave; *3, 4* is the refracted shock waves; *5, 6* is the recompressed refracted detonation waves.

results of calculations of the version with overdriven detonation wave in the second mixture are also presented for the refraction in mixtures $CH_4 + 4O_2$ and $CH_4 + 3O_2$. The table shows that for the refraction in mixtures $CH_4 + 4O_2$ and $CH_4 + 3O_2$, the calculation gives good correlation with the experiment only when there are no chemical reactions in the immediate vicinity of the point of intersection in the refracted wave. This is a confirmation of the experimental fact that refraction in a mixture of any composition takes place momentarily as does the refraction in an inert medium with a decay of the detonation wave into a shock wave and a combustion zone.

10.5 PROPAGATION OF DETONATION WAVES IN NARROW CHANNELS

Interest in the investigation of the propagation of detonation waves in narrow channels is determined by the fact that heat and frictional losses change the state of the gas significantly behind the detonation wave. One of the first attempts to describe the propagation of detonation, taking into account the losses to the walls of the channel, within the limits of one-dimensional theory was made by Zel'dovich [368], who discussed the extended reaction zone with a particular form of kinetic dependence and assessment of heat losses up to the Chapman-Jouguet plane.

The influence of channel diameter, losses, and the features of propagation of detonation in narrow tubes were investigated both experimentally [435–449] and theoretically [368, 434, 450–458]. Heat losses to the wall were investigated in a number of works: in [443] for regimes far from the limits, the friction coefficient and pressure profile were determined from the measurements of heat flow through the wall of the tube behind the Chapman-Jouguet plane. The effect of the boundary layer on the flow behind the wave was discussed in [453]. When the role of heat release is the determining factor during propagation of detonation, heat losses which tend to decrease effective heat release play a large part. As a result, in channels of different diameters detonation limits ensue at some minimal value leading to detonation attenuation. Therefore, of special interest are the detonation regimes near the limits, when the effect of the walls (heat transfer to the wall, friction on the wall) is particularly apparent. Such investigation was conducted for the detonation of gaseous hydrazoic acid HN_3 in tubes of diameter reaching 1 mm at an initial pressure from 1 to 20 mm Hg in [446, 447, 449]. The results of these works will serve as the foundation for the following discussion.

The use of HN_3 for the investigations is expedient for the following reasons:

—High energy release (300 kJ/mole) allows one to conduct the investigations in tubes of small diameter (*d* to 1 mm) when the effect of the walls is particularly pronounced;

—The possibility to eliminate errors which are related to the incorrect determination of the initial concentrations, since detonation occurs as a result of decomposition of HN_3.

Detonation of HN_3 was investigated in the installations of the Center for the investigation of chemistry of combustion and high temperatures in the city of Orleans in calibrated tubes 3.5 m long arranged parallel to each other. The tubes were made of pyrex glass with various internal diameters $d = 1, 2, 5$, and 10 ± 0.01 mm and polished internal surfaces. The tubes were connected to a common initiation chamber of 50 mm diameter and 320 mm long allowing simultaneously obtaining detonation wave in four detonation tubes of small diameter. Detonation in the initiation chamber was caused by an electric charge. Each detonation tube had mounting stations roughly 1 m apart for fitting transducers flush with the internal surface.

In order to determine heat losses to the wall of the tube thin film resistance probes were used, and the pressure behind the detonation wave was measured by piezoelectric transducers.

Velocity measurement was made from the signals of the thin film probes or piezoelectric transduces, and also by registering luminescence of the front of the detonation wave. The use of streak records allowed one to register the appearance of spin and sharp change of velocity upon wave attenuation. The structure of the front of the detonation wave was determined by the soot trace method. The measurement of heat flow, pressure, and registration of soot imprints were made at three positions along the tube length *x* at distances 1.4; 2.4; 3.4 m from the ignition chamber.

Computation of detonation of HN_3 in the absence of and accounting for heat losses was carried out by C. Paillard [449] for the equilibrium state of the gas behind the detonation wave in HN_3 at the initial pressures $p_0 = 1$–5, 10, 15, 20, and 24 mm Hg and initial temperature $T_0 = 298$ K. The following system of chemical reactions was considered

$$HN_3 \rightarrow 0.5\,H_2 + 1.5\,N_2$$
$$H_2 \rightleftarrows 2H$$
$$N_2 \rightleftarrows 2N$$
$$N_2 + H_2 \rightleftarrows 2HN$$
$$N_2 + 2H_2 \rightleftarrows 2NH_2$$
$$N_2 + 3H_2 \rightleftarrows 2NH_3$$
$$3N_2 \rightleftarrows 2N_3$$

The system of equations to be solved included the equations for the balance, equilibrium equations, equation of state, and gasdynamic equations. The system was completed by the Chapman-Jouguet condition on "frozen" speed of sound. The thermodynamic data were taken from handbooks [5, 461].

Given in Table 10.6 are the gas parameters behind the detonation wave in HN_3 in the Chapman-Jouguet plane with total heat release of $H = 300$ kJ/mole. Presented are the temperatures (T, K) pressure (p/p_0), density (ρ_0/ρ), velocity of detonation propagation (D, m/sec), and also the mole fractions of the products of detonation. Furthermore, the table shows the values of the speed of detonation D_∞ for infinitely large tube diameter obtained by extrapolating the experimental values of velocity D_{exp}.

For the investigation of the effect of losses in the energy equation the magnitudes of heat release were varied from 300 to 165 kJ/mole. The results of the calculation of detonation parameters with heat losses for the initial pressures $p_0 = 1$ and 20 mm Hg are given in Table 10.7.

It was noted in [434] that heat losses, which are noticeable in comparison with heat release in the Chapman-Jouguet plane, do not lead to significant

Table 10.6

p_0	T	p/p_0	ρ_0/ρ	D	D_∞
1	3355	49.30	0.5471	2478	–
2	3468	50.60	0.5479	2531	–
3	3537	51.43	0.5480	2535	–
4	3588	52.03	0.5482	2550	2752
5	3629	52.55	0.5478	2562	2754
10	3759	53.94	0.5491	2600	2760
15	3839	54.85	0.5492	2623	2765
20	3898	55.59	0.5487	2639	2770
24	3935	55.88	0.5499	2650	2779

changes in the parameters as a result of the shift of equilibrium state toward non-dissociating products effectively compensating for the influence of heat losses from the reaction zone. In the case of HN_3 a shift of the equilibrium composition is also observed; however, the relative change of detonation velocity and the parameters in the Chapman-Jouguet plane is more pronounced than for mixtures of hydrocarbons and hydrogen with oxygen.

Analysis of the results of experimental investigations and comparison with calculations reveals a number of features of NH_3 detonation in the range of values of the initial parameters (d, p_0) under consideration.

Detonation velocity. When p_0 is decreased in each of the tubes under consideration the presence of a linear dependence of D_{exp} on lg p_0 is observed up to the value of the initial pressure when detonation becomes a spinning detonation. When the pressure p_0 is further decreased, velocity D_{exp} drops according to another more strongly linear dependence of lg p_0 and a velocity 25% lower than D_∞ is recorded near the detonation limits. The magnitude of D_∞, obtained by extrapolation of the dependence of D_{exp} on the inverse diameter $1/d$ for $1/d \to 0$, characterizes the propagation of detonation in the absence of the effect of the tube wall. The value of D_∞ is found to be considerably higher than the velocity D with complete heat release. One of the reasons for this could be the non-monotonous character of heat release with respect to time [368, 459].

In order to compare the experimental data with calculations, a comparison of the values of the deviation of D_{exp} from D_∞ was conducted with the velocity difference obtained in the calculation for complete (300 kJ/mole) and decreased magnitudes of heat release (see Table 10.7). This allowed one to compare the value of heat release decrease with the heat losses recorded during the experiment, and reveal the magnitude of heat losses which determine the drop in detonation velocity by the value D_∞-D_{exp}.

The experimentally obtained parameters were compared with the calculation for the initial pressures that only exceed the value p_0 and correspond to the

Compound							
H_2 (10)	N_2	NH(10^4)	NH_2(10^5)	NH_3(10^7)	N_3(10^{10})	H	N(10^3)
0.437	0.6260	0.0969	0.0134	0.0279	0.0256	0.3300	0.332
0.459	0.6294	0.154	0.0264	0.0656	0.0657	0.3206	0.408
0.531	0.6315	0.202	0.0393	0.108	0.114	0.3149	0.461
0.558	0.6331	0.245	0.0521	0.154	0.169	0.3105	0.502
0.579	0.6343	0.284	0.0647	0.202	0.229	0.3072	0.537
0.650	0.6385	0.447	0.127	0.471	0.586	0.2959	0.655
0.694	0.6410	0.581	0.187	0.771	1.01	0.2888	0.734
0.726	0.6429	0.702	0.246	1.09	1.50	0.2873	0.797
0.748	0.6442	0.786	0.293	1.37	1.90	0.2801	0.835

Table 10.7

H	p/p_0	T	D	Compound	
				H_2(10)	N_2
		p_0 = 1 mm Hg			
300	49.3	3355	2478	0.437	0.6260
285	47.7	3279	2453	0.553	0.6330
270	46.3	3211	2394	0.676	0.6404
255	45.0	3147	2355	0.804	0.6481
240	43.6	3086	2316	0.937	0.6562
225	42.3	3026	2279	1.075	0.6644
210	41.0	2967	2241	1.216	0.6729
195	39.7	2908	2203	1.361	0.6836
180	38.3	2845	2164	1.508	0.6905
165	36.9	2779	2125	1.657	0.6994
		p_0 = 20 mm Hg			
300	55.6	3898	2639	0.726	0.6429
285	53.8	3813	2595	0.843	0.6501
270	52.1	3731	2551	0.965	0.6575
255	50.5	3651	2508	1.092	0.6652
240	48.9	3572	2465	1.221	0.6731
225	47.3	3492	2422	1,353	0.6811
210	45.6	3410	2379	1.489	0.6892
195	44.0	3326	2334	1.625	0.6975
180	42.2	3236	2288	1.761	0.7057
165	40.5	3141	2240	1.897	0.7138

transition to spin and to the above mentioned breaks in the dependence of D_{exp} on lg p_0. This means that only the region of existence of detonation with a multi-headed front was discussed.

Pressure measurements at various locations from the beginning of the tube (x = 1.4; 2.4; 3.4 m) in a tube of 10 mm diameter revealed that the pressure distribution was different at various distances [447], and for x = 3.4 m a stationary profile was produced. Measurements made in various tubes revealed a strong dependence on the diameter of the tube. Oscillograms obtained by a transducer located at distance x = 3.4 m showed that the pressure in the widest tube with 10 mm diameter and in the 5 mm diameter tube was only slightly less than the Chapman-Jouguet pressure. In 2 and 1 mm diameter tubes, on the other hand, the peak pressure was noticeably lower than the Chapman-Jouguet pressure. The time for reaching maximum pressure rises with decreasing diameter of the tube, and is equal to 1 and 2 μsec for a tube of 10 mm diameter and 10 μsec for a tube of 2 mm diameter. Irrespective of the diameter, two distinctive regions are revealed in the pressure profiles: a region where the pressure gradient is considerable, and a region of small pressure drop commencing after longer time interval in narrow tubes. Comparison was made of experimental pressure profiles, determined for p_0 = 10 mm Hg, d = 10 mm, x = 3.4 m, and calculated profiles obtained for three cases: isentropic flow,

Compound				
NH(10^5)	NH_2(10^6)	NH_3(10^8)	H	N(10^4)
		p_0 = 1 mm Hg		
0.969	0.134	0.28	0.3300	3.32
0.826	0.146	0.40	0.3115	2.28
0.701	0.155	0.55	0.2918	1.59
0.595	0.161	0.72	0.2714	1.13
0.499	0.164	0.92	0.2501	0.789
0.414	0.164	1.1	0.2280	0.564
0.338	0.161	1.4	0.2054	0.393
0.271	0.155	1.7	0.1824	0.269
0.210	0.146	2.0	0.1587	0.178
0.158	0.134	2.3	0.1349	0.113
		p_0 = 20 mm Hg		
7.02	2.46	10.9	0.2837	7.79
5.99	2.52	13.7	0.2649	5.82
5.10	2.55	16.9	0.2455	4.27
4.27	2.53	20.4	0.2252	3.09
3.53	2.48	24.2	0.2045	2.22
2.88	2.39	28.5	0.1834	1.56
2.29	2.26	33.1	0.1618	1.07
1.76	2.10	38.1	0.1399	7.10
1.32	1.90	43.6	0.1181	4.51
0.933	1.67	49.6	0.0965	2.68

steady, and unsteady flows. The isentropic solution did not agree at any part with the experimental values. Steady and unsteady solutions were computed with the assumption that the friction coefficient is constant. In large tubes (10 mm) for time $t < 20$ μsec good correlation was observed between the experimental profiles and the steady and unsteady profiles which resemble each other. For $t > 20$ μsec discrepancy between the theory and the experiment was large, particularly for the unsteady solution.

Downstream of the front the degree of pressure drop in the experiment is less than the drop predicted by the steady theory. Deviation is caused by the fact that in this region the assumptions made in the computations are incorrect, since the adiabatic index and gas composition change with temperature. For a tube with 2 mm diameter the experimental profile differs considerably from the observed profile in large-diameter tubes: the peak pressure is significantly wider and its magnitude much lower than the Chapman-Jouguet pressure.

Heat flow to the wall was calculated from the temperature change of each thin-film sensor, the temperature determined from the oscillograph traces of voltage change at the tips of the sensors. At elevated pressures and for wider tubes the signal rises quickly, reaches a plateau then drops slowly, whereas in narrow tubes, a spike is observed in the first microseconds followed by a sharp drop of the signal. At low pressures two types of signals are observed near the limits: either with a flat maximum, or with a sharp maximum and quick drop of the signal down to the value observed under normal conditions, which corresponds to the passage of various parts of the front of spinning detonation near the sensor.

The calculated heat flow $q(t)$, which rises with tube diameter d, usually passes through a maximum during time t_{max} which is shorter the larger the diameter of the tube: for tube $d = 1$ mm $t_{max} = =5$ μsec instead of $t_{max} = 1$–2 μsec for $d = 2.5$ and 10 mm at $F_0 = 10$ mm Hg.

The total quantity of heat Q absorbed by the sensor up to the time t is computed by integrating the function $q(t)$ starting from some chosen moment θ which is close to the corresponding moment of initial temperature jump.

The measurement of the magnitude of heat losses to the wall conducted in the work allowed one to evaluate the time of passage of the reaction zone (from the shock front to the Chapman-Jouguet plane) near the fixed section of the detonation tube. This estimate is considered high, since it was carried out without directly accounting for the losses due to friction. Integration of the rate of heat transfer and subsequent transformations with the assumption that density changes insignificantly behind the front lead to the deduction of the dependences of the quantity of heat lost to the wall per unit area Q on time for $p_0 = 10$ and 20 mm Hg and tubes of diameters 10, 5, and 2 mm. When the diameter of the tube is decreased, heat losses decrease in correspondence with the decreasing rate of heat transfer, which was determined with the help of a thin-film resistance sensor.

The obtained dependence of heat losses allowed one to compare them with the losses adopted in the calculations and therefore to assess the time of passage (and consequently the length) of the reaction zone during HN_3 detonation in narrow tubes. In [449] an assessment of the duration t of the passage of the reaction zone has been carried out as follows: from the magnitude of the deviation of the experimental velocity from D_∞ and the dependence of the decrease of detonation velocity with decreasing heat release the magnitude of heat losses was determined. Comparison of the latter with the experimental dependences of heat losses on time gives the time required for the passage of the reaction zone. For tubes with diameter 10 and 5 mm time t amounts to 5.9 and 8.5 μsec for an initial pressure of 10 mm Hg, and 3.3. and 5 μsec for an initial pressure of 20 mm Hg, respectively.

In the case of detonation in a tube with 2 and 1 mm diameter, the attempts to assess the time for the passage of the reaction zone were not successful, since the decrease of detonation velocity corresponds to large heat losses which are not realized during the time of observation.

Significant heat losses must influence the character of the dependence of the flow parameters behind the Chapman-Jouguet plane on time. To verify the effect of heat losses, calculations of flow parameters were conducted from the experimental pressure profile. These parameters included the stagnation temperature relative to the wall. From the change of this temperature, heat losses were computed and compared with the experimental dependence of Q on time.

Calculations were based on the assumption that the Chapman-Jouguet computation parameters (taking heat losses into account) are implemented in the plane corresponding to the obtained time t. Calculations were conducted for sufficiently large tube diameters and initial pressures when a regime of detonation propagation with multi-headed front was deliberately created. Then, at a moment $t_1 = \sim t, p_{1'\mathrm{exp}} = p$, the flow Mach number relative to the detonation front is equal to $M_B = 1$; for the remaining parameters see Tables 10.6, 10.7. The flow parameters during some moment $t_2 > t$ are determined as follows:

Pressure p_2—from an experimental dependence;

Flow Mach number M_{2B} relative to the front from the dependence

$$(\gamma M_{2B}^2 + 1) p_2 = (\gamma M_{1B}^2 + 1) p_1$$

Temperature for the moment of time t_2

$$t_2 \; T_1 (\gamma M_{1B}^2 + 1)/M_{1B}^2 = T_2 (\gamma M_{2B}^2 + 1)/M_{2B}^2$$

Following the determination of the flow Mach number M_{CT}, relative to the wall of the channel, we determine the deceleration temperature T_{st} for t_1 and t_2.

Using the calculated values of the specific heat c_p and molecular mass m of the detonation products, and also the mean value of the density ρ, we obtain the value of heat losses ΔQ for a mass of gas trapped between the sections corresponding to moments t_1 and t_2:

$$\Delta Q = (T_{st1} - T_{st2}) \, c_p d\bar{\rho}/4\overline{m}$$

Furthermore, calculation is conducted in the same manner taking time t_2 as an initial value and so forth. Comparison of the values of ΔQ with the experimental change of Q indicated satisfactory agreement of the calculation with the experimental curve of heat losses to the wall for $p_0 = 10$ mm Hg and $d = 10$ mm, which is an indication of their determining role in the variation of the parameters behind the detonation front [449].

The present calculation assumes that the parameters change only as a result of heat losses. This is justified near the front and away from the point of detonation initiation, which plays an important role in the process of detonation propagation in narrow tubes.

10.6 REFLECTION OF AN UNSTEADY SYSTEM: SHOCK WAVE-COMBUSTION ZONE

The reflection of detonation waves, which are formed when a shock wave is reflected in a burning mixture, has been discussed previously. The pressure on the reflecting surface, in this case, increases 2–3 times in comparison with the pressure behind the incident detonation wave. The reflection of a shock wave in a combustible mixture in unsteady shock wave-combustion zone systems is of great interest [462–464]. Such systems appear during refraction and diffraction of shock waves, and also during the formation of detonation in tubes. With this, it is possible to increase the pressure on the reflecting surfaces which is related to the unsteady process of ignition of the gas heated beforehand in the shock wave.

From the investigation of the induction period of the combustible mixture by heating the gas in the reflected wave from the end, it was shown in [402, 403] that ignition behind the shock wave, in mixtures far from their limits, does not begin from just one plane behind the front of the shock wave but from a number of ignition spots. Explosions of moderate gas volumes take place at these spots. The detonation fronts from the explosions catch up with the reflected wave and, by merging with it, form a multi-headed detonation wave. The process of formation of detonation behind the shock wave takes place similarly when the low-pressure channel of the shock tube is filled with the combustible mixture [372]. When the residence time of the gas at high temperature behind the wave becomes comparable with the induction period, hot spots appear in the gas flow behind the shock wave followed by microexplosions, the leading front of which overtakes the shock wave front and merges with it. Pressure transducers recorded sharp pressure oscillations in the flow.

The unsteady shock wave-combustion zone system is always formed upon appearance of detonation in tubes. The process of flame propagation in a tube and transition of combustion into detonation has been studied in sufficient detail with the help of high-speed photography by the trace method, and also with the help of measurements of pressure and gas velocities between the shock wave and combustion front [372, 465–467].

Investigations showed that upon ignition of the combustible mixture in a tube, the flame front propagates, accelerating gradually, and, by acting like a piston, causes the formation of one or several shock waves one after the other ahead of it. Depending on the mixture composition, tube diameter, and roughness of tube walls, various laws of front acceleration and, accordingly, distribution of flow parameters are feasible between the front of the first shock wave and the flame front. Transition to detonation takes place as a result of explosion of a hot gas volume which is heated by the shock wave and trapped between the shock wave and the combustion zone. Explosion can start at various places depending on the distribution of the parameters, for example, in the gas between the shock wave and the combustion zone. Part of the created wave is detonation wave, which travels forwards along the tube behind the shock wave and overtakes it, while the other part travels backwards, forming a retonation wave. Here, the detonation wave, as pressure and velocity measurements indicate, is always formed as an overdriven wave with noticeable pressure rise behind it. Pressures of 80 atm have been recorded in a stoichiometric mixture of oxygen with hydrogen in [466], behind the detonation wave at the moment of its formation. This pressure is 4 times higher than the pressure behind the Chapman-Jouguet wave in the same mixture. The velocity of the wave then decreases quickly and the pressure drops to the value in the Chapman-Jouguet wave. If one assumes that the length of the tube is such that the formation of detonation occurs near the end of the tube, then at the moment of reflection of the Chapman-Jouguet wave from far end wall, the pressure at that end wall increases by 12 times.

Such reflection regimes are described in [462], in which investigation of the transition of detonation waves from a tube with 94 mm diameter into a tube with 305 mm and 10 m long has been conducted. The mixture was ignited in the narrow tube where the formation of detonation wave took place, then the wave egressed into the wide tube where decay of the detonation wave and, in some cases, subsequent restoration of detonation occurred depending on the composition of the mixture. Detonation did not form in mixture $CH_4 + 2O_2 + 4N_2$ in the wide tube, and the pressure at the end of the tube reached 332 atm; the calculated pressure upon reflection of the Chapman-Jouguet detonation amounted to 51 atm. The hydrogen-air mixture, containing 20% hydrogen and 20% air, was ignited by detonation in the narrow tube, filled with $2H_2 + O_2$. The recorded pressure at the end of the tube was 105 atm, which is 3 times greater than the pressure of reflection of steady detonation. The detonation wave in a mixture of 8.6% acetylene and 91.4% air did not collapse in the wide tube but propagated at a constant velocity of 1800 m/sec. When the mixture composition was changed slightly to 6.8% acetylene and 93.2% air, the tube end was destroyed. Streak records revealed that the detonation wave was formed in the narrow tube and decayed in the wide tube, the flame velocity increasing from 487 m/sec near the transition region to 880 m/sec at a distance of 2.1 m from the closed end. According to approximate estimates, the pressure which caused flange break off was about 6 times higher than the pressure of the Chapman-Jouguet wave upon reflection. Apparently, when the mixture composition was changed, the ignition process behind the shock wave in the unsteady system occurred at the end of the tube. In [463], experiments are described on obtaining strong shock waves when the chamber is filled with a high-pressure mixture of $2H_2 + O_2$ (20%) and He (80%). In tests with thick diaphragms the combustion of the mixture took place at constant volume. In this case, shock waves with Mach numbers in the range 11–15 were observed in the low-pressure chamber filled with air. When weaker diaphragms were employed the pressure in the high-pressure chamber decreased; however, the Mach numbers of shock waves in air increased, and waves with Mach number equal to 17.5 were obtained in some regimes. The reason for such acceleration, in these cases, was that the formation of the unsteady system in the high-pressure chamber near the diaphragm took place in different ways. Thus, the unsteady interactions of shock waves can lead to significant redistribution of loads inside the channels.

REFERENCES

1. Landau, L. D. and Lifshits, E. M. Mekhanika sploshnykh sred. (Continuum mechanics). Moscow, Gostekhizdat, 1953, 788 pp.
2. Zel'dovich, Ya. B. and Raizer, Yu. P. Fizika udarnykh voln i vysokotemperaturnykh gidrodinamicheskikh yavleniy (Physics of shock waves and high-temperature hydrodynamic phenomena). Moscow, Nauka, 1963, 688 pp.
3. Tikhonov, A. N. and Samarsky, A. A. Uravneniya matematicheskoy fiziki (Equations of mathematical physics). Moscow, Nauka, 1977, 735 pp.
4. Abramovich, G. A. Prikladnaya gazovaya dinamika (Applied gasdynamics). Moscow, Nauka, 1972, 888 pp.
5. Glushko, V. M., et al., ed. Termodinamicheskiye funktsii individual'nykh veshchestv (Thermodynamic functions of individual substances). Moscow, Izd-vo AN SSSR, 1962, Vol. 1, 916 pp.
6. Bazhenova, T. V. and Gvozdeva, L. G. Hestatsionarnyye vzaimodeystviya udarnykh voln (Unsteady interactions of shock waves). Moscow, Nauka, 1977, 274 pp.
7. Predvoditelev, A. S., Stupochenko, E. V., Pleshanov, A. S., Samuylov, E. V., and Rozhdestvensky, I. B. Tablitsy termodinamicheskikh funktsiy vozdukha dlya temperatur ot 200 do 6000 K i davleniy ot 0.00001 do 100 atm (Tables of thermodynamic functions of air for temperatures from 200 to 6000 K and pressures from 0.00001 to 100 atm). Moscow, V(Ts) AN SSSR, 1962, 268 pp.
8. Predvoditelev, A. S., Stupochenko, E. V., Pleshanov, A. S., Samuylov, E. V., and Rozhdestvensky, I. B. Tablitsy termodinamicheskikh funktsiy vozdukha dlya temperatur ot 6000 do 12,000 K i davleniy ot 0.001 do 1000 atm (Tables of thermodynamic functions of air for temperatures from 6000 to 12,000 K and pressures from 0.001 to 1000 atm). Moscow, Izd-vo AN SSSR, 1958, 301 pp.
9. Predvoditelev, A. S., Stupochenko, E. V., Pleshanov, A. S., Samuylov, E. V., and Rozhdestvensky, I. B. Tablitsy termodinamicheskikh funktsiy vozdukha dlya temperatur ot 12,000 do 20,000 K i davleniy ot 0.001 do 10,000 atm (Tables of thermodynamic functions of air for temperatures from 12,000 to 20,000 K and pressures from 0.001 to 10,000 atm). Moscow, Izd-vo AN SSSR, 1959, 205 pp.

10. Predvoditelev, A. S., Stupochenko, E. V., Pleshanov, A. C., Samuylov, E. V., and Rozhdestvensky, I. B. Termodinamicheskiye funktsii vozdukha dlya T = 1000–12,000 K i davleniy ot 0.001 do 1000 atm (grafiki funktsiy) Thermodynamic functions of air for T = 1000–12,000 K and pressures from 0.001 to 1000 atm (graphs of functions)). Moscow, Nauka, 1969, 55 pp.
11. Kuznctsov, N. M. Termodinamicheskikh funktsii i udarnyye adiabaty vozdukha pri vysokykh temperaturakh (Thermodynamic functions and shock adiabatic curves of air at high temperatures). Moscow, Mashinstroyeniye, 1965, 464 pp.
12. Pleshanov, A. S. Sostav, termodinamicheskiye i fazodinamicheskiye svoystva azota dlya temperatur ot 1000 do 12,000 K i davleniy ot 10^{-4} do 10^3 atm (Composition, thermodynamic, and gasdynamic properties of nitrogen for temperatures from 1000 to 12,000 K and pressures from 10^{-4} to 10^3 atm). V: Fizicheskaya gazodinamika i teploobmen (In: Physical gasdynamics and heat transfer). Moscow, Izd-vo AN SSSR, 1962, pp. 36–42.
13. Pleshanov, A. S. and Zaytsev, S. G. Sostav, termodinamicheskiye i fazodinamicheskiye svoystva uglekislovo gaza dlya temperatur ot 1000 do 12,000 K i davleniy ot 10^{-2} do 10^3 atm (Composition, thermodynamic, and gasdynamic properties of carbon dioxide for temperatures from 1000 to 12,000 K and pressures from 10^{-2} to 10^3 atm). V: Fizicheskaya gazodinamika i teploobmen (In: Physical gasdynamics and heat transfer). Moscow, Izd-vo AN SSSR, 1962, pp. 15–24.
14. Predvoditelev, A. S., Stupochenko, E. V., Rozhdestvensky, I. B., and Samuylov, E. V. Tablitsy termodinamicheskikh i gazodinamicheskikh velichin potoka vozdukha za pryamym skachkom uplotneniya dlya skorostey nabegayushchevo potoka do 4500 m/s (Tables of thermodynamic and gasdynamic quantities of air flow behind a normal shock wave for free-stream velocities up to 4500 m/sec). Moscow, V(Ts) AN SSR, 1962, 132 pp.
15. Predvoditelev, E. V., Stupochenko, E. V., Rozhdestvensky, I. B., and Samuylov, E. V. Tablitsy termodinamicheskikh i gazodinamicheskikh velichin potoka vozdukha za pryamym skachkom uplotneniya (Tables of thermodynamic and gasdynamic quantities of air flow behind a normal shock wave). Moscow, Izd-vo AN SSSR, 1959, 78 pp.
16. Bazhenova, T. V., Stanyukovich, A. K., and Uvarov, V. M. Parametry plazmy za pryamym skachkom uplotneniya pri vkhode kosmicheskikh tel v plotniye sloi atmosfery (Plasma parameters behind normal shock wave when space bodies enter the dense layers of the atmosphere). V: Nizkotemperaturnaya plazma v kosmose i na Zemle (In: Low-temperature plasma in space and on earth). Moscow, VAGO AN SSSR, 1977, pp. 247–255.
17. Bazhenova, T. V. and Predvoditeleva, O. A. Znacheniya parametrov vozdukha za pryamym skachkom uplotneniya i za otrazhennoy volnoy pri ravnovesnoy i zamorozhennoy dissotsiatsii (The values of parameters of air behind normal and reflected shock waves under equilibrium and frozen dissociation). V: Fizicheskaya gazodinamika i teploobmen (In: Physical gasdynamics and heat transfer). Moscow, Izd-vo AN SSSR, 1961, pp. 15–24.
18. Belotserkovsky, O. M., ed. Obtekaniye zatuplennykh tel sverkhzvukovym potokom gaza (Flow of supersonic stream of gas over a blunt body). Moscow, V(Ts) AN SSSR, 1966, 338 pp.
19. Lyubimov, A. N. and Rusanov, V. V. Techeniya gaza okolo tupykh tel (Flow of gas near blunt bodies). Moscow, Nauka, 1970, Part 1, 287 pp., Part 2, 379 pp.
20. Lunev, V. V., Magomedov, K. M., and Pavlov, V. G. Giperzvykovoye obtekaniye prituplennykh konysov s uchetom ravnovesnykh fizikokhimicheskikh prevrashcheniy (Hypersonic flow over blunt cones accounting for physico-chemical conversions). Moscow, V(Ys) AN SSSR, 1968, 203 pp.
21. Ginzburg, V. L. Rasprostraneniye elektromagnitnykh voln v plazme (Propagation of electromagnetic waves in plasma). Moscow, Nauka, 1968, 396 pp.
22. Shkarovskiy, I., Johnston, T., and Bachinskiy, M. Kinetika chastits plazmy (Kinetics of plasma particles). Moscow, Atomizdat, 1969, 360 pp.
23. Bazhenova, T. V., Dzagurov, L. Yu., and Lobastov, Yu. S. Vliyaniye zavisimosti secheniy soudareniy ot skorosti elektronov na vzaimodeystviye radiovoln s vozdushnoy plazmoy (Effect of the dependence of collisions cross section on the velocity of electrons on the interaction of radio waves with air plasma). Teplofizika visokikh temperatur (Thermophysics of high temperatures). 1979, Vol. 17, No. 5, pp. 922–927.

24. Gurevich, A. V. and Shvartsburg, A. B. Nelineynaya teoria rasprostraneniya radiovoln v ionosfere (Nonlinear theory of propagation of radio waves in the ionosphere). Moscow, Nauka, 1973, 250 pp.
25. Baker, W. Explosions in air. Austin and London, University of Texas Press, 1973.
26. Stupochenko, E. V., Losev, S. A., and Osipov, A. I. Relaksatsionnyye protsessy v udarnykh volnakh (Relaxation processes in shock waves). Moscow, Nauka, 1965, 484 pp.
27. Losev, S. A. and Polyansky, V. A. Fiziko-khimicheskiy sostav i termodinamicheskiye kharakteristiki neravnovesnovo techeniya za frontom udarnoy volny v vozdukhe (Physico-chemical composition and thermodynamic characteristics of nonequilibrium flow behind a shock-wave front in air). Moscow, Tr. NII mekhaniki MGU, 1970, No. 3, pp. 103–105.
28. Biberman, L. M., Mnatsakanyan, A. Kh., and Yakubov, I. T. Ionizatsionnaya relaksatsiya za udarnymi volnami (Ionized relaxation behind shock waves). UFN, 1970, Vol. 102, Issue 3, pp. 431–462.
29. Kondratiyev, V. N., et al. Spravochnik konstant khimicheskikh reaktsiy (Handbook on constants of chemical reactions). Moscow, Nauka, 1970, 351 pp.
30. Blackman, V. Kolebatel'naya relaksatsiya v kislorode i azote (Vibrational relaxation in oxygen and nitrogen). J. Fluid Mech., 1956, Vol. 1, Part 5, pp. 61–85.
31. White, D. and Millikan, R. Kolebatel'naya relaksatsiya v vozdukhe (Vibrational relaxation in air). J. Chem. Phys., 1963, Vol. 39, No. 1, p. 98.
32. Guenoche, H. and Billiotte, M. B. Sur la mesure du temps de relaxation de vibration derriere une onde de choc. C. r. Acad. sci., A, 1968, Vol. 293, pp. 2663–2670.
33. Vasiliyev, L. A., Yershov, I. V., and Semenov, S. S. Eksperimental'noye issledovaniye neravnovecnykh protsessov za udarnoy volnoy v vozdukhe i azote tenevym metodom (Experimental investigation of nonequilibrium processes behind a shock wave in air and nitrogen by the shadow method). DAN USSR, 1969, Vol. 186, No. 5, pp. 1041–1044.
34. Griffith, W., Brickle, D., and Blackman, V. Structure of shock waves in polyatomic gases. Phys. Rev., 1956, Vol. 102, pp. 1200–1207.
35. Gaydon, A. and Hurl, I. Udarnaya truba v khimicheskoy fizike vysokikh temperatur (The shock tube in high temperature chemical physics). London, Chapman and Hall, 1963.
36. White, D. and Millikan, R. Oxygen vibrational relaxation in O_2–He and O_2–Ar mixtures. J. Chem. Phys., 1963, Vol. 39, No. 7, pp. 1803–1805.
37. Appleton, J. P. and Steinberg M. Vacuum-ultraviolet absorption of shock-heated vibrationally excited nitrogen. J. Chem. Phys., 1967, Vol. 46, pp. 1521–1529.
38. Appleton, J. P. Shock-tube study of the vibrational relaxation of nitrogen using vacuum-ultraviolet light absorption. J. Chem. Phys., 1967, Vol. 47, pp. 3231–3235.
39. Generalov, N. A. and Losev, S. A. K issledovaniyu yavleniy vozbuzhdeniya kolebaniy i raspada molekul O_2 pri vysokikh temperaturakh (On the investigation of the phenomena of vibrational excitation and decay of O_2 molecules at high temperatures). DAN SSSR, 1961, Vol. 161, No. 5, pp. 1072–1075.
40. Millican, R. and White, D. Systematics of vibrational relaxation. J. Chem. Phys., 1963, Vol. 39, pp. 3209–3215.
41. Losev, S. A. and Yalovik, M. S. Kolebatel'naya relakxatsiya i dissotsiatsiya azota pri vysokikh temperaturakh (Vibrational relaxation and dissociation of nitrogen at high temperatures). Khimiya vysokikh energiy, 1970, Vol. 4, No. 3, pp. 202–205.
42. Taylor, R. L. and Bitterman, S. Survey of vibrational relaxation data for processes important in He–Co_2 laser system. Rev. Mod. Phys., 1962, Vol. 41, pp. 26–32.
43. Bershader, W., Bird, P., and Keifer, J. Density gradient measurements of O_2 dissociation in shock waves. J. Chem. Phys., 1971, Vol. 55, No. 8, pp. 4017–4026.
44. Bazhenova, T. V. Otsenka vremeni relaksatsii dissotsiatsii uglekislovo gaza po opytam v udarnoy trube (Evaluation of relaxation time of disassociation of carbon dioxide from experiments in a shock tube). DAN SSSR, 1962, Vol. 146, No. 3, pp. 554–556.
45. Weihs, D. and Manheimer-Timnat, V. Carbon dioxide dissociation relaxation times measured in a shock tube. AIAA J., 1969, Vol. 7, No. 4, pp. 764–766.
46. Anisimov, S. I., Kuznetsov, N. M., and Nagotov, N. F. Struktura udarnykh voln v vozdukhe

s uchetom kinetiki khimicheskikh reaktsiy i vozbuzhdeniya kolebaniy molekul azota (Structure of shock waves in air taking into account the kinetics of chemical reactions and vibrational excitation of nitrogen molecules). Teplofizika vysokikh temperatur, 1964, Vol. 2, No. 3, pp. 343–377.

47. Feldman, A. Eksperimental'noye izucheniye kinetiki khimicheskikh reaktsiy v vozdukhe pri bysokikh temperaturakh v udarnoy trube (Experimental study of kinetics of chemical reactions in air at high temperatures in a shock tube). J. Fluid Mech., 1957, Vol. 3, pp. 3–8.
48. Bazhenova, T. V., Dushin, V. K., and Kotlyarov, A. D. Opredeleniye konstanty skorosti dissotsiatsii vody (Determination of the velocity constants of dissociations of water). V: Kinetika khimicheskikh reaktsiy (In: Kinetics of chemical reactions). Chernogolovka, OI(Kh)F AN SSSR, 1977, pp. 97–101.
49. Bazhenova, T. V., Gvozdeva, L. G., Lobastov, Yu. C., Naboko, I. M., and Predvoditeleva, O. A. Udarnyye volny v real'nykh gazakh (Shock waves in real gases). Moscow, Nauka, 1968, 198 pp.
50. Epstein, P. Thermodynamics, Pasadena, 1936.
51. Clark, J. and MacChesny, M. Dynamics of real gases. London. Chapman and Hall.
52. Wegener, P., ed. Nonequilibrium flows. New York, Dekker, 1970, Part 2, 254 pp.
53. Broer, J. Characteristics of equation of motion of a reacting gas. J. Fluid Mech., 1958, Vol. 4, pp. 276–285.
54. Stakhanov, I. P. and Stupochenko, E. V. Ob uravneniyakh relaksatsionnoy gidrodinamiki (On the equations of relaxation hydrodynamics). DAN SSSR, 1960, Vol. 134, No. 4, pp. 782–785.
55. Stakhanov, I. P. and Stupochenko, E. V. O strukture liniy Makha v relaksiruyushchikh sredakh (On the structure of Mach lines in relaxation media). DAN SSSR, 1960, Vol. 134, No. 5, pp. 901–904.
56. Moore, F. K. and Gibson, W. E. Propagation of weak disturbances in a gas subject to relaxation effects. J. Aero/space Sci., 1960, Vol. 27, No. 2, pp. 15–21.
57. Ting I Li. Recent developments in the investigation of the dynamics of flow equilibrium dissociated gas. ARS Journ., 1961, No. 2, pp. 170–182.
58. Wood, W. and Kirckwood, W. Sound velocity in reacting system. J. Chem. Phys., 1957, Vol. 27, pp. 596–601.
59. Wood, V. and Parker, F. Structure of centered expansion wave in relaxing gas. Phys. of Fluids, 1958, Vol. 1, No. 3, pp. 230–241.
60. Spens, D. A. Unsteady propagation of a shock wave in relaxing gas. Proc. Roy. Soc., 1961, Vol. A264, pp. 221–233.
61. Vasiliyev, L. A. Tenevyye metody. Direct shadow methods, Moscow, Nauka, 1968, 336 pp.
62. Kharitonov, A. I., Sukhorukikh, V. S., and Fokeyev, V. P. Ustanovka Krantsa-Shardina s fotoregistratorom (The Cranz-Schardin system with photorecorder). V: Fizicheskiye metody issledovaniya prozrachnykh neodnorodnostey (In: Physical methods of investigation of transparent discontinuities). Moscow, MDNTP, 1984, pp. 43–44.
63. Gvozdeva, L. G. and Zhilin, Yu. V. P'ezoelektracheskiy datchik davleniya (Piezoelectric pressure transducer). Patent 529385 (USSR), Published in B. I., 1976, No. 35, p. 104.
64. Constantinescu, I. and Dumitrescu, L. Z. Compensation of errors introduced by connecting ducts at pressure measurements in short-duration aerodynamic facilities. In: Shock tube research. London, Chapman and Hall, 1971, pp. 16.1–16.6.
65. Polyakov, Yu. A. Issledovaniye nestatsionarnovo teploobmena pri otrazhenii udarnykh voln (Investigation of unsteady heat transfer with reflection of shock waves). Teplofizika vysokikh temperatur, 1965, Vol. 3, No. 6, pp. 35–41.
66. Polyakov, Yu. A. and Mitkina, E. A. Tonkoplenochniy termometr soprotevleniya (Thin-film resistance thermometer). PTE, 1961, No. 4, pp. 140–142.
67. Hanson, R. Study of gas-solid interaction. In: Shock tube research. London, Chapman and Hall, 1971, pp. 58.1–58.7.
68. Vallrath, K. and Thomer, O., eds. High-speed physics. St. Louis, 1967.
69. Zhilin, Yu. V. Vliyaniye teplofizicheskikh svoystv materiala stenki na pokazaniya plenochnovo termometra soprotivleniya (Effect of thermophysical properties of wall material on the readings of film resistance thermometer). Prepr. IVTAN, No. 2–004, Moscow, 1976, 25 pp.

70. Shtemenko, L. S. Vozniknoveniye udarnykh voln v udarnoy trube v protsesse raskrytiya diafragmy (The formation of shock waves in a shock tube during the opening of the diaphragm). Vestn. MGU, Ser. 3, Fizika, astronomiya, 1982, Vol. 23, No. 3, pp. 27–34.
71. Tkachenko, B. K. Osobennosti vysokotemperaturnykh techeniy v udarnykh trubakh i vazbros eksperimental'hykh dannykh (Specific features of high-temperature flows in shock tubes and scatter of experimental data). V: Vysokotemperaturnaya gazodinamika, udarnyye truby i undarnyye volny (In: High-temperature gasdynamics, shock tubes, and shock waves). Minsk, ITMO im. A. V. Likova, AN USSR, 1983, pp. 46–51.
72. Freeman, N. C. On the stability of plane shock waves. J. Fluid Mech., 1957, Vol. 2, pp. 397–411.
73. Lighthill, M. J. The diffraction of a blast. Proc. Roy. Soc., London A, 1949, Vol. 198, No. 1055, pp. 454–468.
74. Lapworth, K. C. An experimental investigation of the stability of plane shock waves. J. Fluid Mech., 1959, Vol. 6, Part 3, pp. 469–475.
75. Liepman, H. W. and Bowman, R. M. Shape of shock fronts in shock tubes. Phys. Fluids, 1964, Vol. 7, pp. 2013–2018.
76. Weynants, R. R. An experimental investigation of shock-wave diffraction over compression and expansion corners. Toronto, 1968, 55 pp.
77. Ryazin, A. P. Ionizatsionnaya neustoychivost' udarnoy volny v ksenone (Ionized instability of a shock wave in xenon). Letters in ZhTF, 1980, Vol. 6, pp. 516–522.
78. Tumakayev, G. K. Sistematizatsiya anomal'nykh yavleniy v udarno-nagretykh ognoatomhykh gazakh. Priroda ikh vozniknoveniya i razvitiya (Systematizing of anomalous phenomena in shock-treated monatomic gases. Nature of their appearance and development). V: Vysokotemperaturnaya gazodinamika, udarnyye truby i udarnyye volny (In: High-temperature gasdynamics, shock tubes, and shock waves). Minsk, ITMO im. A. V. Likova, AN BSSR, 1983, pp. 154–160.
79. Demiyanov, Yu. A. Formirovaniye pogranichnovo sloya na plastine s dvizhushchimsya skachkom uplotneniya (Formation of boundary layer on a plate with moving shock wave). PMM, 1957, Vol. 21, Issue 3, pp. 368–374.
80. Demiyanov, Yu. A. Vliyaniye pogranichnovo sloya na kharakter techeniya gaza v trube na dvizhushchesya udarnoy volnoy (Effect of boundary layer on the character of gas flow in a tube behind a moving shock wave). PMM, 1957, Vol. 21, Issue 4, pp. 473–477.
81. Hartunian, R. A. and Marrone, P. Fluids viscosity of dissociated gases from shock-tube heat-transfer measurements. Phys. Fluids, 1961, Vol. 4, No. 5, pp. 535–543.
82. Bazhenova, T. V. Izmeneniye skorosti potoka za skachkom v udarnoy trube (Variation of flow velocity behind a shock in a shock tube). V: Fizicheskaya razodinamika i teploobmen (In: Physical gasdynamics and heat transfer). Moscow, Izd-vo AN SSSR, 1961, pp. 31–35.
83. Mirels, H. Test time in low-pressure shock tubes. Phys. of Fluids, 1963, Vol. 9, No. 5, 1201 pp.
84. Mirels, H. Test time in low-pressure shock tubes. Phys. of Fluids, 1963, Vol. 6, No. 9, pp. 1201–1220.
85. Mirels, H. Shock-tube test time limitation due to turbulent-wall boundary layer. AIAA Journ., 1964, Vol. 2, No. 1, pp. 114–126.
86. Breeze, J. C. and Ferriso, C. C. Duration of laminal flow in a shock wave boundary layer. Phys. Fluids, 1964, Vol. 7, No. 7, pp. 1071–1073.
87. Mirels, H. and Mullen, J. F. Small perturbation theory for shock tube attenuation and non-uniformity. Phys. Fluids, 1964, Vol. 7, No. 8, pp. 1208–1218.
88. Demiyanov, Yu. A. and Kireyev, V. T. K analizu odnomernykh nestatsionarnykh techeniy gaza s uchetom teploprovodnosti i vyazkosti (On the analysis of one-dimensional unsteady gas flows accounting for thermal conductivity and viscosity). Izv. AN SSSR, MZhG, 1965, No. 2, pp. 3–7.
89. Jeanmaire, P. and Brocher, E. Influence of turbulent boundary layer on shock tube test time. AIAA Journ., 1965, Vol. 3, No. 5.
90. Mirels, H. and King, W. Series solutions for shock tube laminar boundary layer and test time. AIAA Journ., 1966, Vol. 4, No. 5.

91. Mirels, H. Correlation formulas for laminar shock tube boundary layer. Phys. Fluids, 1966, Vol. 9, No. 7, pp. 1265–1272.
92. Mirels, H. Flow nonuniformity in shock tubes operating at maximum test times. Phys. Fluids, 1966, Vol. 9, No. 10, pp. 1907–1912.
93. Zaytsev, S. G., Lazareva, E. V., and Chebotareva, E. I. Issledovaniya pogranichnovo sloya za udarnoy volnoy v Ar i N_2 (Investigation of boundary layer behind a shock wave in Ar and N_2). V: Svoystva gazov pri vysokikh temperaturakh (In: Properties of gases at high temperatures). Moscow, Nauka, 1967, pp. 142–150.
94. Ackroyd, J. On the laminar compressible boundary layer induced by the passage of a plane shock wave over a flat wall. Proc. Cambridge Philos. Soc., 1967, Vol. 63, No. 3, pp. 889–907.
95. Sheer, R. E. and Nagamatsu, H. T. Methods for distinguishing type of hypersonic boundary layer in shock tunnel. New York, 1968, 9 pp. (AIAA paper, No. 50).
96. Michel, R. and Mentre, P. Approximate study of the boundary layer and of the heat transfer at the wall of a shock tube (Abstr.). Phys. Fluids, 1969, Vol. 12, No. 5, pp. 181–182.
97. Hsu Chen-chi. Boundary-layer growth behind a plane wave. AIAA Journ., 1969, Vol. 7, No. 9, pp. 1810–1812.
98. Mirels, H. Boundary-layer growth effects in shock tubes. In: Shock tubes research. London, Chapman and Hall, 1971, pp. 6.1–6.30.
99. Walker, J. D. A. and Dennis, S. C. R. The boundary layer in a shock tube. J. Fluid Mech., 1972, Vol. 56, No. 1, pp. 19–47.
100. Bazhenova, T. V., Naboko, I. M., and Nemkov, R. G. Experimental study of the boundary layer effects on the distribution of flow paraments behind a shock wave in a shock tube. In: Shock tubes. Toronto, 1969, pp. 60–67.
101. Light, G. G. Test gas properties behind a decelerating shock wave in a shock tube. Phys. Fluids, 1973, Vol. 16, No. 5, pp. 624–628.
102. Shsuke, L. and Akio, H. Shock attenuation in a shock tube due to boundary layer. Trans. Jap. Soc. Aeronaut. and Space Sci., 1973, Vol. 15, No. 30, pp. 193–207.
103. Bazhenova, T. V., Naboko, I. M., Yeremin, A. V. and Kochnev, V. A. Test time behind reflected wave in a shock tube with nozzle. In: Recent developments in shock tube research. Stanford, 1973, pp. 474–484.
104. Yeremin, A. V., Naboko, I. M. and Opara, A. I. Polnoye rabocheye vremya v udarnoy trube pri issledovanii istecheniya iz otverstiya v tortse (Complete working time in a shock tube when investigating the flow from an opening in the tube end). Teplofizika vysokikh temperatur, 1973, Vol. 11, No. 4, pp. 823–831.
105. Dumitrescu, L. Z., Brun, R., and Sides, Y. Computations of gas parameters and compensation of attenuation effects in real shock tube flows. In: Recent developments in shock tube research. Stanford, 1973, pp. 439–448.
106. Dumitrescu, L. Z. An attenuation-free shock tube. Phys. Fluids, 1972, Vol. 15, pp. 207–212.
107. Zeitoun, D. and Brun, R. Methode de calcul approche de la couche limite laminaire hors d'equilibre vibratoire. J. Heat Mass Transfer, 1973, Vol. 16, pp. 427–435.
108. Brabbs, T. A. and Belles, F. E. Experimental study of effects of laminar boundary layers on chemical kinetic measurements in a shock tube. In: Shock tube research. London, Chapman and Hall, 1971, pp. 24.1–24.10.
109. Vasiliyeva, R. V., Zuyev, A. C., and Mirshanov, D. N. O raspredelenii tolkayemovo gaza v udarnoy trube i struktura probki za sil'nymi udarnymi volnami (On the distribution of the driven gas in a shock tube and the structure of the contact region behind shock waves). ZhTF, 1979, Vol. 49, Issue 2, pp. 419–426.
110. Zeitoun, D., Brun, R., and Valetta, M. J. Shock-tube flow computation including the diaphragm and boundary layer effects. In: Shock tubes and waves., Proc. 12th Intern. Symp., 1980, pp. 180–186.
111. Zuyev, A. D. and Mirshanov, D. N. Primeneniye rentgenovskoy diagnostiki dlya izucheniya probki udarnoczhatovo gaza i oblasti kontaktnovo peremeshivaniya (The use of X-ray diagnostics for the study of contact region of shock-compressed gas and region of contact mixing). Teplofizika vysokikh temperatur, 1981, Vol. 19, No. 5, pp. 1111–1114.

112. Deckker, B. E. L. Boundary layer on a shock tube wall and at a leading edge using schlieren. In: Intern. symp. flow visualization. Bochum, 1980, pp. 555–559.
113. Merzkirch, W. Boundary layer visualization in a shock tube In: Recent developments in theoretical and experimental fluid mechanics. Berlin, Springer-Verlag, 1979, pp. 137–143.
114. Savrov, S. D., Ageyev, I. M., and Dolgikh, E. A. K voprosu o mekhanizme perenosa izlucheniya v argonovoy i ksenonovoy plasme na fronte udarnoy volny (On the mechanism of radiation transport in argon and xenon plasma at the shock-wave front). ZhTF, 1981, Vol. 51, Issue 8, pp. 1707–1709.
115. Griffiths, V. Reflection of a shock wave from the end of the shock tube. JAS, 1956, Vol. 23, No. 1, pp. 19–22.
116. Mark, H. The interaction of a reflected shock wave with the boundary layer in a shock tube. JAS, 1957, Vol. 24, No. 4, pp. 304–306.
117. Strehlow, R. and Cohen, A. Limitations of the reflected shock technique for studying fast chemical reactions and its application to the observation of relaxation in nitrogen and oxygen. J. Chem. Phys., 1959, Vol. 30, No. 1, pp. 257–264.
118. Tonies, J. P. and Greene, E. F. Dissociation of nitrogen in shock waves. J. Chem. Phys., 1955, Vol. 23, pp. 1366–1373.
119. Skinner, G. B. Limitation of the reflected shock tube technique for studying fast chemical reactions. J. Chem. Phys., 1959, Vol. 31, p. 268–269.
120. Byron, S. and Rott, N. On the interaction of the reflected shock wave with the laminar boundary layer on the shock tube walls. In: Proc. of the heat transfer and fluid mechanics institute. Stanford, 1961, pp. 38–54.
121. Holder, D. W., Stuart, C. M., and North, R. Y. Reflected shock tunnel. Brit. Aeronaut. Res. Counc. Rep., 1961, No. 22, pp. 891–901.
122. Dumitrescu, L. Z., Popescu, C., and Brun, R. Experimental studies of the shock reflection and interaction in a shock tube. In: Shock tubes: Proc. 7th Intern. shock tube symp. Toronto, 1969, pp. 751–770.
123. Davies, L. and Wilson, J. L. Influence of reflected shock and boundary layer interaction on shock-tube flow. Phys. Fluids, 1969, Vol. 12, No. 5, Part 2, pp. 37–43.
124. Sanderson, R. Interpretation of pressure measurements behind a reflected shock wave in rectangular shock tube. AIAA Journ., 1969, Vol. 7, No. 7, pp. 1370–1372.
125. Honda, M., Takayama, K, Onodera, O., and Kohama, Y. Motion of reflected shock waves in shock tube. In: Modern developments in shock tube research: Proc. 10th Intern. Symp. Kyoto, 1975, pp. 320–327.
126. Matsuo, K., Kawagoe, S., and Kage, K. The interaction of a reflected shock wave with the boundary layer in a shock tube. Bull. JSME, 1974, Vol. 17, No. 110, pp. 1039–1046.
127. Nikolayev, G. N. Eksperimental'noye issledovaniye formirovaniya udarnoy volny, otrazhennoy ot stenki s otverstiyem (Experimental investigation of the formation of a shock wave reflected from a wall with holes). Izv. AN SSSR, MZhG, 1974, No. 2, pp. 176–181.
128. Takano, Y., Miyoshi, S., and Akamatsu, T. Reflection processes of ionizing shocks in argon on end wall of a shock tube. In: Shock tubes and waves: Proc. 12th Intern. shock tube symp. Jerusalem, 1980, pp. 187–196.
129. Zhilin, Yu. V., Bazhenova, T. V., and Gvozdeva, L. G. Bifurcation of the shock wave upon reflecting from the end wall of the shock tube. Arch. Mech., 1978, Vol. 30, No. 4–5, pp. 675–681.
130. Taylor, J. R. and Hornung, H. G. Real gas and wall roughness effects on the bifurcation of the shock reflected from the end wall of a tube. In: Shock tubes and waves: Proc. 13th Intern. shock tube symp. Albany, New York, 1981, pp. 262–270.
131. Fokeyev, V. P., Gvozdeva, L. G., and Kalachev, A. A., et al. Study of shock wave reflection in rectangular channels by means of Cranz-Schardin system in parallel beams. In: Opt. meth. in dynamics of fluids and solids: Proc. IUTAM. Prague, 1985, pp. 59–64.
132. Fokeyev, V. P., Kharitonov, A. I., and Kalachev, A. A. Issledovaniye otrazheniya udarnykh voln v pryamougol'nykh kanalakh (Investigation of the reflection of shock waves in rectangular channels). V: Kineticheskiye i gazodinamicheskiye protsessy v neravnovesnykh sredakh (In:

Kinetic and gasdynamic processes in nonequilibrium media). Tr. vtoroy shk.-konf., Izd-vo MGU, 1984, pp. 24–25.

133. Sukhorukikh, V. S., Kharitonov, A. I., and Sharov, Y. L. The Crantz-Schardin method in the parallel beams. In: Proc. 15th Intern. congr. on high speed photography and photonics. San Diego, SPIE, 1983, pp. 534–535.

134. Ikui, T., Matsuo, K., and Nagai, M. The mechanism of pseudo-shock waves. Bull. JSME, 1974, Vol. 17, No. 108, pp. 731–739.

135. Matsuo, K., Kage, K, and Kawagoe, S. The interaction of a reflected shock wave with the contact region in a shock tube. Bull. JSME, 1975, Vol. 18, No. 121, pp. 681–688.

136. Gel'fand, B. E., Gubin, S. A., Kogarko, S. N., and Popov, D. E. Issledovaniye osobennostey rasprostraneniya i otrazheniya voln davleniya v poristoy srede (Investigation of the features of propagation and reflection of pressure waves in a porous medium). ZhPMTF, 1975, No. 6, pp. 74–77.

137. Gvozdeva, L. G. and Faresov, Yu. M. O vzaimodeystvii udarnoy volny so stenkoy, pokrytoy poristym szhimayemym materialom (On the interaction of a shock wave with a wall covered by a porous compressible material). Letters in ZhTF, 1984, Issue 19, pp. 1153–1156.

138. Gel'fand, B. E., Gubanov, A. V., and Timofeyev, E. I. Vzaimodeystviye vozdushnykh udarnykh voln s poristym ekranom (Interaction of air shock waves with a porous screen). Izv. AN SSSR, MZhG, 1983, No. 4, pp. 79–84.

139. Rudinger, G. Some effects of finite particle volume on the dynamics of gas particle mixture. AIAA Journ., 1965, Vol. 3, No. 7.

140. Campbell, I. J. and Pitcher, A. S. Shock waves in a liquid containing gas bubbles. Proc. Roy. Soc., London A, 1958, Vol. 243, No. 1235, pp. 534–545.

141. Parkin, B. R., Gilmor, F. R., and Brode, G. L. Shock waves in bubbly water. Memorandum RM-2795-PR (Abridged), Oct. 1961.

142. Wallis, G. One-dimensional two-phase flow. New York, McGraw Hill Book Company, 1968.

143. Van Wyngaarden, L. One-dimensional flow of liquids containing small gas bubbles. Annual review of fluid mech., Palo Alto, Vol. 4, 1972, pp. 369–396.

144. Bachelor, G. K. Compression waves in gas-bubble suspension in a liquid. In: Fluid Dyn. Trans., Warzawa, PWN, 1968, No. 4.

145. Gubaidullin, A. A., Ivandayev, A. I., Nigmatullin, R. I., and Khabeyev, I. C. Volny v zhidkostyakh s puzyr'kami (Waves in liquids with bubbles). V: Itogi nauki i tekhniki. Mekhanika zhidkosti i gaza (In: Review of science and technology. Mechanics of liquids and gases). Moscow, 1982, Vol. 17, pp. 93–102.

146. Gvozdeva, L. G. and Faresov, Yu. M. O raschete parametrov statsionarnykh udarnykh voln v poristoy szhimayemoy srede (On the computation of the parameters of steady shock waves in a porous compressible medium). ZhTF, 1985, Vol. 55, Issue 4, pp. 773–775.

147. Kuznetsov, N. M., Timofeyev, E. I., Polenov, A. N., and Gubanov, A. K. O skorosti rasprostraneniya slabykh vozmyshcheniy v vode s puzyr'kami para (On the velocity of propagation of weak perturbations in water with vapor bubbles). DAN SSSR, 1983, Vol. 273, No. 6, pp. 1355–1358.

148. Stanyukovich, K. P. Neustanovivshchiyesya dvizheniya sploshnoy sredy (Unsteady motion of a continuous medium). 2-e izd., pererab. i dop., Moscow, Nauka, 1971, 854 pp.

149. Mach, E. Uber einige mechanische Wirkungen des elektrischen Funkens. Akad. Wiss. Wien, 1878, Vol. 77, 819 pp.

150. von Neuman, J. Collected works. London, Pergamon Press, 1943, Vol. 6, pp. 239–299.

151. Zel'dovich, Ya. B. Teoriya udarnykh voln i vvedeniye v gazodinamiku (Theory of shock waves and introduction to gasdynamics). Moscow, Izd-vo AN SSSR, 1945, 186 pp.

152. Courant, R. and Friedrichs, K. Supersonic flow and shock waves. Wiley-Interscience, New York, 1948.

153. Liepmann, H. W. and Roshko, A. Elements of gas dynamics. London, Chapman and Hall, 1957.

154. Pack, D. C. The reflection and diffraction of a shock wave. J. Fluid Mech., 1964, Vol. 18, pp. 549–576.

155. Polachek, H. and Ziger, R. I. Interaction of shock waves. In: Fundamentals of gasdynamics. Princeton, New Jersey, 1958.
156. Oertel, H. Stossrohre. Wien, Springer, 1965, 982 pp.
157. Gubkin, K. E. Otrazheniye udarnykh voln (Reflection of shock waves). V: Mekhanika v SSSR za 50 let (In: Mechanics in the USSR in 50 years). Moscow, Nauka, 1970, pp. 269–311.
158. Reichenbach, G. Shock waves in gases. In: High Speed Physics. St. Louis, Springer, 1967.
159. Arutyunian, G. M. and Karchevskiy, L. V. Otrazhennyye udarnyye volny (Reflected shock waves). Moscow, Mashgiz, 1973, 376 pp.
160. Baum, F. A., Orlenko, L. P., and Stanyukovich, K. P. Fizika vzryva (Physics of explosions). Moscow, Nauka, 1975, 800 pp.
161. Griffith, W. C. Shock wave interactions. J. Fluid Mech., 1981, Vol. 106, pp. 81–101.
162. Bazhenova, T. V., Gvozdeva, L. G., and Nettleton, M. A. Unsteady interactions of shock waves. Progr. Aerospace Sci., 1984, Vol. 21, No. 4, pp. 249–331.
163. Bleakney, W. and Taub, A. H. Interaction of shock waves. Rev. Mod. Phys., 1949, Vol. 21, pp. 584–605.
164. Fletcher, C. A., Taub, A. H., and Bleakney, W. The Mach reflection of shock waves of nearly glancing incidence. Rev. Mod. Phys., 1951, Vol. 23, pp. 271–286.
165. White, D. R. An experimental survey of shock waves. Proc. 2nd Midwest Conf. Fluid Mech., 1952, Vol. 21, No. 3, pp. 252–263.
166. Griffith, W. and Brickle, D. E. The diffraction of strong shock waves. Phys. Rev., 1953, Vol. 89, No. 2, pp. 451–453.
167. Bleakney, W. Review of significant observations of the Mach reflection of shock waves. Proc. Symp. Appl. Math., 1954, Vol. 5, pp. 41–47.
168. Smith, W. R. Mutual reflection of two shock waves of arbitrary strengths. Phys. Fluids, 1959, Vol. 2, pp. 533–541.
169. Zaslavskiy, B. N. and Safarov, R. A. O podobii techeniy, voznikayushchikh pri otrazhenii slabykh udarnykh voln ot zhestkoy stenki i svobodnoy poverkhnosti (On similarity of flows resulting from the reflection of weak shock waves from surfaces). Fizika goreniya i vzryva, 1973, Vol. 9, No. 4, pp. 579–585.
170. Zaslavskiy, B. N. and Safarov, R. A. O makhovskom otrazhenii slabykh udarnykh voln ot zhestkoy stenki (On the Mach reflection of weak shock waves from a solid wall). ZhPMTF, 1973, No. 5, pp. 26–33.
171. Henderson, L. F. and Siegenthaler, A. Experiments on the diffraction of weak blast waves: The von Neumann paradox. Proc. Roy. Soc., London A, 1980, Vol. 369, No. 1739, pp. 537–555.
172. Henderson, L. F. and Gray, P. M. Experiments on the diffraction of strong blast waves. Proc. Roy. Soc., London A, 1981, Vol. 377, No. 1771, pp. 363–378.
173. Gvozdeva, L. G. and Predvoditeleva, O. A. Eksperimental'noye issledovaniye makhovskovo otrazheniya udarnykh voln pri skorosti 1000-3000 m/s v uglekislom gaze, azote, i vozdukhe (Experimental investigation of Mach reflection of shock waves at a speed of 1000-3000 m/sec in carbon dioxide, nitrogen, and air). DAN SSSR, 1965, Vol. 163, No. 5, pp. 1088–1091.
174. Gvozdeva, L. G. and Predvoditeleva, O. A. Osobennosti makhovskovo otrazheniya udarnykh voln, dvizhushchikhsya v uglekislom gaze, v azote so skorostyami poryadka 2000 m/s (Features of Mach reflections of shock waves travelling in carbon dioxide and nitrogen at a speed of about 2000 m/sec). V: Issledovaniye po fizicheskoy gazodinamike (In: Investigation in physical gasdynamics). Moscow, Nauka, 1966, 183 pp.
175. Gvozdeva, L. G., Predvoditeleva, O. A., and Fokeyev, V. P. Dvoynoye makhovskoye otrazheniye sil'nykh udarnykh voln (Double Mach reflection of strong shock waves). Izv. AN SSSR, MZhG, 1968, No. 1, pp. 12–19.
176. Merritt, D. L. Mach reflection at a cone. AIAA Journ., 1968, Vol. 6, No. 6, pp. 1208–1209.
177. Weinants, R. R. An experimental investigation of shock wave diffraction over compression and expansion corners. Toronto, 1968, 94 pp.
178. Gvozdeva, L. G., Bazhenova, T. V., Predvoditeleva, O. A., and Fokeyev, V. P. Mach reflection of shock waves. Astron. acta, 1969, Vol. 14, pp. 503–508.

179. Gvozdeva, L. G., Bazhenova, T. V., Predvoditeleva, O. A., and Fokeyev, V. P. Pressure and temperature at the wedge surface in Mach reflection of strong shock waves. Astron. acta, 1970, Vol. 15, pp. 503–510.
180. Semenov, A. N., Sishikova, M. P., and Berezkina, M. K. Eksperimental'noye izucheniye osobennostey makhovskovo otrazheniya v udarnoy trube (Experimental study of the features of Mach reflection in a shock tube). ZhTF, 1970, Vol. 15, Issue 5, pp. 795–803.
181. Bertrand, B. F. Measurements of pressure in Mach reflection of strong shock waves in a shock tube. Mem. Rep. Ballist. Res. Lab., 1972, No. 2196.
182. Law, C. K. and Glass, I. I. Diffraction of strong shock waves by a sharp compressive corner. CASI Trans., 1971, Vol. 4, No. 1, pp. 2–15.
183. Makarevich, G. A., Predvoditeleva, O. A., and Lisenkova, G. C. Otrazheniye udarnoy volny ot klina (Reflection of a shock wave from a wedge). Teplofizika vysokikh temperatur, 1974, Vol. 12, No. 6, pp. 625–628.
184. Gvozdeva, L. G., Predvoditeleva, O. A., and Fokeyev, V. P. Issledovaniye davleniya i teplovykh potokov v stenku klina pri makhovskom otrazhenii udarnykh voln (Investigation of pressure and heat flow in the wall of a wedge during Mach reflection of shock waves). V: Teploobmen i fizicheskaya gazodinamika (In: Heat transfer and physical gasdynamics). Moscow, Nauka, 1974, pp. 156–164.
185. Gvozdeva, L. G., Bazhenova, T. V., and Fokeyev, V. P. The regions of various forms of Mach reflection and its transition to regular reflection. Astron. acta, 1975, Vol. 3, pp. 131–140.
186. Semenov, A. N., Sishikova, M. P., and Berezkina, M. K. Svoystva makhovskovo otrazheniya pri vzaimodeystvii udarnykh voln s nepodvizhnym klinom (Properties of Mach reflections during the interaction of shock waves with a stationary wedge). Fizika goreniya i vzryva, 1975, Vol. 11, No. 4, pp. 596–608.
187. Ikui, T., Matsuo, K., Akoi, T., and Kondoh, N. Investigation of Mach reflection of a shock wave. 1. Configurations and domains of shock reflection. Bull. JSME, 1982, Vol. 25, No. 208, pp. 1513–1520.
188. Gvozdeva, L. G. and Fokeyev, V. P. Eksperimental'noye issledovaniye neregulyarnovo otrazhniya udarnykh voln ot poverkhnosti klina (Experimental investigation of irregular reflection of shock waves from the surface of a wedge). Fizika goreniya i vzryva, 1976, Vol. 12, No. 2, pp. 260–269.
189. Gvozdeva, L. G. and Fokeyev, V. P. Perekhod ot makhovskovo otrazheniya k regulyarnomu v oblasti sushchestvovaniya razlichnykh form makhovskovo otrazheniya (Transition from Mach to regular reflection in the region of existence of various forms of Mach reflections). Fizika goreniya i vzryva, 1977, Vol. 13, No. 1, pp. 102–109.
190. Henderson, L. F. and Lozzi, A. Experiments on transition of Mach reflection. J. Fluid Mech., 1975, Vol. 68, Part 1, pp. 139–155.
191. Ben-Dor, G. and Glass, I. I. Domains and boundaries of nonstationary oblique shock-wave reflections. 1. Diatomic gas. J. Fluid Mech., 1979, Vol. 92, Part 3, pp. 459–496.
192. Henderson, L. F. and Lozzi, A. Further experiments on transition to Mach reflection. J. Fluid Mech., 1979, Vol. 94, Part 3, pp. 541–559.
193. Syshchikova, M. P. and Krassovskaya, I. V. Some properties of regular and irregular interaction of shock waves. Arch. Mech., 1979, Vol. 31, No. 1, pp. 135–145.
194. Hornung, H. G. and Kychakoff, G. Transition from regular to Mach reflection of shock waves in relaxing gases. In: Shock tube and shock wave res.: Proc. 11th Intern. shock tube symp., Seattle, 1977, pp. 296–302.
195. Berezkina, M. K. and Plotnikov, N. V. Primeneniye sul'fat-litiyevovo preobrazovatelya impul'snovo davleniya v gazodinamicheskom eksperimente (The use of lithium-sulphate convertor of impulse pressure in gasdynamics experiments). Fizika goreniya i vzryva, 1982, Vol. 18, No. 6, pp. 63–68.
196. Hornung, H. G., Oertel, H., and Sandemen, R. J. Transition of Mach reflection of shock waves in steady and pseudosteady flow with and without relaxation. J. Fluid Mech., 1979, Vol. 60, Part 3, pp. 541–560.

197. Henderson, L. F. On the Whitham theory of shock-wave diffraction at concave corners. J. Fluid Mech., 1980, Vol. 99, Part 4, pp. 801–811.
198. Ben-Dor, G. and Glass, I. I. Domains and boundaries of non-stationary oblique shock-wave reflections. 2. Monatomic gas. J. Fluid Mech., 1980, Vol. 96, Part 4, pp. 735–756.
199. Berezkina, M. K., Sishikov, M. P., and Semenov, A. N. Vzaimodeystviye dvukh posledovatel'nykh udarnykh voln s klinom (Interaction of two consecutive shock waves with a wedge). ZhTF, 1982, Vol. 52, Issue 7, pp. 1375–1378.
200. Deschambault, R. L. and Glass, I. I. An update on non-stationary oblique shock-wave reflection: Actual isopycnics and numerical experiments. J. Fluid Mech., 1983, Vol. 131, pp. 27–57.
201. Ludloff, G. F. Aerodynamics of blast waves. In: Problems in mechanics. Adv. Appl. Mech., Acad. Press, New York, 1953, Vol. 3.
202. Oertel, H. Regular shock reflection used for studying oxygen dissociation. In: Modern developments in shock tube research. Kyoto, 1975, pp. 605–612.
203. Oertel, H. Oxygen vibrational and dissociation relaxation behind regular reflected shocks. J. Fluid Mech., 1976, Vol. 74, Part 3, pp. 477–495.
204. Oertel, H. Oxygen dissociation relaxation behind oblique reflected and stationary oblique shocks. Arch. Mech., 1978, Vol. 30, No. 2, pp. 123–133.
205. Arutyunian, G. M., Belokon, V. A., and Karchevskiy, L. V. O vliyanii pokazatelya adiabaty na otrazheniye udarnykh voln (On the effect of the adiabatic index on the reflection of shock waves). ZhPMTF, 1970, No. 4, pp. 62–66.
206. Henderson, J. F. On the confluence of three shock waves in a perfect gas. Aeronaut. Quart., 1964, Vol. 15, No. 2, pp. 181–197.
207. Kawamura, R. and Saito, H. Reflection of shock waves. 1. Pseudo-stationary case. J. Phys. Soc. Jap., 1956, Vol. 11, No. 5, pp. 584–592.
208. Tzu Sion Shao. Numerical solution of plane viscous shock reflections. J. Comput. Phys., 1967, Vol. 1, No. 1, pp. 367–381.
209. Pantazapol, D., Bellet, J. C., and Soustre, J. Sur les conditions d'apparition de l'éffet de Mach dans la réflexion d'une onde de ohoc oblique en écoulement supersonique. C. r. Acad. sci. A, 1972, Vol. 275, pp. 225–228.
210. Walker, D. K., Dewey, J. M., and Scotten, L. N. Observation of density discontinuities behind reflected shocks close to the transition from regular to Mach reflexion. J. Appl. Phys., 1982, Part 4, No. 3, pp. 1398–1400.
211. Hornung, H. G. and Robinson, H. L. Transition from regular to Mach reflection of shock waves. 2. The steady-flow criterion. J. Fluid Mech., 1982, Vol. 123, pp. 155–165.
212. Hornung, H. G. and Taylor, J. R. Transition from regular to Mach reflection of shock waves. 1. The effect of viscosity in the pseudo-steady case. J. Fluid Mech., 1982, Vol. 123, pp. 143–154.
213. Makomaski, A. H. Some effects of surface roughness on two-dimensional Mach reflection of moving plane shock waves in air. Canad. Aeronaut. and Space J., 1966, pp. 109–111.
214. Schultz-Grunow, F. Production and inhibition of Mach stems. In: Modern develop. in shock tube research: Proc. 10th shock tube symp. Kyoto, 1975, pp. 356–362.
215. Takayama, K., Ben-Dor, G., and Gotoh, J. Regular to Mach reflection transition in truly non-stationary flows: Influence of surface roughness. AIAA J., 1981, Vol. 19, No. 9, pp. 1238–1240.
216. Glass, I. I. Research frontiers at hypervelocities. CASI J., 1967, Vol. 13, No. 8, pp. 347–367.
217. Jones, D. M., Martin, P. M., and Thornhill, C. K. A note on the pseudo-stationary flow behind a strong shock diffracted or reflected at a corner. Proc. Roy. Soc., London A, 1951, Vol. 209, No. 1097, pp. 238–248.
218. Whitham, G. Linear and nonlinear waves. Interscience Publ., New York, John Wilby and Sons, 1977.
219. Oshima, K., Sugaya, K., Yamamoto, M., and Totoki, T. Diffraction of a plane shock wave around a corner. Univ. Press, Tokyo, 1965.

220. Skews, B. W. The shape of the diffracting shock wave. J. Fluid Mech., 1967, Vol. 29, Part 2, pp. 297–304.
221. Skews, B. W. The perturbed region behind a diffracting shock wave. J. Fluid Mech., 1967, Vol. 29, Part 4, pp. 705–719.
222. Bazhenova, T. V., Gvozdeva, L. G., Komarov, V. S., and Sukhov, B. G. Diffraction of strong shock waves in a shock tube. In: Shock tube research. London, Chapman and Hall, 1971, pp. 54/2.
223. Bazhenova, T. V., Gvozdeva, L. G., Komarov, V. S., and Sukhov, B. G. Pressure and temperature change on the wall surface in strong shock wave diffraction. Astron. Acta, 1972, Vol. 17, pp. 659–666.
224. Bazhenova, T. V., Gvozdeva, L. G., Komarov, V. S., and Sukhov, B. G. Issledovaniye difraktsii sil'nykh udarnykh voln na vypuklykh uglakh (Investigation of diffraction of strong shock waves at convex corners). Izv. AN SSSR, MZhF, 1973, No. 4, pp. 122–134.
225. Bazhenova, T. V., Gvozdeva, L. G., Komarov, V. S., and Sukhov, B. G. Techeniye relaksiruyushchevo gaza, voznikayushcheye pri vykhode udarnoy volny v rasshryayushchiysya kanal (Flow of relaxing gas resulting from the expansion of a shock wave into a converging channel). Teplofizika vysokikh temperatur 1973, Vol. 11, No. 6, pp. 1203–1213.
226. Bazhenova, T. V., Gvozdeva, L. G., and Zhilin, Yu. V. Change in the shape of a diffracting shock wave at a convex corner. Acta astron., 1979, Vol. 6, pp. 401–412.
227. Bleackney, W., White, D. R., and Griffith, W. C. Measurement of diffraction of shock waves and resulting loading of structures. J. Appl. Mech., 1950, Vol. 17, pp. 439–445.
228. Emrich, R. and Reichenbach, H. Photographic study of early stages of vortex formation behind an edge. In: Shock tubes. Toronto, 1970, pp. 740–750.
229. Sloan, S. A. and Nettleton, M. A. A model for the axial decay of a shock wave in a large abrupt area change. J. Fluid Mech., 1975, Vol. 71, pp. 769–784.
230. Sloan, S. A. and Nettleton, M. A. A model for the decay of a wall shock in a large abrupt area change. J. Fluid Mech., 1978, Vol. 88, Part 2, pp. 259–272.
231. Nettleton, M. A. Shock attenuation in a "gradual" area expansion. J. Fluid Mech., 1973, Vol. 60, pp. 209–223.
232. Edwards, D. H., Fearnley, P., and Nettleton, M. A. Shock diffraction with 90° bends. J. Fluid Mech., 1983, Vol. 132, pp. 257–270.
233. Zhmakin, A. N., Komissaruk, V. A., Mende, N. P., Fursenko, A. A., and Tsimbalov, V. V. O rasprostranenii udarnoy volny v okrestnosti sopryazheniya ploskikh kanalov (On the propagation of a shock wave in the vicinity of interface of plane channels). Prepr. FTI AN SSSR, No. 950, Leningrad, 1985, 18 pp.
234. Chisnell, R. The motion of a shock wave in a channel with application to cylindrical spherical shock waves. J. Fluid Mech., 1957, Vol. 2, pp. 286–298.
235. Heilig, W. H. Diffraction of a shock wave by a cylinder. Phys. Fluids, 1969, Vol. 12, No. 5, Suppl 1, Part 2, pp. 154–157.
236. Heilig, W. H. A result concerning the transition from regular reflection to Mach reflection of strong shock waves. In: Shock tube and shock wave research. Seattle, Univ. of Washington Press, 1978, pp. 288–295.
237. Golubinskiy, A. I. and Sokolov, K. B. Issledovaniye raspredeleniya davleniya po poverkhnosti ploskikh i tsilindricheskikh tel pri padenii na nikh vzryvnoy udarnoy volny (Investigation of the distribution of pressure across the surface of plane and cylindrical bodies upon impact of a blast shock wave). Moscow, 1970, 17 pp.
238. Sishikova, M. P., Semenov, A. N., and Berezkina, M. K. Otrazheniye udarnoy volny ot krivolineynoy vognutoy poverkhnosti (Reflection of shock wave from a curvilinear concave surface). Letters in ZhTF, 1976, Vol. 2, Issue 2, pp. 61–66.
239. Lozzi, A. and Henderson, L. F. Experiments on reflections of plane shock waves on cylindrical surfaces. In: Modern developments in shock tube research. Kyoto, 1975, pp. 363–368.
240. Gvozdeva, L. G., Lagutov, Yu. P., and Fokeyev, V. P. Perekhod ot regulyarnovo otrazheniya k makhovskomu pri vzaimodeystvii udarnykh voln s tsilindricheskimi poberkhnostyami (Transition from regular to Mach reflection upon interaction of shock waves with cylindrical surfaces). Letters in ZhTF, 1979, Vol. 5, Issue 13, pp. 812–816.

241. Gvozdeva, L. G., Bazhenova, T. V., Lagutov, Yu. P., and Fokeyev, V. P. Interaction of shock waves with cylindrical surfaces. Arch. Mech., 1980, Vol. 32, No. 5, pp. 693–702.
242. Gvozdeva, L. G., Lagutov, Yu. P., and Fokeyev, V. P. Perekhod ot makhovskovo otrazheniya k regulyarnomu pri vzaimodeystvii sil'nykh udarnykh voln s tsilindricheskimi poverkhnostyami (Transition from Mach to regular reflection upon interaction of strong shock waves with cylindrical surfaces). Izv. AN SSSR, MZhG, 1982, No. 2, pp. 132–138.
243. Itoh, S. and Itaya, M. On the transition between regular and Mach reflection. In: Shock tubes and waves. Jerusalem, 1980, pp. 314–323.
244. Itoh, S., Okazaki, N., and Itaya, M. On the transition between regular and Mach reflection in truly non-stationary flows. J. Fluid Mech., 1981, Vol. 108, pp. 383–400.
245. Ben-Dor, G., Takayama, K., and Kawauchi, T. The transition from regular to Mach reflection and from Mach to regular reflection in truly nonstationary flows. J. Fluid Mech., 1980, Vol. 100, Part 1, pp. 147–160.
246. Takayama, K., Gotoh, J. and Ben-Dor, G. Influence of surface roughness on the shock transition in quasistationary and truly nonstationary flows. In: Shock tubes and waves. Albany, New York, 1982, pp. 326–334.
247. Takayama, K. and Sasaki, M. Effects of radius of curvature and initial angle on the shock transition over concave or convex walls. Rep. Inst. High Speed Mech., Tohoku Univ., Ser. B, 1983, Vol. 46, No. 353, pp. 1–30.
248. Zhigalko, E. F. Otrazheniye udarnoy volny ot vognutoy stenki (Reflection of a shock wave from a concave wall). V: Voprosy aerogazodinamiki sverkhzvukovykh prostranstvennykh techeniy (Problems of aerogasdynamics of supersonic space flows). Novosibirsk, 1979, pp. 20–44.
249. Znamenskaya, I. A., Ryazin, A. P., and Shugayev, F. V. Ob osobennosti raspredeleniya parametrov gaza na nachal'noy stadii otrazheniya udarnoy volny ot sfery i tsilindra (On the features of distribution of gas parameters in the initial stage of reflection of a shock wave from a sphere and a cylinder). Izv. AN SSSR, MZhG, 1979, No. 3, pp. 103–110.
250. Nosenko, N. I., Sisoyev, N. N., Shugayev, F. V. Nachal'naya stadiya otrazheniya ploskoy udarnoy volny ot tsilindra, sfery i ellepsoida vrashcheniya (Initial stage of reflection of a plane shock wave from a cylinder, sphere, and ellipsoid of revolution). Izv. AN SSSR, MZhG, 1980, No. 2, pp. 94–100.
251. Shugayev, F. V. Vzaimodeystviye udarnykh voln s vozmushcheniyami (Interaction of shock waves with perturbations). Moscow, Izd-vo MGU, 1983, 96 pp.
252. Ginzburg, I. P. and Markov, Yu. S. Eksperimental'noye issledovaniye otrazheniya udarnoy volny ot dvugrannovo klina (Experimental investigation of reflection of a shock wave from a two-face wedge). V: Gazodinamika i teploobmen (In: Gasdynamics and heat transfer). Leningrad, Izd-vo LGU, 1973, No. 3, pp. 74–81.
253. Grudnitskiy, V. G. and Prokhorchuk, Yu. A. Raschet vzaimodeystviya udarnoy volny s zatuplennym telom (Calculations of interaction of a shock wave with a blunt body). Numerical methods in continuum mechanics, 1975, Vol. 6, No. 4, pp. 42–55.
254. Lyakhov, V. N. Nestatsionarnykh nagruzki pri difraktsii udarnoy volny (Unsteady loads upon diffraction of a shock wave). Izv. AN SSSR, MZhG, 1975, No. 4, pp. 123–129.
255. Lyakhov, V. N. Vzaimodeystviye udarnykh voln umerennoy intensivnosti s tsilindrom (Interaction of shock waves of moderate intensity with a cylinder). Izv. AN SSSR, MZhG, 1979, No. 2, pp. 113–119.
256. Lyakhov, V. N. Approksimatsiya rezul'tatov matematicheskovo modelirovaniya vozdeystviya udarnoy volny na tsilindr (Approximation of the results of mathematical simulation of the action of a shock wave on a cylinder). Izv. AN SSSR, MZhG, 1980, No. 4, pp. 179–184.
257. Voynovich, P. A., Popov, F. D., and Fursenko, A. A. Chislennoye modelirovaniye otrazheniya udarnoy volny ot krivolineynoy vognutoy poverkhnosti (Numerical simulation of reflection of a shock wave from a curvilinear convex surface). Letters in ZhTF, 1978, Vol. 4, Issue 6, pp. 313–316.
258. Murdoch, J. Shock-wave interaction with two-dimensional bodies. AIAA Journ., 1975, Vol. 13, No. 9, pp. 1139–1140.
259. Sandeman, J., Leitch, A., and Hornung, H. The influence of relaxation on transition to Mach reflection in pseudosteady flow. In: Shock tubes and waves. Jerusalem, 1980, pp. 298–307.

260. Takayama, K. and Ben-Dor, G. A reconsideration of the hysteresis phenomenon in the regular Mach reflection transition in truly nonstationary flows. In: Shock tubes and waves. Sydney, 1983, pp. 135–143.
261. Dewey, J. M., Walker, D. K., Lock, G. D., and Scotten, L. N. The effects of radius of curvature and initial angle on transition from regular to Mach reflection of weak shocks reflected from curved surfaces. In: Shock tubes and waves. Sydney, 1983, pp. 144–149.
262. Heilig, W. and Reichenbach, H. About the trend of pressure in shock reflection processes. In: Shock tubes and waves. Sydney, 1983, pp. 151–158.
263. Dumitrescu, L. Z. and Preda, A. Some new results concerning the diffraction of a shock wave around a convex corner. In: Modern developments in shock tube research. Kyoto, 1975, pp. 369–377.
264. Gvozdeva, L. G., Kharitonov, A. I., Sukhorukikh, V. S., and Lagutov, Yu. P. Primeneniye golograficheskovo interferometra dlya issledovaniya difraktsii udarnykh voln (The use of holographic interferometer for the investigation of diffraction of shock waves). V: II vsesoyuz. konf. po metodam aerofiz. issled. (In: II All-Union conference on the methods of aerophysical investigations). Novosibirsk, 28 May–1 June, 1979, Minsk, 1979, Vol. 3, pp. 78–81.
265. Lagutov, Yu. P. O forme udarnoy volny, difragiruyushchey na zakrutlennom ugle (About the form of a shock-wave diffracting at a rounded corner). Izv. AN SSSR, MZhG, 1983, No. 3, pp. 169–173.
266. Richtmyer, R. and Morton, K. Difference methods for initial-value problems. Interscience publishers, New York, 1967.
267. Roache, P. Computational fluid dynamics. Albuquerque, Hermosa Publ., 1976.
268. Kestenboym, Kh. S., Roslyakov, S. S., and Chudov, L. A. Tochechnyy vzryv (Point explosion). Moscow, Nauka, 1974, 254 pp.
269. Balakin, V. B. and Bulanov, V. V. Chislennoye resheniye zadachi o vzaimodeystvii udarnoy volny s tsilindrom v sverkhzvukovom potoke (Numerical solution of the problem of interaction of a shock wave with a cylinder in supersonic flow). Inzh.-fiz. zhurn., 1971, Vol. 21, No. 6, pp. 1033–1039.
270. Worming, R. F., Kutler, P., and Lomax, H. Noncentral difference schemes of the second- and third-order accuracy for the solution of hyperbolic-type nonlinear equations. AIAA Journ., 1973, Vol. 11, No. 2.
271. Lyakhov, V. N., Zitserman, V. Yu., and Kostilevich, E. K. O vliyanii teplofizicheskikh svoystv rabochevo tela na parametry nestatsionarnovo techeniya (About the effect of thermophysical properties of the working medium on the parameters of unsteady flow). V: Metody rascheta teplofizicheskikh svoystv veshchestv (In: Methods of computation of thermophysical properties of substances). Moscow, IVTAN, 1983, pp. 43–62.
272. Loytsyanskiy, L. G. Mekhanika zhidkosti i gaza (Mechanics of liquids and gases). Moscow, Nauka, 1973, 847 pp.
273. Kutler, P. and Sakell, L. Three-dimensional shock-and-shock interaction problem. AIAA Journ., 1975, Vol. 13, No. 10, pp. 1360–1367.
274. Vyakhov, V. N. Sglazhivaniye i iskusstvennaya vyazkost' pri raschetakh dvumernykh nestatsionarnykh techeniy s razryvami (Smoothing and artificial viscosity during the computation of two-dimensional flows with discontinuities). V: Chislennyye metody mekhaniki sploshnoy sredy (In: Numerical methods in continuum mechanics). Novosibirsk, 1974, Vol. 5, Issue 3, pp. 69–74.
275. Ivanov, M. Ya, Koretskiy, V. V., and Kurochkina, N. Ya. Issledovaniye svoystv raznostnykh skhem skvoznovo scheta vtorovo poryadka approksimatsii (Investigation of the properties of a difference scheme of the shock capturing method of the second order of approximation). V: Chislennyye metody mekhaniki sploshnoy sredy (In: Numerical methods in continuum mechanics). Novosibirsk, 1980, Vol. 11, No. 2, pp. 41–63.
276. Sommerfeld, A. Optika (Optics). Moscow, Izd-vo inostr. lit., 1953, 487 pp.
277. Smirnov, V. I. Kurs vysshey matematiki (Course in higher mathematics). Leningrad, Moscow, Gostekhteoretizdat, 1949, Vol. 3, Part 2, 672 pp.
278. Bargman, V. On nearly glacing reflection of shocks. Office Sci. Res. and Develop. Rep., 1945, No. 5177.

279. Sagomonyan, A. Ya. Prostranstvennyye zadachi neustanovivshevosya dvizheniya szhimayemoy zhidkosti (Spatial problems of unsteady flow of compressible liquid). Moscow, Izd-vo MGU, 1962, 80 pp.
280. Ryzhov, O. S. and Khritianovich, S. A. O nelineynom otrazhenii slabykh udarnykh voln (About the nonlinear reflection of weak shock waves). PMM, 1958, Vol. 22, Issue 5, pp. 586–599.
281. Grib, A. A., Ryzhov, O. S., and Khristianovich, S. A. Teoriya korotkikh voln (Theory of short waves). ZhPMTF, 1960, No. 1, pp. 63–74.
282. Kleyner, B. G. and Shindyapin, G. P. Ob odnom klasse tochnykh chastnykh resheniy uravneniya korotkikh voln (About one class of exact special solutions of short waves). PMM, 1970, Vol. 34, Issue 6, pp. 1150–1158.
283. Shindyapin, G. P. Chislennoye resheniye zadachi neregulyarnovo otrazheniya slaboy udarnoy volny ot zhestkoy stenki v ideal'nom gaze (Numerical solution of the problem of irregular reflection of weak shock wave from a solid wall in an ideal gas). Zhurn. vichisl. matematiki i mat. fiziki, 1980, Vol. 20, No. 1, pp. 249–254.
284. Lyakhov, V. N. and Ryzhov, O. S. O zakone podobiya pri nelineynom otrazhenii udarnoy volny ot zhestkoy stenki (About the law of similarity in nonlinear reflection of a shock wave from a solid wall). Izv. AN SSSR, MZhG, 1977, No. 3, pp. 116–123.
285. Shindyapin, G. P. O neregulyarnom otrazhenii slabykh undarnykh voln ot zhestkoy stenki (About irregular reflection of weak shock waves from a solid wall). ZhPMTF, 1964, No. 2, pp. 22–28.
286. Shindyapin, G. P. O regulyarnom otrazhenii slabykh udarnykh voln ot zhestkoy stenki (About regular reflection of weak shock waves from a solid wall). PMM, 1965, Vol. 29, Issue 1, pp. 114–121.
287. Shindyapin, G. P. Neregulyarnoye vzaimodeystviye slabykh udarnykh voln raznoy intensivnosti (Irregular interaction of weak shock waves of different intensity). PMM, 1974, Vol. 38, Issue 1, pp. 105–114.
288. Dulov, V. G. and Rayzberg, B. A. Nachal'naya stadiya obrazovaniya strui (Initial stage of jet formation). Izv. vuzov. Aviats. tekhnika, 1961, No. 4, pp. 30–33.
289. Lighthill, M. J. The diffraction of blast. Part 2. Proc. Roy. Soc., London A, 1950, Vol. 200, No. 1063, pp. 554–565.
290. Ting, L. and Ludloff, H. F. Aerodynamics of blast. JAS, 1951, Vol. 18, No. 2, pp. 143–144.
291. Arutyunyan, G. M. O vzaimodeystvii udarnoy volny s klinom (About the interaction of a shock wave with a wedge). DAN ArmSSR, 1968, Vol. 46, No. 4, pp. 160–169.
292. Zhigalko, E. F. Lineynoye priblizheniye k difraktsii i otrazheniyu udarnoy volny (Linear approximation to diffraction and reflection of a shock wave). Vestn. LGU, 1969, No. 13, Issue 3, pp. 94–104.
293. Zhigalko, E. F. Priblizhennoye lokal'no-nelineynoye resheniye zadachi o vzaimodeystvii udarnoy volny s zhestkoy stenkoy (Approximate local nonlinear solution of the problem on the interaction of a shock wave with a solid wall). V: Gazodinamika i teploobmen (In: Gasdynamics and heat transfer). Leningrad, Izd-vo LGU, 1973, No. 3, pp. 63–74.
294. Bogatko, V. I. and Kolton, G. A. O regulyarnom otrazhenii sil'noy udarnoy volny ot poverkhnosti klina (About regular reflection of a strong shock wave from the surface of a wedge). Izv. AN SSSR, MZhG, 1983, Vol. 2, pp. 92–96.
295. Kireyev, V. T. and Tikhomirov, N. A. Regulyarnoye vzaimodeystviye udarnykh voln pri nalichii kolebatel'noy relaksatsii (Regular interaction of shock waves in the presence of vibrational relaxation). Izv. AN SSSR, MZhG, 1976, No. 4, pp. 90–96.
296. Titarenko, V. V. Asimptoticheskoye issledovaniye nekotorykh avtomodel'nykh vzaimodeystviy udarnykh voln (Asymptotic investigation of some self-similar interactions of shock waves). V: Vychislitel'nyye metody i programmirovaniye (In: Computational methods and programming). Izd-vo Sarat. Un-ta, 1977, Issue 1, pp. 60–72.
297. Titarenko, V. V. K zadache o vnezapnom vdviganii klina (On the problem about the sudden sliding of a wedge). ZhPMTF, 1981, No. 4, pp. 155–156.
298. Ludloff, H. F. and Friedman, M. B. Difference solution of shock diffraction problem. JAS, 1955, Vol. 22, No. 2, pp. 139–140.
299. Rusanov, V. V. Raschet vzaimodeystviya nestatsionarnykh udarnykh voln s prepyatstviyami

(Calculation of interaction of unsteady shock waves with obstacles). Zhurn. vichisl. matematiki i mat. fiziki, 1961, Vol. 1, No. 2, pp. 267–269.

300. Schneyer, G. P. Numerical simulation of regular and Mach reflections. Phys. Fluids, 1975, Vol. 18, pp. 1119–1124.
301. Belotserkovskiy, O. M. and Davidov, Yu. M. Metod krupnykh chastits v gazovoy dinamike (Method of large particles in gasdynamics). Moscow, Nauka, 1982, 391 pp.
302. Richtmyer, R. D. Progress report on the Mach reflection calculation. New York, 1961, 63 pp.
303. Shankar, V., Kutler, P., and Anderson, D. Diffraction of a shock wave by a compression corner. II. Single Mach refraction. AIAA Journ., 1978, Vol. 18, No. 1, pp. 4–5.
304. Beloshitskiy, A. V., Krikonov, V. V., Lipnitskiy, Yu. M., and Lyakhov, V. N. Issledovaniye razlichnykh gazodinamicheskikh techeniy s pomoshch'yu yavnykh raznostnykh skhem skvoznovo scheta (Investigation of various gasdynamic flows with the help of explicit difference schemes of the shock capturing method). Tr. NII mekhaniki MGU, 1973, No. 30, pp. 168–175.
305. Lipnitskiy, Yu. M. and Lyakhov, V. N. Chislennoye resheniy zadachi difraktsii udarnoy volny na kline (Numerical solution of the problem of shock wave diffraction on a wedge). Izv. AN SSSR, MZhG, 1974, No. 6, pp. 88–94.
306. Lyakhov, V. N. Matematicheskoye modelirovaniye makhovskovo otrazheniya udarnykh voln v sredakh s razlichnymi pokazatelyami adiabaty (Mathematical simulation of Mach reflection of shock waves in mediums with various adiabatic indexes). Izv. AN SSSR, MZhG, 1976, No. 3, pp. 90–94.
307. Lyakhov, V. N. K voprosu ob otsenke davleniya pri nestatsiounarnom otrazhenii udarnoy volny (On the question about the evaluation of pressure in unsteady reflection of a shock wave). Izv. AN SSSR, MZhG, 1977, No. 2, pp. 100–106.
308. Lipnitskiy, Yu. M. and Lyakhov, V. N. Vzaimodeystviye udarnoy volny s klinom v sverkhzvukovom potoke (Interaction of a shock wave with a wedge in supersonic flow). Uchen. zap. TsAGI, 1976, Vol. 7, No. 4, pp. 115–119.
309. Ter-Minasyants, S. M. Difraktsiya ploskoy volny na kline, dvizhushchemsya so sverkhzvukovoy skorost'yu (Diffraction of a plane wave on a wedge moving at supersonic speed). PMM, 1971, Vol. 35, Issue 2, pp. 238–247.
310. Kutler, P. and Shankar, V. Diffraction of a shock wave by a compression corner. I. Regular reflection. AIAA Journ., 1977, Vol. 15, No. 2, pp. 197–203.
311. Old and Bird. Simulation of regular reflection and Mach reflection by the Monte Carlo method. AIAA Journ., 1977, Vol. 15, No. 5.
312. Fry, M., Titsworth, J., Kulh, A., et al. Shock capturing using flux-corrected transport algorithms with adaptive gridding. In: Proc. 13th Intern. symp. on shock tubes and waves. Albany, New York, 1981, pp. 376–384.
313. Book, D., Boris, J., Kulh, A., et al. Simulation of complex shock reflections from wedges in inert and reactive gaseous mixtures. In: Proc. 7th Intern. conf. on numer. meth. in fluid dyn. 1981, pp. 84–90.
314. Woodward, P. and Collela, P. High resolution difference schemes for compressible gas dynamics. Lect. Notes Phys., 1981, Vol. 141, pp. 434–441.
315. Woodward, P. and Colella, P. The numerical simulation of two-dimensional fluid flow with strong shocks. J. Comput. Phys., 1984, Vol. 54, pp. 115–173.
316. Roslyakov, G. S. Vzaimodeystviye ploskikh skachkov odnovo napravleniya (Interaction of plane shocks travelling in the same direction). V: Chislennyye metody v gazovoy dinamike (In: Numerical methods in gasdynamics). Sb. rabot VTs MGU. Moscow, Izd-vo MGU, 1965, Issue 4, pp. 28–51.
317. Godunov, S. K., Zabrodin, A. V., Prokopov, G. P., and Cherkashin, V. A. O chislennom reshenii mnogomernykh zadach gazovoy dinamiki (About the numerical solution of multidimensional problems of gasdynamics). V: Teoreticheskiye osnovy i konstruirovaniye chislennykh algoritmov zadach matematicheskoy fiziki (In: Theoretical principles and construction of numerical algorithms for problems of mathematical physics). Moscow, Nauka, 1979, pp. 201–216.

318. Tarnavskiy, G. V., Khonichev, V. I., and Yakovlev, V. I. Difraktsiya udarnoy volny na pryamom ugle i na vykhode iz ploskovo kanala (Diffraction of a shock wave at a right angle and at the outlet from a plane channel). Izv. CO AN SSSR, Ser. tekh. nauk, 1974, Issue 2, No. 8, pp. 56–65.
319. Tarnavskiy, G. V., Khonichev, V. I., and Yakovlev, V. I. O rezul'tatakh rascheta difraktsii udarnoy volny na pryamom ugle s uchetom real'nykh svoystv gaza (About calculation results of shock-wave diffraction at a right angle accounting for real gas properties). Izv. AN SSSR, MZhG, 1976, No. 5, pp. 126–129.
320. Rudakova, G. M. and Shashkin, A. P. K raschetu nekotorykh ploskikh nestatsionarnykh techeniy gaza (On the calculation of some plane unsteady gas flows). Izv. CO AN SSSR, Ser. tekh. nauk, 1975, Issue 1, No. 3, pp. 70–77.
321. Gridnev, N. P. and Katsnel'son, S. S. Ispol'zovaniye raznostnykh skhem 3-vo poryadka tochnosti dlya rascheta gazodinamichkikh i magnitogidrodinamicheskikh techeniy (Use of difference schemes of the 3rd order of accuracy for the calculation of gasdynamic and magnetohydrodynamic flows). V: 6-ya Mezhdunar. konf. po chislennym metodam v gidrodinamike (In: 6th International conference on numerical methods in hydrodynamics). Tbilisi, 1978, Vol. 2, pp. 70–75.
322. Vasiliyev, E. I. Nestatsionarnoye istecheniye strui v zatoplennoye prostranstvo (Unsteady flow of a jet in a submerged space). Izv. AN SSSR, MZhG, 1984, No. 1, pp. 42–46.
323. Bloor, M. I. and Evans, R. A. Use of vortex discretisation in establishing weak shock diffraction patterns. AIAA Journ., 1979, Vol. 17, No. 10, pp. 1046–1049.
324. Demiyanov, A. Yu. and Panasenko, A. V. Chislennoye resheniye zadachi difraktsii ploskoy udarnoy volny na vypuklom ugle (Numerical solution of the problem of diffraction of a plane shock wave at a convex corner). Izv. AN SSSR, MZhG, 1981, No. 5, pp. 103–109.
325. Demiyanov, A. Yu. Lipnitskiy, Yu. M., Lyakhov, V. F., Panasenko, A. V., and Yatsuk, V. F. Rezul'taty sistematicheskikh chislennykh raschetov vzaimodeystviya udarnykh voln s vypuklym uglom (Results of systematic numerical calculations of the interaction of shock waves with a convex corner). Letters in ZhTF, 1981, Vol. 7, Issue 24, pp. 1481–1484.
326. Tkachenko B. K. Effekt stenki v zadachakh po difraktsii udarnykh voln (The effect of walls in problems on shock-wave diffraction). V: Aerofizika i prikladnaya matematika (In: Aerophysics and applied mathematics). Moscow, 1981, pp. 75–76.
327. Taylor, T. D. and Lin, T. C. Numerical model for muzzle blast flow fields. AIAA Journ., 1981, Vol. 19, No. 3, pp. 346–349.
328. Girb, A. A. and Iskhakov, V. A. K zadache ob istechenii udarnoy volny iz truby v svobodnoye prostranstvo (On the problem about the flow of a shock wave from a tube into a free space). Vestn. LGU, 1981, No. 13, pp. 75–79.
329. Graham, J. M. R. and Hiller, R. Shock wave diffraction at a sharp edge and the effect of baffles in a shock tube. In: Shock tubes and waves. Albany, New York, 1982, pp. 402–410.
330. Sugimura, T., Taki, S., and Fusiwara, T. Formation of two-dimensional non-steady detonation in a channel with backward-facing step. In: Shock tubes and waves. Albany, New York, 1982, pp. 411–419.
331. Lyakhov, V. N. K voprosu o vykhode udarnoy volny v rasshiryayushchuyusya chast'kanala (On the question about the flow of a shock wave into an expanding section of a channel). Izv. AN SSSR, MZhG, 1978, No. 6, pp. 191–193.
332. Shokin, Yu, I. Metod differentsial'novo priblizheniya (Method of differential approximation). Novosibirsk, Nauka, 1979, 221 pp.
333. Zamishlyayev, B. V. and Yakovlev, Yu. S. Dinamicheskiye nagruzki pri podvodnom vzryve (Dynamic loads from an underwater blast). Leningrad, Sudostroyeniye, 1967, 292 pp.
334. Rakhmatullin, Kh. A., Saatov, Ya. U., Sabodash, P. F., and Fillipov, I. G. Dvumernyye zadachi po neustanovivshemusya dvizheniyu szhimayemykh sred (Two-dimensional problems of unsteady flow of compressible media). Tashkent, Fan, 1969, 182 pp.
335. Kireyev, V. T. Ob otrazhenii sil'noy udarnoy volny ot sfery i tsilindra (About the reflection of a strong shock wave from a sphere and a cylinder). Izv. AN SSSR, MZhG, 1969, No. 3, pp. 53–61.

336. Sokolov, V. B. Otrazheniye sil'noy udarnoy volny ot ellipsoida vrasheniya i ellipticheskovo tsilindra (Reflection of a strong shock wave from an ellipsoid of revolution and an elliptical cylinder). Izv. AN SSSR, MZhG, 1974, No. 4, pp. 173–174.
337. Poluboyarinov, A. K. O dvizhenii udarnovo fronta, otrazhennovo ot zatuplennovo tela (About the travel of a shock front reflected from a blunt body). Izv. AN SSSR, MZhG, 1971, No. 2, pp. 70–77.
338. Moran, J. P. and Moorhem, W. K. Diffraction of a plane shock by an analytic blunt body. J. Fluid Mech., 1969, Vol. 38, No. 1, pp. 127–136.
339. Bryson, A. E. and Gross, R. N. Diffraction of strong shocks by cones, cylinders and spheres. J. Fluid Mech., 1961, Vol. 10, No. 1, pp. 1–16.
340. Whitham, G. B. A new approach to problems of shock dynamics. Part 1. Two-dimensional problems. J. Fluid Mech., 1957, Vol. 2, No. 2, pp. 145–171.
341. Whitham, G. B. A new approach to problems of shock dynamics. Part 2. Three-dimensional problems. J. Fluid Mech., 1959, Vol. 5, No. 3, pp. 369–386.
342. Sishikova, M. P., Berezkina, M. K., and Semenov, A. N. Formirovaniye techeniya okolo modeli v udarnoy trube (Flow formation near a model in a shock tube). V: Aerofizicheskiye issledovaniya sverkhzvukovykh techeniy (In: Aerophysical investigations of supersonic flows). Moscow, Leningrad, Nauka, 1967, pp. 7–12.
343. Zhmayeva, E. A. and Kharitonov, A. I. Formirovaniye golovnovo skachka okolo zatuplennykh tel, pomeshchennykh v udarnoy trube (Formation of a leading shock near a blunt body in a shock tube). Izv. AN SSSR, MZhG, 1971, No. 6, pp. 131–136.
344. Braslavets, V. A., Zhmayeva, E. A., and Kharitonov, A. I. K voprosu ob izmenenii plotnosti, davleniya i otkhoda pri vozdeystvii udarnoi volny na nepodvizhnoye zatuplennoye telo (On the question about the density change, pressure, and reflection upon the interaction of a shock wave with a stationary blunt body). Izv. AN SSSR, MZhG, 1974, No. 3, pp. 166–170.
345. Heilig, W. Theoretische und experimentelle Untersuchungen zur beugung von stosswellen an Kugeln und Zylindren: Doctor-Ing. Diss. Karlsruhe, Univ. Karlsruhe, 1969, 115 pp.
346. Patz, G. Ausbildung der Kopfwelle stumpfer Korper in der Strömung hinter dem laufenden Stoβ: Doctor-Ing. Diss. Karlsruge, Univ. Karlsruhe, 1976, 81 pp.
347. Bogoslovskiy, N. E. Issledovaniye nestatsionarnovo obtekaniya tel potokom, dvizhushchimsya za udarnoy volnoy (Investigation of unsteady flow around a body by a stream moving behind a shock wave). Izv. AN SSSR, MZhG, 1966, No. 3, pp. 24–30.
348. Ivanov, A. N. and Chernyavskiy, S. Yu. Issledovaniye difraktsii udarnoy volny na tsilindre (Investigation of shock-wave diffraction on a cylinder). Uchen. zap. TsAGI, 1970, Vol. 1, No. 6, pp. 131–133.
349. Kolgan, V. P. and Fonarev, A. S. Ustanovleniye obtekaniya pri padenii udarnoi volny na tsilindr i sferu (Establishment of flow when a shock wave impinges on a cylinder or a sphere). Izv. AN SSSR, MZhG, 1972, No. 5, pp. 98–103.
350. Godunov, S. K., Zabrodin, A. V., et al. Chisclennoye resheniye mnogomernykh zadach gazovoy dinamiki (Numerical solution of multi-dimensional problems of gasdynamics). Moscow, Nauka, 1976, 400 pp.
351. Lipnitskiy, Yu. M., Lyakhov, V. N., and Fel'dshteyn, V. A. Vzaimodeystviye uprutoy obolochki s volnoy sil'novo razryva v gaze (Interaction of an elastic shell with a wave of a strong discontinuity in gas). V: IX Vsesoyuz. konf. po teorii obolochek i plastin, dek. 1973 (In: IX All-Union conference on the theory of shells and plates, Dec. 1973). Leningrad, Sudostroyeniye, 1975, pp. 198–199.
352. Lipnitskiy, Yu. M., Lyakhov, V. N., and Fel'dshteyn, V. A. Nestatsionarnoye vzaimodeystviye udarnoy volny s uprugoy tsilindricheskoy obolochkoy (Unsteady interaction of a shock wave with an elastic shell). Uchen. zap. TsAGI, 1976, Vol. 7, No. 1, pp. 80–88.
353. Vol'mir, A. S. Obolochiki v potoke zhidkosti i gaza (Shells in flows of liquids and gases). Moscow, Nauka, 1976, 416 pp.
354. Mark, A. and Kutler, P. Shock diffraction computations over complex structures. In: Proc. of 13th Intern. symp. on shock tubes and waves. Albany, New York, 1982, pp. 772–780.
355. Howarth, L., ed. Recent developments in fluid dynamics, high-speed flow. 1956, Part 2, 682 pp.

356. Vinogradov, R. I., Zhukovskiy, M. I., and Yakubov, I. R. Gazogidravlicheskaya analogiya i eye prakticheskoye primeneniye (Gas-hydraulic analogy and its practical application). Mashinostroyeniye, 1978, 152 pp.
357. Merritt, D. L. and Aronson, P. M. Study of blast-bow wave interactions in a wind tunnel. New York, 1965, 74 pp. (AIAA paper, No. 5).
358. Ruetenik, J. R. and Lemcke, B. Study of blast-bow wave interactions in a shock tube and shock tunnel. New York, 1966, 48 pp. (AIAA paper, No. 409).
359. Shugayev, F. V. and Lisin, Yu. G. Issledovaniye vzaimodeystviya ploskoy udarnoy volny s zatuplennym telom, nakhodyashchimsya v sverkhzvukovom potoke (Investigation of the interaction of plane shock wave with a blunt body in supersonic flow). Inzh.-fiz. zhurn., 1971, Vol. 21, No. 3, pp. 37–40.
360. Beans, E. W. A definition of the time duration of pressure phase resulting from intersecting shock waves. AIAA J., 1965, Vol. 3, No. 12, pp. 168–169.
361. Taylor, T. D. and Hudgins, H. E. Interaction of blast wave with blunt body travelling at supersonic speeds. AIAA J., 1968, Vol. 6, No. 2, pp. 196–204.
362. Arutyunyan, G. M. K raschetu davleniya v kriticheskoy tochke pri padenii udarnoy volny na telo, dvizhushcheyesya so sverkhzvukovoy skorost'yu (On the calculation of pressure at the critical point when a shock wave hits a body travelling at supersonic speed). Izv. AN SSSR, MZhG, 1972, No. 6, pp. 94–101.
363. Braslavets, V. A., Zhmayeva, E. A., Krupenya, B. I., Makarevich, G. A., Nevskiy, A. G., and Shimarev, S. K. Davleniye v kriticheskoy tochke pri vstrechnom vzaimodeystvii udarnoy volny s letyashchim telom (Pressure at the critical point due to head-on interaction of a shock wave with a flying body). Izv. AN SSSR, MZhG, 1974, No. 5, pp. 166–169.
364. McNamara, W. Flame computer code for the axisymmetric interaction of a blast wave with a shock layer on a blunt body. J. Spacecraft and Rockets, 1967, Vol. 4, No. 6, pp. 790–795.
365. Riyes, D. T. Vzaimodeystviye nestatsionarnykh udarnykh voln s ploskimi giperzvukovymi techeniyami u zatuplennykh tel (Interaction of unsteady shock waves with plane hypersonic flows at blunt bodies). V: 13-y mezhdunar. kongr. po teoret. i prikl. mekhanike (In: 13th International congress on theoretical and applied mechanics). Sb. annot. Moscow, 1972, 77 pp.
366. Lyakhov, V. N. Nestatsionarnoye vzaimodeystviye udarnoy volny s zatuplennym telom v serkhzvukovom potoke (Unsteady interaction of a shock wave with a blunt body in supersonic flow). Izv. AN SSR, MZhG, 1979, No. 6, pp. 70–76.
367. Zel'dovich, Ya. B. K teorii rasprostraneniya detonatsii v gazoobraznykh sistemakh (On the theory of propagation of detonation in gaseous systems). ZhETF, 1940, No. 10, pp. 542–568.
368. Zel'dovich, Ya. B. and Kompaneyets, A. S. Teoriya detonatsii (Theory of detonation). Moscow, Gostekhizdat, 1955, 268 pp.
369. Zel'dovich, Ya. B. Izbrannyye trudy: Khimicheskaya fizika i gidrodinamika (Selected works: Chemical physics and hydrodynamics). Moscow, Nauka, 1984, 374 pp.
370. Lewis, B. and Elbe, G. Combustion, flame, and explosions in gases. Cambridge, 1938.
371. Sokolik, A. S. Samovosplameneniye, plamya i detonatsiya v gazakh (Self-ignition, flame, and detonation in gases). Moscow, Izd-vo AN SSSR, 1960, 427 pp.
372. Salamandra, G. D., Bazhenova, T. V., Zaytsev, S. G., Soloukhin, R. I., Naboko, I. M., and Sevastiyanova, I. K. Nekotoryye metody issledovaniya bystroprotekayushchikh protsessov i ikh primeneniye k izucheniyu formirovaniya detonatsionnoy volny (Some methods of investigation of rapid processes and their application to the study of detonation-wave formation). Moscow, Izd-vo AN SSSR, 1960, 92 pp.
373. Voytsekhovskiy, B. V., Mitrofanov, V. V., and Topchiyan, M. E. Struktura front detonatsii v gazakh (Structure of detonation front in gases). Novosibirsk: Izd-vo AN SSSR, 1960, 168 pp.
374. Shchelkin, K. M. and Troshin, Ya. K. Gazodinamika goreniya. Combustion gasdynamics. Moscow, Nauka, 1963, 255 pp.
375. Soloukhin, R. I. Udarnyye volny i detonatsiya v gazakh (Shock waves and detonation in gases). Moscow, Fizmatgiz, 1963, 175 pp.

376. Shchetinkov, E. S. Fizika goreniya gazov. (Physics of combustion of gases). Moscow, Nauka, 1965, 739 pp.
377. Penner, S. S. and Williams, F. A., eds. Detonation and two-phase flow. New York, Acad. Press, 1962.
378. Dremin, A. N., Savrov, S. D., Trofimov, V. S., and Shvedov, K. K. Detonatsionnyye volny v kondensirovannykh sredakh (Detonation waves in condensed media). Moscow, Nauka, 1970, 163 pp.
379. Williams, F. A. Combustion theory. Palo Alto, London, Addison Wesley Publ. Corp., 1971.
380. Freiwald, G. Fast chemical reactions, combustion, and detonation in gases. In: Highspeed Physics. St. Louis, 1967.
381. Salamandra, G. D. Fotograficheskiye methody issledovaniya bystroprotekayushchikh protsessov (Photographic methods of investigation of rapid processes). Moscow, Nauka, 1974, 200 pp.
382. Jost, V. Explosions und Verbrennungs-vergänge in gasen. Berlin, Springer-Verlag, 1939.
383. Bartlma, F. Gazdynamik. Wien, der Verbrennung, 1975.
384. Wood, W. W. and Fickett, W. Investigation of the Chapman-Jouget hypothesis. Phys. Fluids, 1963, Vol. 6, No. 5, pp. 648–652.
385. Wood, W. W. and Salsburg, Z. V. Investigation of the steady state of sustained detonation and shock waves. In: Hypersonic flow. London, 1960.
386. Shchelkin, K. N. Dva sluchaya neustoychivovo goreniya (Two cases of unsteady combustion). ZhETF, 1959, Vol. 36, Issue 2, p. 600.
387. Vasiliyev, A. A. and Nikolayev, Yu. A. Model' yacheyki mnogofrontovoy gazovoy detonatsii (Simulation of a cell of multi-front gas detonation). Fizika goreniya i vzryva, 1976, Vol. 12, No. 5, pp. 744–755.
388. Uliyanitskiy, V. Yu. O roli 'vspyshki' i soudareniya poperechnykh voln v formirovanii mnogofrontovoy struktury detonatsionnykh voln v gazakh (On the role of 'flash' and collision of transverse waves in the formation of a multi-front structure of detonation waves in gases). Fizika goreniya i vzryva, 1981, Vol. 17, No. 2, pp. 127–133.
389. Subbotin, V. A. Stolknoveniya poperechnykh detonatsionnykh voln v gazakh (Collision of transverse detonation waves in gases). Fizika goreniya i vzryva, 1975, Vol. 11, No. 3, pp. 486–491.
390. Mitrofanov, V. V. and Subbotin, V. A. O mekhanizme detonatsionnovo sgoraniya v gazakh (About the mechanism of detonation combustion in gases). V: Goreniye i vzryv (In: Combustion and explosion). Moscow, Nauka, 1977, pp. 447–453.
391. Taki, S. and Fujiwara, T. Numerical analysis of two-dimensional nonsteady detonations. AIAA Journ., 1978, Vol. 16, No. 1, pp. 73–77.
392. Subbotin, V. A. Dva tipa struktury poprechnykh voln v mnogofrontovoy detonatsii (Two types of structures of transverse waves in multi-front detonation). Fizika goreniya i vzryva, 1975, Vol. 2, No. 1, pp. 96–102.
393. Manzhaley, V. N. and Subbotin, V. A. Eksperimental'noye issledovaniye ustoichivosti pereszhatoy detonatsii v gaze (Experimental investigation of the stability of recompressed detonation in gases). Fizika goreniya i vzryva, 1976, Vol. 12, No. 6, pp. 935–942.
394. Manzhaley, V. N. O predele sushchestvovaniya mezhfrontovoy struktury v pereszhatykh detonatsionnykh volnakh v gazakh (About the limit of existence of inter-front structure in recompressed detonation waves in gases). Fizika goreniya i vzryva, 1979, Vol. 15, No. 4, pp. 107–114.
395. Manzhaley, V. N. Ob ustoychivosti sil'nykh udarnykh voln v gazakh (About the stability of strong shock waves in gases). Fizika goreniya i vzryva, 1980, Vol. 16, No. 5, pp. 149–150.
396. Vasiliyev, A. A., Gavrilenko, T. P., Mitrofanov, V. V., Subbotin, V. A., and Topchiyan, M. E. O polozhenii tochki perekhoda cherez zvukovuyu skorost' za frontom detonatsii (About the location of the point of transition through sound speed behind the detonation front). Fizika goreniya i vzryva, 1972, Vol. 8, No. 1, pp. 98–104.
397. Vasiliyev, A. A., Gavrilenko, T. P., and Topchiyan, M. E. Usloviya Chepmena-Zhuge dlya real'nykh detonatsionnykh voln (The Chapman-Jouget conditions for real detonation waves). Fizika goreniya i vzryva, 1973, Vol. 9, No. 2, pp. 304–309.

398. Edwards, D. H., Jones, A. T., and Phillips, D. E. The location of the Chapman-Jouget surface in multiheaded detonation. J. Phys. D: Appl. Phys., 1976, Vol. 9, p. 1331.
399. Uliyanitskiy, V. Yu. Issledovaniye galopiruyushchevo rezhima gazovoy detonatsii (Investigation of the galloping regime of gas detonation). Fizika goreniya i vzryva, 1981, Vol. 17, No. 1, pp. 118–124.
400. Gvozdeva, L. G. Eksperimental'noye issledovaniye difraktsii detonatsionnykh voln v stekhiometricheskoy smesi metana s kislorodom (Experimental investigation of detonation-wave diffraction in a stoichiometric mixture of methane with oxygen). ZhPMTF, 1961, No. 5, pp. 53–56.
401. Gvozdeva, L. G. and Predvoditeleva, O. A. Issledovaniye troynykh konfiguratsiy detonatsionnykh voln v gazakh (Investigation of triple configurations of detonation waves in gases). Fizika goreniya i vzryva, 1969, Vol. 5, No. 4, pp. 451–461.
402. Zaytsev, S. G. and Soloukhin, R. I. K voprosu o vosplamenenii adiabaticheski nagretoy gazovoy smesi (On the question about the ignition of an adiabatically heated gas mixture). DAN SSSR, 1958, Vol. 22, No. 6, pp. 1039–1045.
403. Zaytsev, S. G. Opredeleniye i raschet nekotorykh kharakteristik protsessa vosplameneniya gazovoy smesi (Determination and computation of some characteristics of the ignition process of a gas mixture). V: Tr. III Vsesoyuz. soveshch. po teorii goreniya (In: Transactions of III All-Union meeting on the theory of combustion). Moscow, Izd-vo AN SSSR, 1960, Vol. 1, pp. 214–217.
404. Gel'fand, B. E., Gubin, S. A., Mikhalkin, V. N., and Shargatov, V. A. Raschet parametrov ploskikh udarnykh voln v vozdukhe pri detonatsii gazovykh smesey (Calculation of the parameters of plane shock waves in air upon detonation of gas mixtures). Khim. fizika, 1984, Vol. 3, No. 6, pp. 879–884.
405. Gel'fand, B. E., Gubin, S. A., Mikhalkin, V. N., and Shargatov, V. A. K raschetu techeniy s detonatsionnymi volnami (On the calculation of flows with detonation waves). Khim. fizika, 1984, Vol. 3, No. 3, pp. 435–442.
406. Eysen, K., Gross, R., and Rivlin, T. Theoretical computation of gas detonation. Combustion and Flame, Vol. 4, No. 2, pp. 137–147.
407. Zel'dovich, Ya. B., Kogarko, S. M., and Simonov, N. I. Eksperimental'noye issledovaniye sfericheskoy detonatsii (Experimental investigation of spherical detonation). ZhTF, 1957, Vol. 26, Issue 8, pp. 1744–1752.
408. Kogarko, S. M. O vozmozhnosti detonatsii gazovykh smesey v konusnykh trubkakh (About the possibility of detonation of gas mixtures in conical tubes). Izv. AN SSSR, Ser. Khim., 1956, No. 4, pp. 419–426.
409. Gvozdeva, L. G. Difraktsiya detonatsionnykh voln (Diffraction of detonation waves). V: Fizicheskaya gazodinamika, teploobmen i termodinamika gazov vysokikh temperatur (In: Physical gasdynamics, heat transfer, and thermodynamics of high-temperature gases). Moscow, Izd-vo AN SSSR, 1962, pp. 131–139.
410. Mitrofanov, V. V. and Soloukhin, R. I. O difraktsii mnogofrontovoy volny (About the diffraction of a multi-front wave). DAN SSSR, 1964, Vol. 159, No. 5, pp. 1003–1006.
411. Soloukhin, R. I. and Ragland, K. W. Ignition processes in expanding detonation. Combust. and Flame, 1969, Vol. 13, pp. 295–351.
412. Maksimik, B. Ya., Sukhov, V. V., and Kharkin, A. P. Izmeneniye gazodinamichkikh parametrov detonatsionnoy volny pri rasprostranenii v geometricheski izmenyaemom obeme (Change of gasdynamic paraments of a detonation wave during propagation in geometrically varying volume). Khim. tekhnologiya, 1977, No. 4, pp. 54–55.
413. Troshin, Ya. B. Mekhanizm vozniknoveniya dopolnitel'nykh troynykh konfiguratsii na poverkhvosti raskhodyashcheysya detonatsionnoy volny (Mechanism of occurrence of additional triple configurations on the surface of a diverging detonation wave). DAN SSSR, 1978, Vol. 240, No. 3, pp. 644–647.
414. Edwards, D. H., Nettleton, M. A., and Thomas, G. O. The diffraction at a planar detonation wave at an abrupt area change. J. Fluid Mech., 1979, Vol. 95, No. 1, pp. 79–93.
415. Edwards, D. H., Thomas, G. O., and Nettleton, M. A. Diffraction of a planar detonation in

various fuel oxygen mixtures at an area change. Progr. Astron. and Aeronaut., 1981, Vol. 75, pp. 341–357.

416. Vasiliyev, A. A. and Grigoriyev, V. V. Kriticheskiye usloviya rasprostraneniya gazovoy detonatsii v rezkorasshiryayushchikhsya kanalakh (Critical condition of propagation of gas detonation in abruptly diverging channels). Fizika goreniya i vzryva, 1980, Vol. 16, No. 5, pp. 117–125.
417. Moen, I. O., Murray, S. B., Bjerketvedt, D., et al. Diffraction of detonation from tubes into a large fuel-air explosive cloud. In: 19th Intern. symp. on combust., Pittsburgh, 1982, pp. 635–644.
418. Urtiew, P. A. and Tarver, C. M. Effects of cellular structure on the behavior of gaseous detonation waves under transient conditions. Progr. Astron. and Aeronaut., 1981, Vol. 75, pp. 370–384.
419. Gubin, S. A., Kogarko, S. M., and Mikhalin, V. N. Eksperimental'nyye issledovaniya gazovoy detonatsii v konicheskikh trubakh (Experimental investigations of gas detonation in conical tubes). Fizika goreniya i vzryva, 1982, Vol. 18, No. 5, pp. 111–117.
420. Matsui, H. and Lee, J. H. On the measure of the reactive detonation hazards of gaseous fuel-oxygen and air mixtures. In: 17th Intern. symp. on combust., Pittsburgh, 1978, pp. 1269–1279.
421. Vasiliyev, A. A. Issledovaniye kriticheskovo initsiirovaniya gazovoy detonatsii (Investigation of the critical initiation of gas detonation). Fizika goreniya i vzryva, 1983, Vol. 21, No. 1, pp. 121–131.
422. Vasiliyev, A. A. O kriticheskom diametre detonatsii gazovykh smesey (About the critical detonation diameter of gas mixtures). Fizika goreniya i vzryva, 1982, Vol. 14, No. 3, pp. 98–104.
423. Levin, V. A. and Markov, V. V. Vykhod detonatsionnoy volny v rasshiryayushchiysya kanal (Discharge of a detonation wave into diverging channel). Izv. AN SSSR, MZhG, 1981, No. 3, pp. 82–86.
424. Ayvazov, V. B. and Zel'dovich, Ya. B. Obrazovaniye pereszhatoy detonatsionnoy volny v suzhayushcheysya trubke (Formation of recompressed detonation wave in converging tube). ZhETF, 1947, Vol. 17, No. 10, pp. 899–900.
425. Gvozdeva, L. G. Otrazheniye detonatsionnykh voln v gazakh (Reflection of detonation waves in gases). V: Fizicheskaya gazodinamika i svoystva gazov pri vysokikh temperaturakh (In: Physical gasdynamics and properties of gases at high temperatures). Moscow, Nauka, 1964, pp. 157–174.
426. Gavrilenko, T. P., Nikolayev, Yu. A., and Topchiyan, M. E. Issledovaniye pereszhatykh detonatsionnykh voln (Investigation of recompressed detonation waves). Fizika goreniya i vzryva, 1979, Vol. 15, No. 5, pp. 119–123.
427. Edwards, D. H., Walker, J. R., and Nettleton, M. A. On the propagation of detonation waves along wedges. Arch. Combust., 1984, Vol. 4, No. 3, pp. 197–209.
428. Gavrilenko, T. P. and Prokhorov, E. S. Pereszhataya detonatsionnaya volna v gaze (Recompressed detonation wave in a gas). Fizika goreniya i vzryva, 1981, Vol. 17, No. 6, pp. 121–125.
429. Gvozdeva, L. G. Prelomleniye detonatsionnykh voln pri padenii na granitsu razdela gazovykh smesey (Refraction of detonation waves upon incidence on the interface of gaseous mixtures). ZhTF, 1961, Vol. 31, No. 6, pp. 731–739.
430. Jahn, R. I. The refraction of shock waves at a gaseous interface. J. Fluid Mech., 1956, Vol. 1, No. 5, pp. 457–489.
431. Meyer, R. F. Refraction of weak shock waves. J. Fluid Mech., 1957, Vol. 3, pp. 309–315.
432. Abd-El-Fattah, A. M., Henderson, L. F., and Lozzi, A. Precursor shock waves at a slow-fast gas interface. J. Fluid Mech., 1976, Vol. 76, pp. 157–179.
433. Abd-El-Fattah, A. M. and Henderson, L. F. Shock waves at a fast-slow gas interface. J. Fluid Mech., 1978, Vol. 80, No. 3, pp. 156–158.
434. Nikolayev, Yu. A. and Topchiyan, M. E. Raschet ravnovesnykh techeniy v detonatsionnykh volnakh v gazakh (Calculations of equilibrium flows in detonation waves in gases). Fizika goreniya i vzryva, 1977, Vol. 13, No. 3, pp. 393–404.

435. Guenoche, H. Recherches sur la détonation et la déflagration dans les mélanges gazeux. Rev. Inst. fr. pétrole, 1949, Vol. 4, No. 1, pp. 15–36; Vol. 4, No. 2, pp. 48–69.
436. Tsukhanova, O. A. Peredacha vzryva cherez kapillyary (Transfer of explosion through capillaries). V: Tr. III Vsesoyuz. soveshch. po teorii goreniya (In: Transactions of III All-Union meeting on the theory of combustion). Moscow, Izd-vo AN SSSR, 1960, Vol. 1, pp. 187–192.
437. Pusch, W. and Wagner, H. G. Einfluss des Rohrdurchmessers auf die Ausbreitung einer Detonation in explosiblen Gasgemischen. T. I. - Ztschr. Elektrochem., 1965, Vol. 69, pp. 503–513.
438. Guhlmann, K, Pusch, W., and Wager, H. G. Einfluss des Rohrdurchmessers auf die Ausbreitung einer Detonation in explosiblen Gasgenischen. T. II. Ztschr. Elektrochem., 1966, Vol. 70, No. 2, pp. 143–148.
439. Sichel, M. and David, T. S. Transfer behind detonations in H_2-O_2 mixtures. AIAA Journ., 1966, Vol. 4, No. 6, pp. 1089–1090.
440. Zakaznov, V. F., Rozlovskiy, A. I., and Strizhevskiy, I. I. Gasheniye detonatsii i osobennost' eye rasprostraneniya v uzkikh kanalakh (Detonation quenching and the features of its propagation in narrow channels). Fizika goreniya i vzryva, 1967, Vol. 3, No. 2, pp. 217–224.
441. Skinner, J. H. Friction and heat transfer effects in the nonsteady flow behind a detonation. AIAA Journ., 1967, Vol. 5, No. 11, pp. 2069–2071.
442. Brochet, C., Guerrand, C., Manson, N., and Veyssière, M. Variation de la pression des gaz derrière le front des détonations dans les mélanges stricts propane-oxygène, influence du diamètre des tubes et de la pression initiale. C. r. Acad. sci., B, 1969, Vol. 268, pp. 361–364.
443. Edwards, D. H., Brown, D. R., Hooper, G., and Jones, A. T. The influence of wall heat transfer on the expansion following a C-J detonation wave. J. Phys. D: Appl. Phys., 1970, Vol. 3, No. 3, pp. 365–376.
444. Paillard, C., Dupré, G., and Combourieu, J. Étude de la détonation de composés endothermiques gazeux. I. Célérités de détonation de l'azoture de chlore dans des tubes cylindriques. Limites de détonation de l'azoture de chlore et de l'azoture d'hydrogène. J. chim. phys., 1973, Vol. 70, No. 5, pp. 811–818.
445. Paillard, C., Dupré, G., and Combourieu, J. Étude de la détonation de composés endothermiques gazeux. II. Propagation et conditions critiques d'extinction dans des tubes capillaires de la flamme de décomposition de l'azoture de chlore gazeux pur ou dilué. J. chim. phys., 1974, Vol. 71, No. 2, pp. 175–181.
446. Paillard, C., Dupré, G., Lisbet, R., Combourieu, J., Fokeyev, V. P., and Gvozdeva, L. G. A study of hydrogen azide detonation with heat transfer at the wall. Acta Astronaut., 1979, Vol. 6, pp. 227–242.
447. Paillard, C., Dupré, G., Lisbet, R., Combourieu, J., Fokeyev, V. P., Gvozdeva, L. G., and Bazhenova, T. V. Pressure and wall heat transfer behind a hydrogen azide detonation wave in narrow tubes. In: Gas dynamics of detonation and explosions. New York, 1981, pp. 134–149.
448. Paillard, C., Dupré, G., Lisbet, R., and Combourieu, J. Influence de la structure d'une onde de détonation sur les vitesses de transfert de chaleur aux parois. In: Colloque Intern. Mallar. Le Chatelier. Bertlot et Vielle. Bordeaux, France, 1981, pp. 449–454.
449. Bazhenova, T. V., Gvozdeva, L. G., Fokeyev, V. P., Paillard, C., Combourieu, J., Dupré, G., and Lisbet, R. Issledovaniye detonatsii HN_3, v blizi peredelov (Investigation of detonation of HN_3 near the boundaries). Fizika goreniya i vzryva, 1985, Vol. 21, No. 1, pp. 120–124.
450. Manson, N. La theorie hydrodynamique et le diamètre limite de propagation des ondes explosives. Ztschr. Elektrochem., 1957, Vol. 61, No. 5, pp. 586–592.
451. Manson, N. and Guenoche, H. Effect of the charge diametre on the velocity of detonation wave in the mixtures. In: 6th Symp. (Intern.) Combust. New Haven, 1956, pp. 631–639.
452. Belles, F. E. Detonability and chemical kinetics: Prediction of limits of detonability of hydrogen. In: 7th Symp. (Intern.) Combust. Leningrad, 1958, pp. 745–751.
453. Fay, J. A. Two-dimensional gaseous detonations, velocity deficit. Phys. Fluids, 1959, Vol. 2, No. 3, pp. 283–289.

454. Fujiwara, T., Sugimura, M., and Inagawa, H. Oxyhydrogen detonations near pressure limit. Mem. Fac. Eng. Nagoya Univ., 1971, Vol. 23, No. 2, pp. 309–315.
455. Tsuge, S. The effect of boundaries on the velocity deficit and the limit of gaseous detonations. Combust. Sci. Tech., 1971, Vol. 3, pp. 195–205.
456. Dove, J. E., Scroggie, B. J., and Semerjian, H. Velocity deficits and detonability limits of hydrogen-oxygen gas detonations at low initial pressures. Acta Astronaut., 1974, Vol. 1, pp. 345–359.
457. Gladilin, A. M. and Karpilovskiy, E. I. Uchet vzaimodeystviya gazovovo potoka so stenkoy stvola detonatsionnoy ustanovki (Consideration of interaction of gaseous flow with the wall of the barrel of detonation apparatus). Fizika goreniya i vzryva, 1980, Vol. 16, No. 3, pp. 148–150.
458. Zhdan, S. A. and Fedenok, V. I. Parametry ravnovesnovo gazovovo potoka v stvole detonatsionnoy ustanovki (Parameters of equilibrium gaseous flow in the barrel of detonation apparatus). Fizika goreniya i vzrvya, 1982, Vol. 18, No. 6, pp. 103–107.
459. Guenoche, H. Le Diuzet, P. and Sedes, C. Influence of the heat release function on the detonation states. In: Gas dynamics of detonation and explosion. New York, 1981, pp. 387–407. Progr. Astronaut. and Aeronaut., Vol. 75.
460. Strehlow, R. A., Liaugminas, R., Watson, R. H., and Eyman, J. R. Transverse wave structure in detonations. In: 11th Symp. (Intern.) Combust. Pittsburgh, 1961, pp. 683–692.
461. Stull, D. R., ed. JANAF Thermochemical tebles. Midland (Mich.), Dow Chemical Co., 1970.
462. Kogarko, S. M. Issledovaniye davleniya v tortse truby pri nestatsionarnom bystrom gorenii (Investigation of pressure at the tube end during unsteady rapid combustion). ZhTF, 1958, Vol. 28, Issue 9, pp. 2041–2045.
463. Herzberg, Z. and Smith, W. Method of excitation of strong shock waves. Journ. of Appl. Phys., 1954, Vol. 25, No. 1, pp. 130–131.
464. Denisov, Yu. N. 'Uprugiye' i 'neuprugiye' otrazheniya skachkov plotnosti v vosplamenyayushchikhsya sredakh ('Elastic' and 'inelastic' reflections of shock waves in combustible media). DAN AN SSSR, 1984, Vol. 179, No. 3, pp. 665–669.
465. Salamandra, G. D., Bazhenova, T. V., and Naboko, I. M. Formirovaniye detonatsii pri sgoranii gaza v trubakh (Formation of detonation during the combustion of a gas in tubes). ZhTF, 1959, Vol. 29, Issue 11, pp. 1354–1359.
466. Urtiew, P. A. and Oppenheim, A. K. Experimental observations of the transition to detonation in an explosive gas. Proc. Roy. Soc., 1966, Vol. A 295, pp. 13–28.
467. Salamandra, G. D. Fotograficheskiye metody issledovaniya bystroprotekayushchikh protsessov (Photographic methods of investigation of rapid processes). Moscow, Nauka, 1974, 200 pp.

INDEX